MCBU
Molecular and Cell Biology Updates

Series Editors:

Prof. Dr. Angelo Azzi
Institut für Biochemie
und Molekularbiologie
Bühlstr. 28
CH-3012 Bern
Switzerland

Prof. Dr. Lester Packer
Dept. of Molecular
and Cell Biology
251 Life Science Addition
Membrane Bioenergetics Group
Berkeley, California 94720
U. S. A.

Adenine Nucleotides in Cellular Energy Transfer and Signal Transduction

Edited by   S. Papa
                 A. Azzi
                 J.M. Tager

Birkhäuser Verlag
Basel · Boston · Berlin

Volume Editors' addresses:

Prof. Dr. Sergio Papa
Prof. Dr. Joseph M. Tager
Institute of Medical Biochemistry
and Chemistry
University of Bari
I-70124 Bari
Italy

Prof. Dr. Angelo Azzi
Institut für Biochemie
und Molekularbiologie
Bühlstrasse 28
CH-3012 Bern
Switzerland

Deutsche Bibliothek Cataloging-in-Publication Data
Adenine nucleotides in cellular energy transfer and signal transduction /
ed. by S. Papa ... -Basel; Boston; Berlin:
Birkhäuser, 1992
(Molecular and cell biology updates)
ISBN 3-7643-2673-5 (Basel ...)
ISBN 0-8176-2673-5 (Boston)
NE: Papa, Sergio (Hrsg.)

© 1992 Birkhäuser Verlag
      P. O. Box 133
      4010 Basel
      Switzerland
      FAX (++41) 61 271 76 66

Printed in Germany on acid-free paper,
directly from the authors' camera-ready manuscripts.
ISBN 3-7643-2673-5
ISBN 0-8176-2673-5

TABLE OF CONTENTS

PROTEIN KINASES

ADENYLATE CYCLASE

PREFACE

Adenine Nucleotides play a major role in cellular metabolism and functions, serving as high-potential phosphate transfer compounds in energy metabolism and as substrates and co-factors for proteins involved in signal transduction.

During the last few years definite advancement has been made in elucidating the molecular and genetic aspects of the enzyme involved in oxidative phosphorylation, the so-called F_oF_1 H^+-ATP synthase. Non-invasive NMR technologies have been developed to monitor in vivo the energy level of tissues based on determination of the concentrations of adenine nucleotides, phosphate and phosphate esters.

Thus it became clear that the capacity of oxidative phosphorylation adapts itself to the ATP demand which changes continuously with the physiological state in various tissues. This is achieved by regulation of the enzyme activity as well as by regulation of its biogenesis.

The reversible phosphorylation of proteins is recognised as a major regulatory mechanism in eukaryotic as well as in prokaryotic cells for cellular signal transduction and control of gene expression, cell growth, differentation and oncogenesis. The same applies to the role played by cAMP.

A further topic of growing interest concerns the discovery of the ATP binding cassette (ABC) superfamily of transport proteins which includes systems of primary importance in medicine such as the multi-drug resistance P glycoprotein, the cystic fibrosis transmembrane conductance regulator (CFTR) and the 70 kd peroxisomal membrane protein.

Finally, much attention is being devoted in many laboratories to the molecular structure and role of ATP-modulated channels.

Experts in these various fields were invited to participate in the meeting in Fasano and to contribute chapters for this volume.

We believe that the book should be of interest to students of basic biomedical sciences as well as to clinicians.

We would like to thank Dr. Helena Kirk and Dr. Carla Del Pesce for their help in organising the material for the book, Dr. Rita Peccarisi and Dr. Nino Villani, who prepared the subject index, and the staff of Birkhäuser for their kind cooperation.

S. Papa
A. Azzi
J.M. Tager

Adenine Nucleotides in Cellular Energy Transfer and Signal Transduction
S. Papa, A. Azzi & J.M. Tager (eds)
© 1992 Birkhäuser Verlag, Basel/Switzerland

MOLECULAR BIOLOGY AND ASSEMBLY OF YEAST MITOCHONDRIAL ATP SYNTHASE

Rodney J. Devenish, Maria Galanis, Theo Papakonstantinou, Ruby H.P.Law, David G. Grasso, Leon Helfenbaum and Phillip Nagley

Department of Biochemistry and Centre for Molecular Biology and Medicine, Monash University, Clayton, Victoria 3168, Australia

SUMMARY: This work concerns the structure, function and assembly of the non-F_1 proteins of the mitochondrial ATP synthase complex (mtATPase). A variety of in vitro and in vivo approaches based on molecular genetic manipulation has been used in our laboratory to study mtATPase in bakers' yeast Saccharomyces cerevisiae. Here we review our recent advances concerning this enzyme complex including: 1) the dramatic enhancement of import of proteins into mitochondria by duplication of the N-terminal cleavable leader sequence; 2) the development of immunochemical procedures to study the assembly into mtATPase of subunits imported into isolated mitochondria; and 3) site-directed mutagenesis of an artificial nuclear gene encoding subunit 8. This molecular genetic manipulation of subunit 8 has demonstrated the role of its C-terminal positively-charged region in the assembly of subunit 8 into mtATPase. Further, the minimal effects on function of positive charges introduced into the hydrophobic domain of subunit 8 questions the assumption that this domain represents a transmembrane stem.

INTRODUCTION

The mitochondrial ATP synthase (mtATPase) complex of eukaryotes is the key energy-transducing enzyme in oxidative phosphorylation, catalysing in vivo the synthesis of ATP driven by energy released by mitochondrial respiration (Senior, 1988). The synthesis of ATP from ADP and P_i, which takes place on the soluble F_1 sector extrinsic to the bilayer of the inner-mitochondrial membrane, is coupled to proton translocation through the proton channel formed by the F_o sector subunits embedded in that membrane. In yeast mtATPase the non-F_1 subunits are classified as belonging to F_o or F_A (Nagley, 1988). F_o comprises three integral membrane subunits, denoted 6,8 and 9, thought to be integral membrane proteins. F_A represents a group of additional proteins, amongst which are those involved in coupling proton translocation to ATP synthesis. The F_A proteins in yeast include three proteins: OSCP (Tzagoloff, 1970) and two proteins of M_r 18,000 and 25,000, designed P18 and P25, respectively.

The non-F_1 proteins in yeast mtATPase are the subject of continuing investigations in our laboratory by approaches based on molecular genetic manipulation to study aspects of the structure, function and assembly of the enzyme complex. Since the mtATPase complex is assembled in a membrane-bound organelle, separated from the cytosolic site of synthesis of the majority of mtATPase subunits that are encoded in nuclear DNA, a full appreciation of the biogenesis of the complex requires understanding not only of gene structure and regulation, but also of protein targeting, membrane translocation and subunit interactions within mitochondria (Hartl & Neupert, 1990; Pfanner et al., 1991; Baker & Schatz, 1991). By means of the allotopic expression strategy (Nagley & Devenish, 1989), we have directly applied the sophisticated nuclear gene expression technology available in yeast to the manipulation of proteins normally encoded by mitochondrial DNA (mtDNA). This involves the functional relocation of mitochondrial genes to the nucleus and delivery of the protein products back to mitochondria under the targeting control of a suitable cleavable N-terminal leader sequence.

In our investigations allotopic expression in vivo has been coupled with in vitro analysis of import and assembly. The assembly assay requires the means of generating mitochondria depleted of a specific subunit (see below). The recent advances of our laboratory presented here will focus on the application of these techniques together with genetic manipulations that either provide enhanced import capability to chimaeric precursors previously demonstrated to be non-importable, or that utilise site-directed mutagenesis to probe subunit structure and function.

DRAMATIC ENHANCEMENT OF PROTEIN IMPORT INTO MITOCHONDRIA BY DUPLICATION OF THE N-TERMINAL CLEAVABLE LEADER SEQUENCE

We have utilised the allotopic expression approach in our studies of two F_o sector subunits 8 (Y8; 48 amino acids) and 9 (Y9; 76 amino acids) normally encoded in yeast mtDNA. For allotopic expression of Y8 and Y9 a series of chimaeric precursor proteins has been constructed that incorporate the 66 amino acid leader (N9L) from Neurospora crassa mtATPase subunit 9, a nuclearly encoded protein (Viebrock et al., 1982). Fusions of N9L to either Y8 or Y9 have been made via a 7 amino acid bridge (constructs denoted N9L/Y8-1 or N9L/Y9-1, respectively) or without any intervening amino acids (i.e. direct fusions denoted N9L/Y8-2 or N9L/Y9-2, respectively) (Gearing & Nagley, 1986; Farrell et al., 1988; Law et al., 1988). Both N9L/Y8-1 and N9L/Y8-2 are imported into mitochondria in vitro and, when expressed in vivo in a yeast host lacking

```
                              |        ••
N9L/Y8-1        SIVNATTRQAFQKRA YSSEISSMPQLVPFYFMNQLTY
                                        |
N9L/Y8-2        SIVNATTRQAFQKRAMPQ LVPFYFMNQLTYGFLLMIT
                                   |
                              |        ••
N9L/Y9-1        SIVNATTRQAFQKRA YSSEISSMQLVLAAKYIGAGIST
                                   |    ••
N9L-D/Y9-1      SIVNATTRQAFQKRA YSSEISSMQLVLAAKYIGAGIST
                                   |
N9L/Y9-2        SIVNATT RQAFQKRAMQLVLAAKYIGAGISTIGLLGAG
                       |
N9L-D/Y9-2      SIVNATT RQAFQKRAMQLVLAAKYIGAGISTIGLLGAG
                                          |
N9L/Y9-3        SIVNATTRQAFQKRAYSS VLAAKYIGAGISTIGLLGAG
                                  |
N9L-D/Y9-3      SIVNATTRQAFQKRAYSS VLAAKYIGAGISTIGLLGAG
```

Fig. 1   Sequences of chimaeric import precursor proteins around the sites of fusion. Each sequence represented starts at residue SER52 of the N9L sequence. Sequences are shown in single letter code. The sequence in bold represents residues of mature <u>N. crass</u> subunit 9. Adjacent residues (with dots) complete the bridge upstream of subunit 8 in N9L/Y8-1 or of subunit 9 in N9L/Y9-1. Underlined sequences are the N terminal regions of subunit 8 or subunit 9. In Y9-3 constructs the three N-terminal amino acids of subunit 9, MQL, have been removed. The arrows indicate the positions of cleavage following import as inferred from radiosequencing of the processed product. Data are compiled from: Gearing & Nagley (1986); Farrell et al. (1988); Galanis et al. (1990; 1991b).

mitochondrially encoded Y8, lead to the functional incorporation of imported Y8 into mtATPase. Allotopic expression of Y8-1, bearing a 7 amino acid N-terminal extension after processing of N9L/Y8-1 in mitochondria (Fig. 1; Gearing & Nagley, 1986), generates a strain having indices of mitochondrial function only slightly less than those of wildtype strains (Nagley et al., 1988). In contrast, allotopic expression of Y8-2 results in only partial restoration of mtATPase function, generating a strain with markedly impaired oxidative metabolism (Grasso et al., 1991). The 3 amino acid N-terminal truncation of Y8 after processing of N9L/Y8-2 (Fig. 1.; Galanis et al., 1990) is considered to interfere with a domain of Y8 important for the function of the F_o sector (Galanis et al., 1991b; see below).

In contrast to N9L/Y8-1, N9L/Y9-1 is imported relatively weakly into isolated mitochondria and N9L/Y9-2 does not import at all (Law et al., 1988). Neither Y9 construct is able to restore the ability of a subunit 9-deficient yeast to grow on non-fermentable substrate following expression <u>in vivo</u>.

Since the failure of N9L/Y9 constructs to rescue may be related to the

general deficiency or failure of import observed in vitro, a novel means
of enhancing protein import into mitochondria was implemented.

We devised a gene manipulation strategy, leading to the tandem
duplication of the N9L cleavable leader peptide, to achieve a solution to
this problem (Galanis et al., 1991a). Thus the Y9 constructs endowed
with a tandemly duplication N9L moiety show remarkable improvement in
import into isolated mitochondria. However, the double leader
constructs, when expressed in vivo, do not lead to complementation of a
subunit 9-deficient mutant, most likely because of aberrant processing of
the precursor to yield non-functional subunit 9 molecules within the
mitochondrial matrix (Galanis et al., 1990). N-terminal sequencing of
^{35}S-methionine labelled processed Y9 isolated from mitochondria after in
vitro import indicated the site of cleavage of the matrix protease in the
precursor. In both N9L-D/Y9-1 and N9L-D/Y9-2 the processed imported Y9
protein carries an N-terminal extension of 7 and 8 residues respectively
(Fig. 1). Apparently, such N-terminal extensions cannot be tolerated for
the assembly or function of Y9 in the mtATPase complex.

One attempt to remedy this situation involved the use of gene
manipulation to generate new constructs, N9L/Y9-3 and N9L-D/Y9-3, in
which the leader (single or double) plus the first three amino acids of
mature N. crassa subunit 9 are directly fused to Y9 lacking the N-
terminal three amino acids of Y9 (Fig. 1; Galanis et al., 1991b). The
design of the fusion point between the N9L and Y9 moieties was aimed at
maintaining an authentic matrix protease cleavage site in the chimaeric
precursor such that following import and subsequent processing a mature
Y9 derivative lacking an N-terminal extension would be generated. Import
into isolated mitochondria was more efficient for the double leader
chimaeric precursor than the single leader chimaeric precursor (data not
shown). Microsequencing of processed Y9 isolated from these mitochondria
indicated a site of matrix protease cleavage such that Y9 is effectively
truncated by three amino acids (Fig. 1). This is the case whether single
or double leader constructs are utilised. In neither case is the
construct able to restore growth of the mutant yeast strain that lacks
endogenous mitochondrially synthesised subunit 9.

It is clear that the nature of the amino acid sequence around the
matrix protease cleavage site influences the conformation of the Y9
chimaeric precursors resulting in aberrant processing. In an attempt to
overcome this problem we have designed a strategy to incorporate variable
amino acid sequence flanking the conserved matrix protease cleavage site
in order to provide a means of generating modified versions of N9L-D/Y9-2

that will be correctly processed and able to rescue subunit 9 defective
strains. This strategy is now being applied by means of a random
localised mutagenesis technique centred upon the matrix protease cleavage
site region of the chimaeric precursor proteins. Two single-stranded
oligonucleotides have been synthesised to enable the generation of a
gene segment with localised variable sequence in the regions -13 to -3
and +1 to +4 (relative to the point of fusion between the leader sequence
and mature Y9) in order to generate variable codons flanking a conserved
ARG-ALA cleavage motif associated with the matrix protease processing
site (Hendrick et al., 1989). Following replacement of the -13 to the +4
segment of the gene encoding the double leader Y9 construct with an
appropriately mutagenised synthetic DNA segment, a population of mutated
N9L/Y9-2 coding regions will be generated that can be tested for
expression _in vivo_ for the ability to rescue subunit 9-deficient mutant
yeast strains.

The double leader strategy has proved very useful for the analysis of
variants of the Y8 protein, particularly in cases where the Y8 variant
fused to a single N9L leader is imported to a very limited extent. C-
terminal truncation mutants of Y8 have been generated by _in vitro_
mutagenesis which are either extremely poor importers or totally
incompetent for import (Nero et al., 1990). However, when the N9L leader
has been duplicated, in every case tested, efficient import of the
precursor _in vitro_ was observed (Galanis et al., 1991 a, 1991 b; see
also Table 1). This has opened the way for further analysis of these
muteins at the biological level, which was in some cases previously
impossible due to the inefficiency of delivery of the protein to its site
of assembly in mitochondria.

Comparison of _in vivo_ expression of single and double leader
constructs has demonstrated that in the case of the Y8-1 (K47→STP)
variant expressed in yeast cells using the expression vector pPD72
(Papakonstantinou et al., 1991) delivery into the mitochondrial matrix of
sufficient quantity of this Y8 variant is key to achieving the functional
assembly of the subunit into the mtATPase complex thus restoring
function. When expressed from the pPD72 vector, the single leader
construct shows no detectable rescue of subunit 8-deficient host cells
in vivo. In contrast, the corresponding double leader construct shows
efficient restoration of mtATPase function (Galanis et al. 1991b). We
have previously shown that expression of the same Y8-1 (K47→STP) variant
in a single leader construct using the vector pLF1 is able to partially
restore function of subunit 8-deficient yeast cells (Grasso et al., 1991);

Table 1   Comparison of single and double leader subunit 8 chimaeric
           precursors.

	Construct	Import[a]	Rescue[b]
Single Leader	N9L/Y8-1	++	++++
	N9L/Y8-2	+	++
	N9L/Y8-1 (K47→STP)	+	++
	N9L/Y8-2 (K47→STP)	±	−
	N9L/Y8-1 (R42→STP)	−	−
	N9L/Y8-2 (R42→STP)	−	−
	N9L/Y8-1 (R37→STP)	−	−
Double Leader	N9L-D/Y8-1	++++	++++
	N9L-D/Y8-2	++++	++
	N9L-D/Y8-1 (K47→STP)	++++	++++
	N9L-D/Y8-2 (K47→STP)	++++	−
	N9L-D/Y8-1 (R42→STP)	++ *	−
	N9L-D/Y8-2 (R42→STP)	++ *	−
	N9L-D/Y8-1 (R37→STP)	++ *	−

Data compiled from: Law et al. (1990); Grasso et al. (1991); Galanis
et al. (1991b); Papakonstantinou et al. (1991).
a.   Radiolabelled chimaeric precursors were incubated with isolated
mitochondria partially depleted of Y8 under active import conditions and
samples processed for gel electrophoresis as described (Law et al.,
1990). ++++, corresponds to almost quantitative import of bound chimaeric
precursor;   fewer plus signs indicate a reduced extent of import;   *
indicates degradation of mature subunit 8 derivative.
b.   Subunit 8 gene constructs on a multicopy expression vector with a
PGK1 promoter were introduced into host cells lacking endogenous
mitochondrially synthesized subunit 8. Ability to restore growth of host
cells on non-fermentable substrate ('rescue') was scored by a plate assay
(Nagley et al., 1988).   ++++, corresponds to a growth rate on ethanol
medium of about 5.5 hours;   fewer plus signs indicate a slower growth
rate.

efficient rescue is observed with a duplicated N9L leader (Table 1).
Presumably the expression level from the pLF1 vector is just high enough
to achieve measurable assembly from the N9L/Y8-1 (K47→STP) variant. We
are currently investigating the differences in expression levels of Y8
variants between the pLF1 and pPD72 vectors.

    As we discuss in more detail below, the defect in the Y8-1(K47→STP)
variant is manifested at the level of an assembly step itself. The
functional defects in cells expressing the single leader construct
N9L/Y8-1 (K47→STP) can be overcome by use of the double leader construct
N9L-D/Y8-1 (K47→STP) (Table 1). By contrast, incomplete restoration of
function is observed in vivo, with N9L/Y8-2, comprising full length Y8
directly fused to N9L (Fig.1), although Y8-2 does assembly at a clearly
measurable rate into mtATPase in vitro (Grasso et al., 1991). It is

notable that this function defect _in vivo_ is unable to be remedied by use of the double leader construct N9L-D/Y8-2 (Table 1), in spite of vastly improved import efficiency. This indicates that the mitochondrially processed Y8-2 derivative, which lacks the first 3 residues of Y8 at its N-terminus (Fig. 1), is defective in its ability to perform as such, rather than in the rate of assembly. This implicates the N-terminal region of Y8 as important in some aspect of energy transduction (Galanis et al., 1991b), which is consistent with the conservation of amino acid sequence in this region amongst fungal subunit 8 homologues (Nagley et al., 1990). Note that the combination of N-terminal and C-terminal truncations that are present in mitochondrially processed N9L-D/Y8-2 (K47→STP) prevents functional assembly _in vivo_ of Y8 (Table 1).

ASSEMBLY OF mtATPase SUBUNITS IMPORTED INTO ISOLATED MITOCHONDRIA

We have recently developed an _in vitro_ system using isolated mitochondria to monitor the assembly into mtATPase of imported components such as allotopically expressed Y8 (Law et al., 1990). Target mitochondria that had been partially depleted of endogenous Y8 by deliberately regulated allotopic expression _in vivo_ (Law et al., 1990), when incubated with radiolabelled N9L/Y8-1, were able to assemble imported Y8, as judged by its immunoadsorption to an appropriate anti-F_1 subunit monoclonal antibody, specific for either subunit β (Hadikusumo et al., 1984), or α subunit (Grasso et al., 1991). This versatile _in vitro_ assembly assay has now been extended to naturally imported F_A subunits of mtATPase, OSCP and P25 (homologue of bovine subunit _b_). Gene cassettes encoding OSCP and P25 have been isolated and used to disrupt the chromosomal copy of the gene thus creating null mutant strains specifically lacking OSCP or P25. Low copy number yeast expression vectors have been used to endow the appropriate null mutant with a regulated version of either the gene cassette encoding OSCP or the gene cassette encoding P25 under the control of the strictly inducible GAL1 promoter. Controlled depletion of the relevant subunit in cells is achieved by turning off the GAL1 promoter under growth conditions where the inducer galactose is no longer present. Such cells yield mitochondria containing mtATPase complexes depleted of OSCP or P25 into which the assembly of radiolabelled subunit can be detected (Devenish et al., 1991). Figure 2 shows the assembly of imported OSCP into mtATPase depleted of endogenous OSCP. Interestingly, P25, in contrast to allotopically expressed Y8 or OSCP, assembles into mtATPase complexes of mitochondria isolated from wild-type cells (data not shown), implying

A. **Approach:**

- Incubate mitochondria with pOSCP for 60 min
- lyse in 0.5% cholate/1% octyl-β-D-glucopyranoside

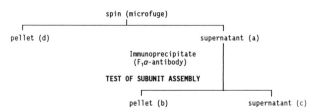

spin (microfuge)

pellet (d)   supernatant (a)

Immunoprecipitate
(F₁α-antibody)

TEST OF SUBUNIT ASSEMBLY

pellet (b)   supernatant (c)

B. <u>Mitochondria of YRDI5 (rho⁺)</u>

Solubilization/Immunoprecipitate

Precursor Import   a   b   c   d

C. <u>Mitochondria of OSCP-depleted transformant (NRG-1)</u>

Solubilization/Immunoprecipitate

Precursor Import   a   b   c   d

NRG-1(Gal)

NRG-1(EtOH)1

NRG-1(EtOH)2

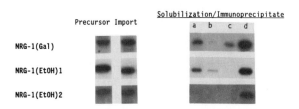

Fig. 2.   Assembly of imported OSCP into mtATPase depleted of endogenous
OSCP.   In panel A is given a summary of the experimental steps for tests
of assembly into mtATPase of OSCP imported into mitochondria <u>in vitro</u>.
Stages (a) through (d) at which protein samples were taken for analysis
by gel electrophoresis are indicated.   Detailed methods are described by
Law et al. (1989).   Cells of strains J69-1B (a wild-type <u>rho</u>⁺ strain) and
NRG-1 (a <u>rho</u>⁺ strain bearing a disrupted <u>ATP5</u> gene [encoding OSCP] and
carrying a low copy number plasmid expressing OSCP under control of the
<u>GAL1</u> promoter [Devenish et al., 1991;   see also text]) were pregrown in
rich medium containing galactose and ethanol to mid-logarithmic phase,
then shifted to rich medium containing ethanol as sole carbon source.
These   cells   were   grown   at   28°C   with   aeration   and   harvested   for
mitochondrial preparation at the late logarithmic phase.   Radiolabelled
OSCP precursor was incubated under import conditions with mitochondria
from J69-1B (panel B) or from NRG-1 which had been grown in galactose or
undergone   either   one   or   two   transfers   into   ethanol   medium   to   deplete
endogenous OSCP (panel C).   Proteins were analysed by gel electrophoresis
directly from untreated mitochondria (Import).   The fate of radiolabelled
OSCP imported into these mitochondria through the fractionation scheme
outlined in panel A is indicated in lanes (a) through (d).

differences exist in the assembly pathway of individual subunits.

C-TERMINAL POSITIVE CHARGED REGION OF SUBUNIT 8 IS REQUIRED FOR ASSEMBLY INTO mtATPase

Site-directed mutagenesis is potentially a powerful tool with which to probe the structure and functions of mtATPase subunits. The effects of changing amino acid residues can be readily assessed by employing both the in vitro import/assembly and in vivo rescue assays. Positively charged residues at the C-terminus represent a highly conserved feature of Y8 homologues in fungi and metazoa (Nagley et al., 1990). We have created a variant of Y8, by site-directed mutagenesis of the artificial nuclear subunit 8 gene, that is truncated at the C-terminal positively charged residue LYS47 (Nero et al., 1990). Using expression vector pLF1, partial restoration of growth is observed for expression of the variant N9L/Y8-1 (K47→STP) (Nero et al., 1990). Using the in vitro assembly assay, Y8-1(K47→STP) was found to be imported but inefficiently assembled into mtATPase complexes of mitochondria depleted of full length Y8 (Grasso et al., 1991). Even the equivalent duplicated N9L construct which is imported almost quantitatively in vitro does not deliver Y8-1 (K47→STP) in a form that can be readily detected to assemble in vitro (Galanis et al., 1991b).

In spite of the inefficient assembly of Y8-1 (K47→STP) in vitro, partial restoration of respiratory growth properties occurs in host cells lacking endogenous Y8 expressing N9L/Y8-1 (K47→STP) on vector pLF1; moreover, almost full restoration of function occurs when N9L-D/Y8-1 (K47→STP) is expressed under similar conditions (Galanis et al., 1991b). These data suggest that the positively charged C-terminus of Y8 may influence the rate of interaction between incoming Y8 and other mtATPase subunits, which can be substantially overcome by improving delivery of Y8-1 (K47→STP) in vivo using a duplicated N9L leader. Since a requisite for the assembly of Y8 is the presence of Y9 in the membrane (Linnane et al., 1985; Hadikusumo et al., 1988) the interactions between Y9 and Y8 may be influenced by the C-terminal positive charges although the interaction with other nuclearly encoded subunits cannot be excluded. Interactions between Y8 and subunit 6 are probable since the latter is not assembled in the absence of Y8 (Linnane et al., 1985; Hadikusumo et al., 1988).

Further truncation of Y8 to delete the other two positive charges, generating fusion proteins such as N9L/Y8-1 (R42→STP) and N9L/Y8-1 (R37→STP) (Nero et al., 1990), leads to constructs unable to be imported.

While the double leader equivalents of these variants are readily imported (Galanis et al. 1991b) no rescue occurs _in vivo_. It might be argued that in the truncation variants created it is not merely the loss of positively charged residues but the loss of additional amino acid residues distal to the truncation point that lead to loss of assembly or functional competence. Accordingly, we have created a set of variants in which the positively charged residues have been converted to the neutral amino acid ILE. At the _in vivo_ level the variants N9L/Y8-1 (R37→ILE) and N9L/Y8-1 (R42→ILE) demonstrate failure to restore growth of cells lacking endogenous Y8 while the variant N9L/Y8-1 (K47→ILE) results in partial restoration of function, reminiscent of N9L/Y8-1 (K47→STP). Assessment of the _in vitro_ import and assembly properties of these variant Y8 proteins reveals that import is relatively efficient but that assembly into mtATPase complexes is apparently deficient (Papakonstantinou et al., 1991). These results suggest the involvement of each of the three charged residues in the assembly process.

PROBING THE PUTATIVE TRANSMEMBRANE STEM REGION OF SUBUNIT 8

We have previously sought to define precisely those amino acid residues making up the transmembrane stem of Y8. Consideration of comparative hydropathy plots for Y8 led to the assignment of residues 14 to 32 as forming the transmembrane stem of Y8 (Nagley et al., 1990). We have carried out additional site-directed mutagenesis of N9L/Y8-1 in order to introduce additional positive charges into the transmembrane stem region of Y8 (Fig. 3). Surprisingly, these substitutions of lysine residues for the original non-polar amino acids make little impact on the assembly and functional properties of mtATPase subunit 8 _in vivo_ (Fig. 3; Papakonstantinou et al., 1991). This observation has raised a key question as to whether subunit 8 is correctly thought of as being anchored in the membrane via a transmembrane stem in order for its C-terminal region to be effective in ensuring correct assembly of the complex. The orientation of subunit 8 with the C-terminal moiety located on the matrix side of the inner mitochondrial membrane was established by chemical modification with the reagent isothionylacetimidate which specifically reacts with primary amines. It was found by Velours and Guerin (1986) that the affinity reagent bound to the ε amino group of the only lysine residue (LYS47) of Y8; this residue contains the only free amino group in Y8 since the N-terminal methionine residue of this polypeptide is blocked by N-formylation. This labelling of Y8 does not preclude a topology of Y8 such that the protein is only partially

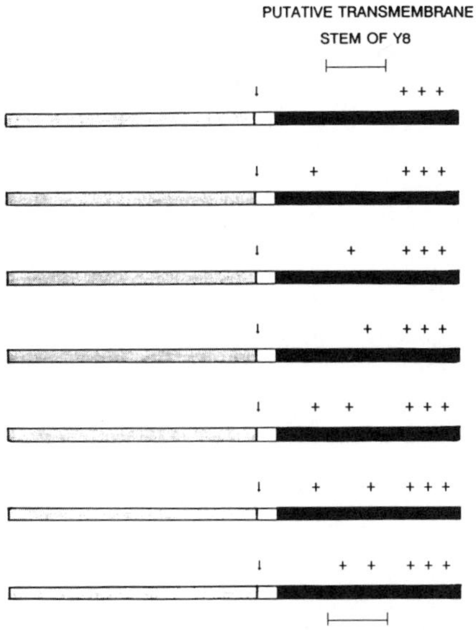

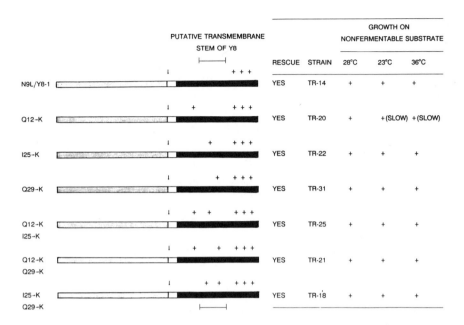

			GROWTH ON NONFERMENTABLE SUBSTRATE		
	RESCUE	STRAIN	28°C	23°C	36°C
N9L/Y8-1	YES	TR-14	+	+	+
Q12-K	YES	TR-20	+	+(SLOW)	+(SLOW)
I25-K	YES	TR-22	+	+	+
Q29-K	YES	TR-31	+	+	+
Q12-K I25-K	YES	TR-25	+	+	+
Q12-K Q29-K	YES	TR-21	+	+	+
I25-K Q29-K	YES	TR-18	+	+	+

Fig. 3 Chimaeric precursors of subunit 8 having extra positively charged residues. ▨, N9L; ■, Y8; □ bridge of 7 residues linking N9L and Y8 in these precursors of the Y8-1 configuration. The distribution of positively charged amino acids throughout the Y8 moeity in each precursor is indicated. Also indicated are the growth properties of strains deficient in mitochondrially encoded subunit 8 bearing genes encoding chimaeric subunit 8 precursors on a yeast expression vector. Data from Papakonstantinou et al. (1991).

embedded in the membrane or even located on the matrix surface of the inner-mitochondrial membrane in a hydrophobic pocket perhaps created by one or more subunits of the mtATPase complex. Hydropathy plots of Y8 variants bearing multiple positive charges show a considerable dimunition in the hydrophobic domain of Y8 considered till now to represent the transmembrane stem (data not shown). The key questions to be resolved by future work are whether subunit 8 obligatorily traverses the membrane, and at the same time, to elucidate the identity of the other mtATPase subunits interacting with the various domains of Y8 (N-terminal, central hydrophobic, and C-terminally charged regions).

ACKNOWLEDGEMENT: This work was supported by the Australian Research Council.

12

REFERENCES

Baker, K.P. and Schatz, G. (1991) Nature 349, 205-208.

Devenish, R.J., Bush, N.C. and Nagley, P. (1991) in preparation.

Farrell, L.B., Gearing, D.P. and Nagley, P. (1988) Eur. J. Biochem. 173, 131-137.

Galanis, M., Law, R.H.P., O'Keefe, L.M., Devenish, R.J. and Nagley, P. (1990) Biochem. Int. 22, 1059-1066.

Galanis, M., Devenish, R.J. and Nagley, P. (1991a) FEBS Lett. 282, 425-430.

Galanis, M., Devenish, R.J. and Nagley, P. (1991b) in preparation.

Gearing, D.P. and Nagley, P. (1986) EMBO J. 5, 3651-3655.

Grasso, D.G., Nero, D., Law, R.H.P., Devenish, R.J. and Nagley, P. (1991) Eur. J. Biochem. 199, 203-209.

Hadikusumo, R.G., Hertzog, P. and Marzuki, S. (1984) Biochem. Biophys. Acta 765, 257-267.

Hadikusumo, R.G., Meltzer, S., Choo, W.M., Jean-Francois, M.J.B., Linnane, A.W. and Marzuki, S. (1988) Biochim. Biophys. Acta 933, 212-222.

Hartl, F.-U. and Neupert, W. (1990) Science 247, 930-938.

Hendrick, J.P., Hodges, P.E. and Rosenberg, L.E. (1989) Proc. Natl. Acad. Sci. 86, 4056-4060.

Law, R.H.P., Farrell, L.B., Nero, D., Devenish, R.J. and Nagley, P. (1988) FEBS Lett. 236, 501-505.

Law, R.H.P., Devenish, R.J. and Nagley, P. (1990) Eur. J. Biochem. 188, 421-429.

Nagley, P. (1988) Trends Genet. 4, 46-52.

Nagley P., Devenish, R.J., Law, R.H.P., Maxwell, R.J., Nero, D. and Linnane, A.W. (1990) In: Bioenergetics: Molecular Biology, Biochemistry and Pathology (Kim, C.H. and Ozawa, T. eds) Plenun Press, New York, pp. 305-325.

Nagley, P. and Devenish, R.J. (1989) Trends Biochem. Sci. 14, 31-35.

Nero, D., Ekkel, S.M., Wang, L., Grasso, D.G. and Nagley, P. (1990) FEBS Lett. 270, 62-66.

Papakonstantinou, T., Nagley, P. and Devenish, R.J. (1991) in preparation.

Pfanner, N., Söllner, T. and Neupert, W. (1991) Trends Biochem. Sci. 16, 63-67.

Senior, A.E. (1988) Physiol. Rev. 68, 177-231.

Tzagoloff, A. (1990) J. Biol. Chem. 245, 1545-1551.

Velours, J. and Guerin, B. (1986) Biochem. Biophys. Res. Commun. 138, 78-86.

Viebrock, A., Perz, A. and Sebald, W. (1982) EMBO J. 1, 565-571.

Adenine Nucleotides in Cellular Energy Transfer and Signal Transduction
S. Papa, A. Azzi & J.M. Tager (eds)
© 1992 Birkhäuser Verlag, Basel/Switzerland

THE MITOCHONDRIAL H^+-ATP SYNTHASE: POLYPEPTIDE SUBUNITS, PROTON TRANSLOCATING AND COUPLING FUNCTION

Sergio Papa, Ferruccio Guerrieri, Franco Zanotti, Karlheinz Altendorf and Gabrielle Deckers-Hebestreit

Institute of Medical Biochemistry and Chemistry, Centre for the Study of Mitochondria and Energy Metabolism, C.N.R., Bari, Italy and Fachbereich Biologie Chemie der Universitat, Osnabruck, FRG

SUMMARY: The F_OF_1 H^+-ATP synthase of mitochondria is made up of 8 conserved subunits homologous to those of prokaryotes and 6 supernumerary subunits. The conserved subunits participate in the hydro-anhydro catalysis, H^+ translocation and energy coupling. This paper concerns the supernumerary subunits whose role is less better known.
Evidence will be presented showing that the supernumerary subunits: ATPase inhibitor protein (IF_1), oligomycin sensitivity conferral protein (OSCP) and F_6 contribute together with F_OI-PVP (thought by some authors to be analogous to subunit b of prokaryotes) and subunit γ of F_1 to the gate function of the pump. In particular the structure-function relationship of F_OI-PVP, its role in the binding of F_1 to F_O, modulation of transmembrane H^+ conduction and its sensitivity to specific inhibitors of the mitochondrial ATP synthase are examined.

INTRODUCTION

The H^+-ATP synthase of coupling membranes (mitochondria plasma membrane of bacteria, thylakoid membrane of chloropasts) is an oligomeric, membrane-bound enzyme that catalyzes ATP synthesis, driven by transmembrane $\Delta\mu H^+$, in aerobic and photosynthetic cells. Like other energy-transfer proteins the H^+-ATP synthase has acquired with evolution supernumerary subunits, [Senior (1988); Papa (1989)]. The enzyme, which in bacteria is

14

composed of eight subunits [Senior (1988)] , in mammalian mitochondria has fourteen subunits. The functional meaning of the additional polypeptides is unknown. They could contribute to transmembrane H^+ translocation, which in prokaryotes is operated by 3 subunits. The supernumerary subunits could be involved in regulation of the enzyme so to adapt its activity to the energy demand, changing, in eukaryotic cells, with their physiological state. Finally supernumerary subunits may act as chaperones [Ellis et al (1991)] for membrane assembly of the enzyme.

TABLE I - SUBUNIT COMPOSITION OF THE F_O, MEMBRANE SECTOR OF H^+-ATP SYNTHASE. Sequences from Walker et al (1991, #1987) and Fearnley et al (*1986) (°see also Zanotti et al 1988).

F_O-E. Coli	F_O-BEEF HEART	
Subunits	Subunits	Sequence
	b	°PVPPLPEHGGY......
a	OSCP	FAKLV...........
	d	#MAGKKIAIKT.......
	a	*MNENLFTSFI.......
b	e	VPPVQVSPLIKL.....
	IF_1	UKELD...........
	F_6	NKELD...........
	A_6L	MPQLD...........
c	c	DIDTAAKFIG.......

The subunits of the H^+-ATP synthase are organized in three sectors (Fig. 1): 1. the catalytic sector or F_1, universally consisting of five subunits; 2. the H^+ translocating, membrane integral F_O sector, which consists of a variable number of subunits (Table I); 3. the stalk, which connects the first two parts and is made up by some F_1 and F_O subunits. The stalk is involved in the coupling and gate function of the H^+-ATP synthase

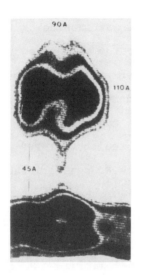

Fig. 1 - THE STRUCTURE OF F_OF_1 ATP SYNTHASE. From Gogol et al (1987).

and appear as the part of the enzyme which has undergone more extensive evolution in eukaryotes.

The present paper is focused on F_O and F_1 subunits which contribute to the coupling and regulatory function in the H^+-ATP synthase complex of mitochondria. Three experimental approaches were used in studying these subunits: (i) enzymatic proteolytic digestion of the complex, characterization of the products and functional analysis of the digested enzyme; (ii) functional analysis of heterologous complexes obtained by cross-reconstitution of F_1 and F_O sectors from bovine-heart mitochondria and E. Coli membranes; (iii) chemical modification of aminoacid residues and its functional impact.

RESULTS AND DISCUSSION

Proteolytic cleavage of subunits.

In "inside out particles" of the inner mitochondrial membrane the

16

F_1 sector, located at the matrix side of the membrane, becomes exposed to the external medium. After removal of F_1 by urea (USMP particles), peripheral segments of F_O subunits, contributing to the binding of F_1, can be digested by proteolytic enzymes [Houstek et al (1988); Zanotti et al (1988)]. Under these conditions trypsin digested the F_OI-PVP polypeptide (which was cleaved to a fragment few Kd smaller) and OSCP [Houstek et al (1988); Zanotti et al (1988); Papa et al (1989)] (F_6 was likely also digested). Addition of soluble F_1 to untreated USMP reconstitutes a membrane associated ATPase activity which is fully inhibited, like the native F_OF_1 complex, by oligomycin.

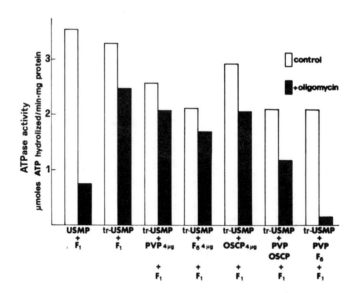

Fig. 2 - ATPase ACTIVITY IN TRYPSINIZED USMP RECONSTITUTED WITH PURIFIED F_1. For trypsin digestion (50µg/mg USMP protein), reconstitution of ATPase activity and other experimental conditions see Houstek et al (1988). Where indicated, after trypsin digestion, isolated F_OI-PVP or OSCP or F_6 were added before F_1. Solid columms: + oligomycin (2µg/mg particle protein).

Oligomycin binds to the F_O sector and inhibits directly H^+ conduction [Pansini et al (1978)]; extension of the inhibitory effect to ATP hydrolysis depends on a functionally correct binding of F_1 to F_O. Correct binding of F_1 to F_O was lost after digestion of USMP by trypsin, as shown by the loss of oligomycin sensitivity of the ATPase activity of reconstituted F_1 (Fig.2). Oligomycin sensitivity of ATP hydrolysis could, now, be restored by adding, to digested USMP reconstituted with soluble F_1, F_OI-PVP and OSCP or F_OI-PVP and F_6 together. Each one of these proteins, added alone, was practically uneffective in restoring oligomycin sensitivity.

Direct aminoacid sequence of the proteolytic product of F_OI-PVP, which remained associated with F_O in the membrane, revealed that digestion took place from the carboxyl end (M214) up to lysine residues preceeding the single cysteine present in the protein at position 197 (Table II) [(Papa et al (1989)]. Thus a short carboxyl-terminal segment of the F_OI,PVP protein extends out of the membrane at the matrix side and contributes to the correct binding of F_1.

TABLE II - PROTEIN SEQUENCE OF PVP-F_OI PROTEIN OF BOVINE MITOCHONDRIAL ATP SYNTHASE. From Walker et al (1991).

10	194
Pro-Val-Pro-Pro-Leu-Pro-Gln-Asn-Gly-Gly......Ile-Ala-Lys-**Cys**-Ile
200	210	214
Ala-Asp-Leu-Lys-Leu-Leu-Ser-Lys-Lys-Ala-Gln-Ala-Gln-Pro-Val-Met

Other experiments showed that digestion of this segment of the F_OI-PVP protein resulted also in inhibition of H^+ conduction by F_O (Fig.3). The residual proton conductivity was oligomycin

18

insensitive. H$^+$ conduction and its oligomycin sensitivity could be restored by adding back the isolated native F$_0$I-PVP protein to digested F$_0$ (Fig.3).

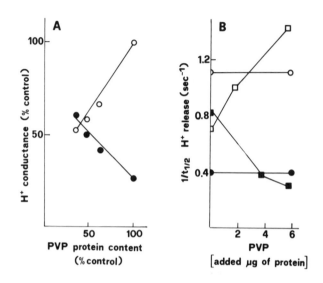

Fig. 3 - RELATIONSHIP BETWEEN H$^+$ CONDUCTION IN USMP AND THE CONTENT OF F$_0$I-PVP PROTEIN IN USMP (panel A) OR LIPOSOMES RECONSTITUTED WITH F$_0$ EXTRACTED FROM TRYPSIN TREATED USMP (panel B). For experimental conditions see Zanotti et al (1988).
Panel A: (O—O) control; (●—●) oligomycin (1µg/mg particle protein.
Panel B: (O—O) F$_0$ extracted from USMP; (□—□) F$_0$ extracted from USMP treated with trypsin (50µg/mg protein); (●—●) F$_0$ extracted from untreated USMP + oligomycin (2µg/mg protein); (■—■) F$_0$ extracted from USMP treated with trypsin (50µg/mg protein) +oligomycin (2µg/mg protein) from Zanotti et al (1988).

Thus the peripheral extension of the F$_0$I-PVP protein, which protrudes out of the matrix side of the membrane and is attached to F$_1$, promotes transmembrane H$^+$ conduction by F$_0$ and its inhibition by oligomycin. It was observed that proteolytic removal of the carboxyl-terminal region of F$_0$I-PVP protein caused also loss of the sensitivity of H$^+$ conduction to DCCD; the binding of this reagent to its specific target in F_O (Glu 97 in subunit c) was, however, unaffected [Guerrieri et al (1991)].

$\underline{F_1-F_O\ hybrydis.}$

Fig. 4 shows that soluble F_1 isolated from E. Coli (EcF$_1$) can reconstitute not only with F_1-depleted E. Coli membranes (UPEc) but also with F_1-depleted submitochondrial vesicles (USMP). The

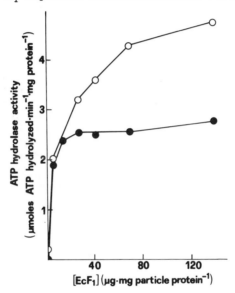

Fig. 4 -RECONSTITUTION OF ATPase ACTIVITY IN UPEc AND USMP BY ADDITION OF PURIFIED EcF$_1$. For the reconstitution of the ATPase activity the membrane particles were incubated with the reported concentration of EcF$_1$ as reported in Guerrieri et al (1989). Symbols: (●——●) UPEc, (O——O) USMP.

binding of EcF$_1$ to UPEc (followed by measuring the ATPase activity appearing in the sedimented membranes) followed saturation kinetics. Half-maximal ATP hydrolase activity could be restored with the addition of about 4 μg EcF$_1$ per mg of membrane protein. USMP also bound EcF$_1$ with saturation kinetics. The reconstituted activity in this case was higher, but the affinity lower. Half-maximal restoration of ATP hydrolase activity required the addition of about 11 µg EcF_1 per mg particles protein. When the reconstituted ATP hydrolase was tested for its sensitivity to oligomycin it could be observed that the activity was sensitive to oligomycin when either EcF_1 or BF_1 were reconstituted with USMP but it was completely insensitive to this inhibitor when EcF_1 or BF_1 were reconstituted with UPEc (Table III).

TABLE III - SENSITIVITY OF RECONSTITUTED ATP HYDROLASE ACTIVITY TO OLIGOMYCIN. For reconstitution experiments, membrane particles were incubated as reported in the legend of Fig. 4 with 120µg of F_1/mg particle protein. The reconstituted particles were preincubated 5min with oligomycin (15µg/mg particle protein)

	ATPase ACTIVITY		
	$\dfrac{\mu mol\ ATP\ hydrolyzed}{min \cdot mg\ particle\ protein}$		
	−	+Oligomycin	%
USMP	0.20	0.06	70
UPEc	−	−	−
USMP+BF_1	5.36	1.76	67
UPEc+EcF_1	2.33	2.33	−
USMP+EcF_1	4.76	1.96	59
UPEc+BF_1	2.68	2.68	−

The addition to the UPEc-BF_1 hybrid of OSCP, F_O I-PVP and F_6, each alone, did not induce significant oligomycin sensitivity of the reconstituted ATP hydrolase activity (Fig. 5). However the addition of the three subunits together restored oligomycin

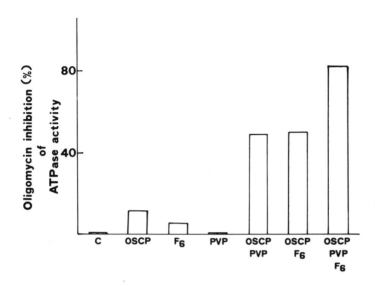

Fig. 5 - OLIGOMYCIN SENSITIVITY OF THE HYDROLASE ACTIVITY OF THE HYBRID FORMED BY UPEc AND BF$_1$. Reconstitution of ATPase activity of UPEc + BF$_1$ was carried out as reported in the legend to Fig. 4. Where indicate isolated OSCP (2μg/mg UP$_{Ec}$ protein), F$_0$I-PVP (4μg/mg UPEc protein), F$_6$ (4 μg/mg UPEc protein) or combinations of same amounts of these proteins were incubated with UPEc for 10 min at room temperature before addition of BF$_1$. OSCP and F$_0$I-PVP were purified as reported in Zanotti et al (1988). F$_6$ was purified as described by Kanner et al (1976). Oligomycin (2μg/mg UPEc protein) was added 5 min before addition of ATP.

sensitivity.

Fig. 6 shows that the valinomycin + K$^+$ induced passive proton translocation in UPEc was, like in USMP, inhibited by oligomycin. Whilst 0.25 μg of oligomycin/mg particle protein were, however, sufficient to cause 50% inhibition of H$^+$ translocation in USMP, 2 μg of oligomycin/mg particle protein were required to obtain 50% inhibition in UPEc. The addition of F$_0$I-PVP to UPEc enhanced the sensitivity of proton conduction to oligomycin.

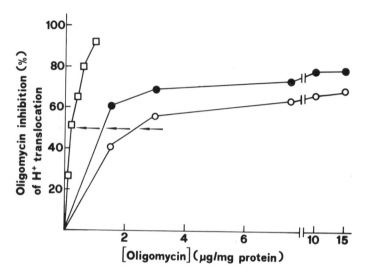

Fig. 6 - TITRATION OF THE INHIBITORY EFFECT OF OLIGOMYCIN ON H⁺ CONDUCTION IN USMP AND UPEc. For measurement of proton conduction mediated by valinomycin pulses (2 μg/mg) see Zanotti et al (1987). Symbols: (O——O) UPEc (1 mg protein/ml) or (☐——☐) USMP (1 mg protein/ml) were incubated for 5 min in KCl (150 mM) in presence or in the absence of the reported concentrations of oligomycin. (●——●) UPEc (1 mg protein/ml) were incubated with F_oI-PVP (4 μg/mg UPEc protein) for 10 min before addition of oligomycin at the concentration reported in the figure. The arrows indicate 50% inhibition.

Chemical modification of residues

Treatment of "inside out" submitochondrial particles (containing both F_O and F_1) with diamide (which oxidizes vicinal dithiols to disulphide bridge), [Kosower et al (1972)] butanedione (which modifies arginine residues) [Riordan (1973)] or ethoxyformic anhydride (a reagent for hystidine residues) [Miles (1977)], caused a dramatic, oligomycin sensitive, enhancement of H⁺ conduction (Fig. 7). This enhancement of H⁺ conduction effected by the three reagents, is symptomatic of opening of the gate of

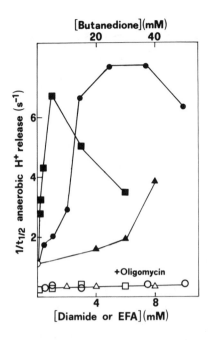

Fig. 7 - TITRATION OF THE EFFECT OF CHEMICAL MODIFIERS OF AMINOACIDS ON H+ CONDUCTION IN ESMP. For experimental conditions and treatment with butanedione see Guerrieri and Papa (1981); with ethoxyformic anhydride (EFA) see Guerrieri et al (1987a) and with diamide see Zanotti et al (1985) .
Symbols: (■——■) EFA; (●——●) diamide; (▲——▲) butanedione.

the H+ channel. To identify the subunits modified by the reagents, particles with various degree of resolution of the $F_0 F_1$ complex were used (Table IV). No stimulatory effect by ethoxyformic anhydride could be observed in particles from which the ATPase inhibitor protein had been removed by sephadex chromatography [Guerrieri et al (1987a)]. The isolated inhibitor protein can be added back to depleted particles and its inhibitory activity on ATP hydrolisis [Guerrieri et al (1987b)] and on proton translocation restored (Fig.8). When the isolated inhibitor protein was treated with ethoxyformic anhydride, its inhibitory

TABLE IV - EFFECT OF AMINOACID REAGENTS ON H^+ CONDUCTION IN SUBMITOCHONDRIAL PARTICLES WITH VARIOUS DEGREES OF RESOLUTION OF $F_O F_1$ COMPLEX. For preincubation with aminoacid reagents and measurement of proton conduction see legend of Fig. 7.

	Anaerobic H^+ release $1/t_{\frac{1}{2}}$ (s^{-1})		
	ESMP	Sephadex	USMP
Control	1.0	2.0	2.0
Diamide 5mM	7.0	-	6.6
EFA 1mM	7.0	2.0	1.7
Butanedione 20 mM	3.0	4.0	5.0

activity was abolished (Fig. 8) [Guerrieri et al (1987a)]. It was shown that under these conditions one out of four hystidine residues in the inhibitor protein was modified by ethoxyformic anhydride. These observations show that a critical hystidine residue in the inhibitor protein is essential for its activity. The protein, being active in modulating both hydro-anhydro catalysis in F_1 and transmembrane H^+ translocation by F_O, qualifies itself as a component of the gate.

The stimulatory effect of diamide and butanedione were preserved in particles deprived of the inhibitor protein as well as of F_1 (Table IV). The following finding allowed to identify the subunits whose modification by diamide results in enhancement of H^+ conduction. The stimulatory effect exerted by diamide on proton translocation in USMP was lost in reconstituted F_O-liposomes (Fig. 9). Immunoblot analysis revealed that subunits of F_1, still present in traces in USMP, were absent in the isolated

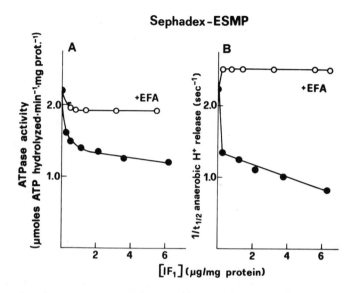

Fig. 8 - EFFECT OF EFA TREATMENT OF PURIFIED F_1 INHIBITOR PROTEIN ON ITS INHIBITION OF ATPase ACTIVITY AND ANAEROBIC RELEASE OF RESPIRATORY PROTON GRADIENT IN SEPHADEX-EDTA-TREATED SUBMITOCHONDRIAL PARTICLES. ATPase activity (A); anaerobic release of proton gradient (B).
Symbols: (●——●), F_1 inhibitor protein; (O——O), F_1 inhibitor protein modified by treatment with 0.5 mM EFA. From Guerrieri et al (1987b).

F_O. The enhancement of H^+ conduction caused in USMP by increasing concentrations of diamide was associated with a progressive decrease of the immunodetected bands of the F_OI-PVP protein and of the γ subunit (Fig. 9). No change in the band of F_OI-PVP was, on the other hand, observed after treatment of F_O-liposomes with an excess of diamide.

The enhancement of H^+ conduction, induced in USMP by diamide treatment, was directly correlated with the decrease of the antigenic bands of the γ subunit and of the F_OI-PVP protein (Fig. 10).

The experiment in Fig. 11 shows that the addition of

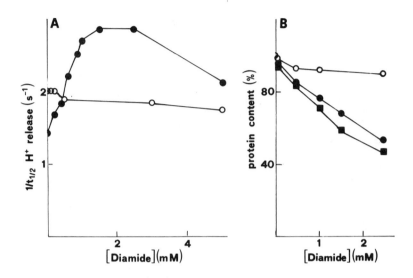

Fig. 9 - TITRATION OF THE EFFECT OF DIAMIDE ON H⁺ CONDUCTION   AND
IMMUNOBLOT   ANALYSIS   OF BANDS OF F_OI-PVP AND γ IN USMP AND F_O-
LIPOSOMES. For experimental conditions see Papa et al (1990).
Panel A: USMP (●——●)   or F_O-liposomes (O——O) were preincubated
for   2   min with diamide at the concentrations   reported   in   the
figure before a measurement of H⁺ conduction
Panel   B: Immunodetected F_OI-PVP   band   in   USMP (●——●) or F_O-
liposomes (O——O) and immunodetected γ band in USMP (■——■).

increasing concentrations of purified γ subunit to F_O-liposomes

caused a   progressive, marked inhibition of passive H⁺ conduction

(cf.[Papa et al (1990)]). Treatment of F_O-liposomes with diamide,

which   caused per se   about   40% inhibition of H⁺   conduction   in

controls,   dramatically   enhanced it in the γ-supplemented F_O-

liposomes (also   in   this   case   H⁺   conduction   was   oligomycin

sensitive).   Immunoblot   analysis (Fig. 11) showed   that   diamide

treatment,   which did not affect the F_OI-PVP band in control F_O-

liposomes,   produced,   when   purified γ subunit   was   added,   a

decrease of the bands of both the F_OI-PVP and of γ subunit.

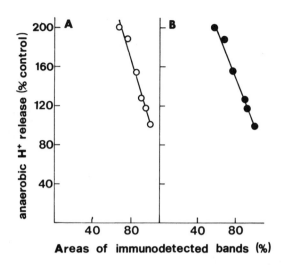

Fig. 10 - RELATIONSHIP BETWEEN % OF ANAEROBIC H$^+$ RELEASE AND AMOUNT OF IMMUNOREACTIVE BANDS F_OI-PVP AND γ PROTEIN AFTER DIAMIDE TREATMENT OF USMP. Immunoblot analysis with a specific serum anti F_OI-PVP (A) or anti-F_1 (B). Semiquantitative analysis was carried out by densitometry of the nitrocellulose sheets at 590nm. Reproduced from Papa et al (1990).
Symbols: (O—O) immunodetected F_OI-PVP band; (●—●) immunodected γ band.

These observations provide evidence showing that in the mitochondrial H$^+$-ATP synthase the γ subunit of F_1 functions as a component of the gate for the H$^+$ channel in F_O. There are three regions rather well conserved in the γ subunit, i.e. those from 18-28, 88-92 and 251-299 [Walker et al (1985)]. The carboxyl-region from 251-299 is relatively rich in acidic and basic residues [Walker et al (1985)]. It is possible that this segment of the γ subunit is in close contact with the carboxyl-terminal region of the F_OI-PVP subunit (Fig. 11). These segments of the two proteins can be supposed to interact through salt-

28

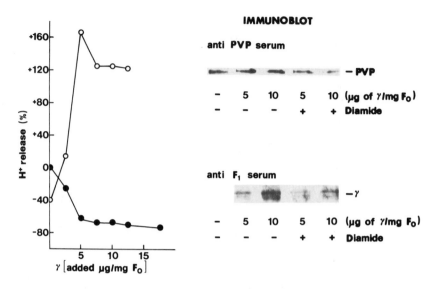

Fig. 11 - CONTROL OF H^+ CONDUCTION IN F_O-LIPOSOMES BY γ SUBUNIT. Symbols: (●——●) isolated γ subunit was added, at the concentrations reported in the figure, to F_O-liposomes. (O——O) F_O-liposomes reconstituted with γ subunit, at the concentration reported in the figure, were treated with diamide (2.5 mM) for 2 min. Reproduced from Papa et al (1990).

bridges, involved in the controlled transfer of H^+ from the catalytic domain in F_1 to the F_O channel. The conserved cysteine-91 in the subunit γ and the cysteine-197 in the F_OI-PVP protein are apparently in vicinal position as indicated by the observation that diamide induces disulphide bridging of the two subunits. This modification apparently alters the gate of the proton channel with dramatic enhancement of H^+ conduction.

CONCLUSIONS

The observations presented provide evidence indicating that the carboxyl-terminal region of the F_OI-PVP protein, residues

214-197, a segment around Cys 91 of the γ subunit, OSCP and F_6 contribute to the stalk of the mitochondrial F_O-F_1, H^+-ATP synthase. F_6, OSCP and the carboxyl-terminal region of F_OI-PVP appear, in fact, essential for the correct functional binding of F_1 to F_O, being all of them necessary for oligomycin sensitivity of the hydrolase activity of the $F_O F_1$ complex. The inhibitor protein, which modulates the catalytic activity of F_1 as well as H^+ conduction by F_O, may also be involved in the functional connection of F_1 with F_O.

Disulfide cross-linking of the F_OI-PVP protein with the γ subunit results in dramatic enhancement of proton conduction by F_O. Removal of F_1 from F_O results in an enhancement of H^+ conduction by F_O which can be depressed by the addition of the isolated γ subunit.

$F_1 F_O$-H^+ATPase

SCHEME - SCHEME PROPOSED FOR THE EFFECT OF DIAMIDE ON ATPase COMPLEX.

The carboxyl-terminal region of $F_O I$-PVP, which seems to be in close contact with the γ subunit (Fig. 11 and scheme), modulates transmembrane H^+ conduction by F_O and its sensitivity to oligomycin. Thus the $F_O I$-PVP protein appears to be an essential component for the functional organization of the transmembrane H^+ channel in F_O.

On the basis of the resemblance of the respective hydropathy plots and possibly and of their three-dimensional structure, Walker et al (1987) have proposed that the mitochondrial $F_O I$-PVP protein is analogous to the b subunit of E. Coli. It should however be noted that there doesn't exist any direct identity in the residue sequences of the bacterial b subunit and mitochondrial $F_O I$-PVP subunit (Table Va). On the other hand the bacterial sequences of the b subunit, exhibit among them a

TABLE Va - HOMOLOGIES OF PROKARYOTES AND EUKARYOTES F_O-SUBUNITS. From National Italian Node EMBnet (C.N.R. Bari).

	c	SUBUNITS a	b ($F_O F_1$-PVP)
		(% identity)	
Escherichia Coli	100	100	100
Vibrio Algynoliticus	49	62	72
Thermophile Bacterium PS3	44	30	28
Human	25	28	--
Bovine	25	24	0
Yeast	29	20	0
Rat	--	18	0

significant sequence identity. The same holds for the mitochondrial F_OI-PVP proteins which, among them, also show significant sequence identity (Table Vb). This identity is of the

TABLE Vb - HOMOLOGIES OF EUKARYOTIC F_O-SUBUNITS. From National Italian Node EMBnet (C.N.R. Bari).

	c	SUBUNITS a	b ($F_O F_1$-PVP)
		(% identity)	
Bovine	100	100	100
Human	97	78	--
Yeast	67	32	23
Rat	--	77	84

same order as that exhibited by the c and a subunit in prokaryotes and eukaryotes. Thus although the F_OI-PVP seems to play an important functional role in the mitochondrial $F_O F_1$-ATP complex, it may not be related to the b subunit of E. Coli, unless in the course of evolution the protein has acquired additional segments and functions.

REFERENCES

Ellis,R.J. and Saskia van der Vies, M. (1991) Annu. Rev. Biochem. 60, 321-347.
Fearnley, I.M. and Walker, J.E. (1986) EMBO Journal 5, 2003,2008.
Gogol, E.P., Lucken, U. and Capaldi, R.A. (1987) FEBS Lett. 219, 274-278.
Guerrieri, F. and Papa, S. (1981) Journal Bioenergetics and Biomembranes 13, 393-409.
Guerrieri, F., Zanotti, F., Che, Y.W., Scarfò, R. and Papa, S. (1987a) Biochimica Biophysica Acta 892, 284-293.
Guerrieri, F., Scarfò, R., Zanotti, F., Che, Y.W. and Papa, S. (1987b) FEBS Lett. 213, 67-72.
Guerrieri, F., Capozza, G., Houstek, J., Zanotti, F., Colaianni, G., Jirillo, E. and Papa, S. (1989) FEBS Lett. 250, 60-66.

Guerrieri, F., Zanotti, F., Capozza, G., Colaianni, G., Ronchi, S. and Papa, S. (1991) Biochim. Biophys. Acta 1059, 348-354.

Houstek, J., Kopecky, J., Zanotti, F., Guerrieri, F., Jirillo, E., Capozza, G. and Papa, S. (1988) Eur J. Biochem 173, 1-8.

Kanner, B.I., Serrano, M., Kandrach, M.A. and Racher, E. (1976) Biochem. Biophys. Res. Commun. 69, 1050-1056.

Kosower, E.M., Corea, W., Kinon, B.J. and Kosower, W.S. (1972) Biochim. Biophys. Acta 264, 39-44.

Miles, E.W. (1977) Methods Enzymology 47, 431-442.

Pansini, A., Guerrieri, F. and Papa, S. (1978) Eur. J. Biochem. 92, 545-551.

Papa, S. (1989) in: "Organelles of Eukaryotic Cells: Molecular, Structure and Interactions (Tager, J.M., Guerrieri, F., Azzi, A. and Papa, S., eds) Plenum Publ. Co. New York and London, pp. 9-26.

Papa, S., Guerrieri, F., Zanotti, F., Houstek, J., Capozza, G. and Ronchi, S. (1989) FEBS Lett. 249, pp. 62-66.

Papa, S., Guerrieri, F., Zanotti, F., Fiermonte, M., Capozza, G. and Jirillo, E. (1990) FEBS Lett. 272, 117-120.

Riordan, J.F. (1973) Biochemistry 12, 3915-3923.

Senior, A.E. (1988) Physiol. Rev. 68, 177-231.

Walker, J.E., Fearnley, J.M., Gay, N.J., Gibson, B.W., Northrop, F.D. Powell, S.J., Runswick, M.J., Sarasta, M. and Tybulewicz, V.L.J. (1985) J. Mol. Biol. 184, 677-701.

Walker, J.E., Runswick, M.J. and Poulter, L. (1987) J. Mol. Biol. 197, 89-100.

Walker, J.E., Lutter, R., Dupuis, A. and Runswich, J. (1991) Biochemistry 30, 5369-5378.

Zanotti, F., Guerrieri, F., Scarfò, R., Berden, J. and Papa, S. (1985) Biochem. Biophys. Res. Commun. 132, 985-990.

Zanotti, F., Guerrieri, F., Che, Y.W., Scarfò, R. and Papa, S. (1987) Eur. J. Biochem. 164, 517-523.

Zanotti, F., Guerrieri, F., Capozza, G., Houstek, J., Ronchi, S. and Papa S. (1988) FEBS Lett. 237, 9-14.

Adenine Nucleotides in Cellular Energy Transfer and Signal Transduction
S. Papa, A. Azzi & J.M. Tager (eds)
© 1992 Birkhäuser Verlag, Basel/Switzerland

STUDIES ON THE STRUCTURE OF THE MITOCHONDRIAL ATP SYNTHASE COMPLEX AND THE MECHANISM OF ATP SYNTHESIS

Youssef Hatefi and Akemi Matsuno-Yagi

Division of Biochemistry, Department of Molecular and Experimental Medicine, The Scripps Research Institute, La Jolla, California 92037

SUMMARY: Described in this article are the membrane topography of subunits b, d, F_6, OSCP and A6L of the bovine mitochondrial ATP synthase complex, and the stoichiometry of these subunits as well as of the ATPase inhibitor protein. Also described are findings regarding the regulation of the kinetics of ATP synthesis by the respiratory chain, the turnover capacity of the bovine mitochondrial ATP synthase complex, and cooperativity in ATP synthesis. The latter involves the binding of ADP to 3 exchangeable sites before rapid ATP synthesis takes place.

The mitochondrial ATP synthase complex is a heteropolymeric enzyme composed of 13 unlike subunits, several of which occur in the enzyme complex in multiple copies. Of particular interest in this regard is that each ATP synthase molecule contains 3 potential catalytic sites. This structural complexity is also reflected in the mechanism of action of the ATP synthase complex. Both energy transduction and transfer by the membrane sector of the enzyme and ATP synthesis and hydrolysis by its catalytic sector are complex processes, which we have only recently begun to understand in some detail. However, since space limitation compels brevity, our remarks will be confined here to certain aspects of the structure of the ATP synthase complex and the mechanism of ATP synthesis.

MEMBRANE TOPOGRAPHY AND STOICHIOMETRY OF THE SUBUNITS OF THE
MITOCHONDRIAL ATP SYNTHASE COMPLX

Purified bovine ATP synthase complex, which catalyzes
oligomycin-sensitive ATP hydrolysis and ATP-^{32}Pi exchange, is
composed of 13 unlike subunits. These are the $\alpha,\beta,\gamma,\delta$ and ϵ
subunits of the catalytic sector F_1; OSCP, F_6, DCCD-binding
protein (subunit c), A6L (subunit 8) and subunits 6, b and d,
which together comprise the membrane sector F_o and the stalk;
and the ATPase inhibitor protein (IF_1). In the bovine ATP
synthase preparations of Walker et al. (1991) there is an
additional polypeptide, which they have designated as subunit e.
All the subunits of our ATP synthase complex, except A6L, were
purified and polyclonal antibodies were raised to each (Hekman &
Hatefi, 1991; Hekman et al., 1991). For A6L two antipeptide
antibodies were raised to synthetic peptides corresponding to
residues 30-43 and 54-66 of A6L (Hekman et al., 1991).
Antibodies to b, c, d, OSCP, F_6, subunit 6 and α/β reacted with
similar antigens in human and rat mitochondria. Only anti-α/β
and anti-subunit 6 cross-reacted with antigens in yeast
mitochondria and in *Paracoccus denitrificans* and *Escherichia
coli* membranes (Hekman & Hatefi, 1991). In bovine, human and
rat mitochondria and in *P. denitrificans* and *E. coli* membranes
there were two proteins of different M_r that reacted with anti-
subunit 6, but in purified bovine ATP synthase complex, yeast
mitochondria and *E. coli* F_oF_1 only one protein reacted (Hekman &
Hatefi, 1991). The nature of the second antigen, which in
bovine mitochondria exhibits a M_r of 26 kDa and is not a
component of the ATP sythase complex, remains to be determined.

Study of the membrane topography of F_o subunits in SMP
(inside-out inner membrane vesicles) and mitoplasts
(mitochondria denuded of outer membrane) showed that b, d, F_6,
OSCP and A6L (including its C-terminal end), but not c and
subunit 6, were accessible on the matrix side to antibodies and

proteolytic enzymes. Sufficient masses of these subunits to recognize antibodies or undergo proteolysis by several different proteases were not exposed on the cytosolic side of the mitochondrial inner membrane. The stoichiometries of several F_0 subunits in SMP were determined by immunoblotting and [^{125}I] protein A binding for quantitation. Per mol F_1, there were in SMP 1 mol each of d, OSCP and IF_1 (as determined in intact mitochondria), and 2 mol each of b and F_6. The stoichiometries of c and subunit 6 could not be determined because subunit 6

Table I

Bovine Mitochondrial ATP Synthase Complex
Subunit Composition and Stoichiometry

Subunit	Molecular Weight	mol/mol of enzyme
α	55,164	3
β	51,595	3
γ	30,141	1
δ	15,065	1
ϵ	5,652	1
6*	24,816	?
b	24,668	2
OSCP	20,967	1
d	18,603	1
F_6	9,006	2
A6L*	7,965	1
c	7,402	?
IF_1	9,578	1

*mtDNA encoded.

transferred poorly to nitrocellulose and antibody to c did not bind protein A. Data regarding molecular weights and stoichiometry of ATP synthase subunits are summarized in Table I, in which the value for the stoichiometry of A6L is from Muraguchi et al. (1990). It should also be mentioned that Penin et al. (1985) have suggested that in pig heart mitochondria there are 2 copies of OSCP per mol of F_1, while Dupuis et al. (1985) have demonstrated with the use of OSCP labeled with $[^{14}C]N$-ethylmaleimide that bovine F_1 has one high-affinity binding site (K_d = 0.08 μM) and two low-affinity binding sites (K_d = 6-8 μM) for OSCP.

KINETIC ASPECTS OF THE MITOCHONDRIAL MECHANISM OF ATP SYNTHESIS

During the past decade, studies on the mechanism of ATP hydrolysis by a number of laboratories have shown that isolated F_1-ATPase catalyzes unisite and trisite ATP hydrolysis. At [ATP] < [F_1], ATP binds very tightly (K_a = 10^{12} M^{-1}) to a single catalytic site per F_1, undergoes reversible hydrolysis and resynthesis on the surface of the enzyme, and leads to slow product release at reported rates of 4 x 10^{-3} s^{-1} (Grubmeyer et al., 1982; Cross et al., 1982; Cunningham & Cross, 1988) or 10-20 times faster (Milgrom & Murataliev, 1987; Fromme & Gräber, 1990; Berden et al., 1991). Addition of ADP and Pi to F_1 also results in enzyme-bound ATP synthesis, as expected from the reversible unisite reaction described (Feldman & Sigman, 1982; Sakamoto & Tonomura, 1983; Yoshida, 1983). At [ATP] >> [F_1], trisite ATP hydrolysis takes place with a turnover number of 500-600 s^{-1} (Cross et al., 1982; Wong et al., 1984). Trisite ATP hydrolysis involves negative cooperativity with respect to [ATP], and positive catalytic cooperativity in the sense that ATP binding to the second and third catalytic sites of F_1 increases the rate of ATP hydrolysis at the first site by 10^4-10^5 fold (Cunningham & Cross, 1988).

These findings prompted us to begin a thorough study of the mechanism of ATP synthesis. Prerequisite for these studies was availability of an enzyme preparation which was stable, reproducible, well-coupled and highly active. Minor modifications of the procedure of Hansen and Smith (1964) provided a preparation of bovine SMP, which was stable at -70°C for >6 months and catalyzed ATP synthesis at rates of 2500-3000 nmol(min. mg of protein)$^{-1}$, an activity that was about one order of magnitude greater than those found in the literature. It became clear later that without such highly active SMP many of the important findings described below would not have been possible.

(a) Regulation of the kinetics of ATP synthesis: It was first found that, contrary to the general experience, kinetic plots of ATP synthesis at 1-1200 μM ADP were not monophasic, but curvilinear and analyzable in terms of at least 3 apparent K_m^{ADP} values. Variable K_m values were also found for Pi. However, unlike ADP, Pi could not be regenerated in the reaction mixture to allow studies of Pi effect below 0.1 mM. At first glance, the curvilinear Eadie-Hofstee plots suggested negative cooperativity with respect to [ADP] and [Pi], but it was soon found that the apparent K_m changes were the result of modulation of the kinetics of ATP synthesis by the coupled rate of respiration (Matsuno-Yagi & Hatefi, 1985; Hekman et al., 1988). At low and high rates of energy production by the respiratory chain, Eadie-Hofstee plots of the kinetics of ATP synthesis at 1-1200 μM ADP were essentially linear, with a single low K_m^{ADP} of 2-4 μM at low membrane potential ($\Delta\psi$) and a single high K_m^{ADP} of 120-160 μM at high $\Delta\psi$, while at intermediate $\Delta\psi$ levels the plots were curvilinear involving the low and the high K_m^{ADP} components mentioned plus intermediate, variable K_m^{ADP} values (Fig. 1). Further scrutiny showed that the transition from low to high apparent K_m^{ADP} was accompanied by the same degree of increase ($\sim$50 fold) in apparent V_{max} for ATP synthesis (Fig. 1 inset). Thus, it became clear that these transitions occurred at constant V_{max}/K_m, indicating that the K_m changes were due mainly to k_{cat} changes. Literature data indicated that similar V_{max} and K_m changes occur in photophosphorylation as a function of light intensity or duration (Vinkler, 1981; Stroop & Boyer, 1987; Junge, 1987; Quick & Mills, 1987; Quick & Mills, 1988; Bizouarn et al., 1991), and possibly even in whole cells as the rate of oxygen consumption is varied with the use of different substrates (From et al., 1990).

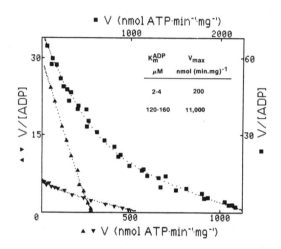

Fig. 1   Effect of attenuation of respiration rate and fractional inactivation of the ATP synthase complexes of SMP on the kinetics of ATP synthesis at 1-1,200 μM ADP. The rate of ATP synthesis was measured at 30°C. Respiratory substrates were 0.5 mM NADH (■,▼) or 30 mM DL-β-hydroxybutyrate plus 1 mM NAD (▲). In ▼, the ATP synthase complexes of SMP were fractionally inactivated by incubating the particles with 20 μM DCCD at 0°C for 210 min. Inhibition of ATPase activity by this treatment was 94% The dots represent the curves calculated from the K_m and V_{max} values obtained by computer-assisted curve fitting of the data points (symbols). The inset shows apparent K_m^{ADP} and V_{max} values at low (upper line) and high (lower line) membrane potential. The value of 11,000 nmol (min. mg)$^{-1}$ was calculated as described in part (b,ii). From Matsuno-Yagi & Hatefi (1985) and Hekman et al. (1988).

Regulation of oxidative phosphorylation is often described in terms of near-cessation of oxygen uptake when ADP is exhausted (respiratory control) or inhibition of F_1 by the ATPase inhibitor protein when respiration ceases and $\Delta\mu_{H+}$ drops. Clearly, however, these mechanisms must come into play under calamitous conditions to prevent futile substrate oxidation or ATP hydrolysis. It seems more probable that the V_{max} and K_m changes discussed above might be the regulatory mechanism by which the kinetics of ATP synthesis are adjusted to the metabolic vicissitudes of the cell.

(b) Turnover capacity of the ATP synthase complex: In the course of the above studies, it was observed that inhibitors of F_1 or F_o inhibited ATP hydrolysis by SMP much more than ATP synthesis. Others had made somewhat similar observations regarding the differential effects of inhibitors on ATP synthesis and hydrolysis, even though their SMP preparations catalyzed ATP synthesis at a fraction of the rate of ATP hydrolysis (Schäfer, 1982; Emanuel et al., 1984; Van der Bend et al., 1985). The conclusions of these laboratories ranged from different affinities of the inhibitors for energized versus nonenergized F_oF_1 to separate sites on F_1 for ATP synthesis and hydrolysis or separate paths for proton conduction in ATP synthesis and hydrolysis. These different interpretations agreed only with the particular inhibitor each laboratory had used, i.e., whether the inhibitor was a covalent modifier or not, and whether it reacted with F_1 catalytic sites or with F_o. We had found the differential effect with all kinds of F_1 or F_o inhibitors, and our experience did not agree with separate F_1 sites or proton paths for ATP synthesis and hydrolysis. Therefore, a different explanation was sought. We had already observed that fractional inactivation of F_oF_1 in SMP elevates steady-state $\Delta\psi$ during oxidative phosphorylation and increases apparent K_m for ADP and Pi. In some experiments, we could inhibit the ATPase activity of SMP by up to 80% with little or no change in apparent V_{max} but tenfold or more increase in apparent K_m^{ADP} (Fig. 2). These results suggested, therefore, that $\Delta\psi$ increase as a result of fractional inactivation of F_oF_1

40

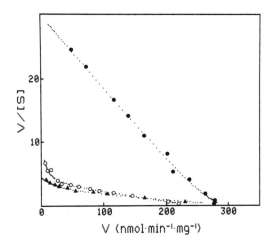

Fig. 2   Eadie-Hofstee plot of the kinetics of ATP
synthesis showing partial inhibition of the ATP synthases
of SMP results in conversion of low K_m (●) to high K_m
(o , ▲) kinetics with little or no change in overall V_{max}.
The respiratory substrate was 30 mM DL-β-hydroxybutyrate
plus 1 mM NAD.   In o and ▲, SMP at 10 mg/ml was treated
with 10 μg/ml of efrapeptin (o) or 2.5 μg/ml of
oligomycin (▲) before assay, and the inhibitions of
ATPase activities of these particles were, respectively,
80% and 60%.   From Matsuno-Yagi & Hatefi (1986).

complexes results in k_{cat} increase for ATP synthesis by the
remaining, uninhibited fraction of the ATP synthase molecules.
This interpretation proved correct and led to several important
conclusions, all derived from the data of Fig. 3.

    In this figure, the abscissa shows percent ATPase activity of
SMP which had been progressively inhibited with increasing
amounts of either N,N'-dicyclohexylcarbodiimide (DCCD) or
tributyltin chloride (TBT-Cl).   The ordinate shows the
reciprocal of the activity of these particles for ATP synthesis
per mol of the *uninhibited fraction* of F_oF_1 molecules present in
the inhibitor-treated SMP preparations.   Respiratory substrate
was NADH in one set of experiments, and succinate in another.
It is seen that as the ATPase activity of SMP was progressively
inhibited and steady-state $\Delta\psi$ was increased (see arrow under

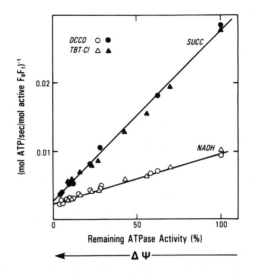

Fig. 3 Increase in the turnover rate of F_oF_1 complexes for ATP synthesis as the population of active F_oF_1 complexes in SMP was decreased by fractional inactivation with DCCD or TBT-Cl. Respiratory substrates were NADH and succinate, as shown, and the extent of fractional inactivation of F_oF_1 complexes was monitored by the decrease in ATPase activity of uncoupled SMP as depicted on the abscissa as well as by [^{14}C]DCCD binding at F_o (Matsuno-Yagi & Hatefi, 1988). The fractional inactivation of F_oF_1 complexes resulted in increase in the steady-state free energy of the system. At high inhibition of ATPase activity (e.g.,≥80%), the steady-state $\Delta\psi$ was very close to static-head $\Delta\psi$, which was about 150 mV. From Matsuno-Yagi & Hatefi (1988).

abscissa), the activity of the *remaining, active* F_oF_1 complexes for ATP sythesis increased and the two lines for NADH and succinate as respiratory substrates converged at the ordinate, where by definition steady-state $\Delta\psi$ approaches the static-head value. The data of Fig. 3 also indicated the following.

i - Since steady-state $\Delta\psi$ increases from right to left on the abscissa of Fig. 3, this figure is like a double reciprocal plot with $\Delta\psi$ as the variable substrate. Thus, it appears that steady-state $\Delta\psi$ acts as a Michaelis substrate in oxidative phosphorylation and is capable of saturating the ATP synthase complexes at static-head.

ii - Since the experiments of Fig. 3 were performed at saturating [ADP] and [Pi], the ordinate intercept at saturating $\Delta\psi$ indicates the turnover number of F_oF_1 for ATP synthesis. This number is 440 s^{-1}, which is the same as V_{max} for ATP hydrolysis by membrane-bound F_1 (400-500s^{-1}.mol F_1^{-1}) (Matsuno-Yagi & Hatefi, 1988), and the same as V_{max} for photophosphorylation (400-420 s^{-1}.mol $CF_oF_1^{-1}$) (Junesch & Gräber, 1985; Matsuno-Yagi & Hatefi, 1988). It was previously difficult to understand why rates of chloroplast photophosphorylation were so much higher than those of oxidative phosphorylation by mitochondria or SMP. However, the reason is now obvious. Photophosphorylation rates were measured under high light intensity, therefore, high $\Delta\mu_{H+}$, whereas oxidative phosphorylation rates were limited by the low rates of substrate oxidation and the low $\Delta\mu_{H+}$ generated. The turnover number of 440 s^{-1} amounts to a rate of 11,000 nmol ATP synthesized/min/mg of SMP protein. Intact bovine heart mitochondria synthesize ATP at rates no more than 700-900 nmol/min/mg of protein (Hatefi & Lester, 1958).

iii - It has long been a matter of debate whether the mitochondrial protonmotive force (pmf) is localized (e.g., involving pairs of energy-yielding and energy-consuming enzyme complexes) (Westerhoff et al., 1984; Slater, 1987) or whether it is delocalized over the entire membrane, as originally envisioned by Mitchell (Mitchell, 1966; Mitchell, 1976). The data of Fig. 3 provide strong support for a delocalized pmf, because fractional inactivation of F_oF_1 complexes by a covalently reacting modifier such as DCCD increases steady-state $\Delta\psi$ as well as the turnover rate of the remaining, active F_oF_1 molecules for ATP synthesis.

iv - Various investigators have attempted to relate the rate of phosphorylation to the rate of coupled respiration (see, for example, Slater, 1987). It is clear from Fig. 3 that what determines the rate of phosphorylation at saturating [ADP] and [Pi] is the magnitude of steady-state $\Delta\psi$ (or $\Delta\mu_{H+}$) and the number of functioning ATP synthase molecules. Steady-state $\Delta\psi$ (or $\Delta\mu_{H+}$) is in turn affected not only by the rate of energy production by the respiratory chain, but also by energy leak of the system, of which intrinsic slip by functioning F_oF_1 complexes is a major component (Zoratti et al., 1986; Slater, 1987; Kamp et al., 1988; Matsuno-Yagi & Hatefi, 1989).

(e) Catalytic site cooperativity in ATP synthesis: The kinetic studies summarized in part (a) were carried out at an ADP concentration range of 1-1200 μM. When the kinetics of oxidative phosphorylation were investigated with SMP depleted of bound ADP/ATP at "exchangeable" sites, using [ADP] below 1μM or [GDP] below its apparent K_m, the results indicated positive cooperativity (Matsuno-Yagi & Hatefi, 1990a). The data were fitted to an equation containing two binding constants, the first binding occurring at low [ADP] or [GDP] with no (or extremely slow) product formation, and the second occurring at higher substrate concentration, which resulted in rapid product formation. Direct binding studies with radioactive ADP and GDP showed that SMP contained a still higher-affinity nucleotide binding site, which could not have been detected by kinetic experiments (Matsuno-Yagi & Hatefi, 1990a). The ADP and GDP dissociation constants for this high-affinity site were calculated from Scatchard plots, and cold chase experiments with unlabeled ADP and GDP showed that this high-affinity site was also an "exchangeable" site. The results are summarized in Fig. 4, in which K_I, K_{II}, and K_{III} (equivalent to apparent K_m) represent the dissociation constants for ADP or GDP at the 3 "exchangeable" nucleotide binding sites. The figure shows that ADP must bind to 3 "exchangeable" sites (presumably the 3 β subunits of F_1) before rapid, physiologically relevant, ATP synthesis takes place. It also shows the apparent K_m^{ADP} and k_{cat} values at low and high protonmotive force as derived from the studies summarized under part (a) above.

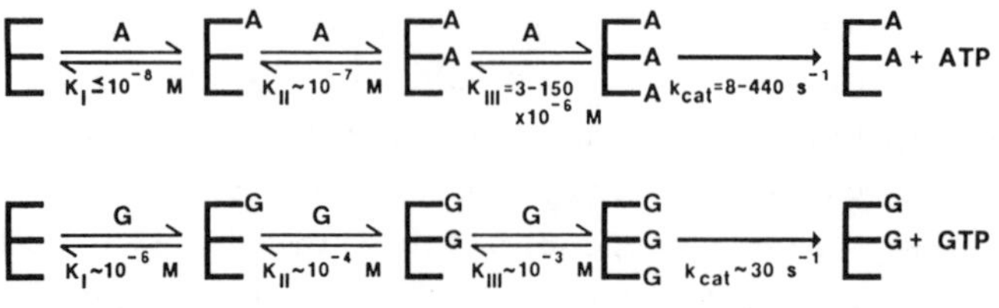

Fig. 4 Scheme showing the magnitude of the dissociation constants (K_I and K_{II}) involved in ADP (A) and GDP (G) binding to SMP (E) before rapid ATP or GTP synthesis ensued with the K_{III}(apparent K_m) values shown. For ATP synthesis the K_{III} and k_{cat} values shown are for the low K_m (low respiration rate, low steady-state $\Delta\psi$) and high K_m (high respiration rate, high steady-state $\Delta\psi$) modes. For GTP synthesis, the K_{II}, K_{III}, and k_{cat} values shown are with NADH as the respiratory substrate. From Matsuno-Yagi & Hatefi (1990a,b).

Subsequent studies indicated that sites I, II, and III need not be occupied by the same nucleotide for product formation to take place (Matsuno-Yagi & Hatefi, 1990b). Thus, when sites I and II were occupied by ADP, GDP added at μM concentrations resulted in GTP formation with an "apparent" K_m^{GDP} of 4 μM. However, the ADP at sites I and II could not be replaced with a nonproductive substrate analog (for example, TNP-ADP or TNP-ATP) which was capable of binding to F_1 "exchangeable" sites. This suggested that sites I and II are not mere regulatory sites, but are catalytically active sites and, once occupied by the same substrate, the three sites might become catalytically equivalent.

It is clear from Fig. 4 and the results summarized in the first paragraph of this section that the kinetics of both ATP synthesis and hydrolysis display negative cooperativity with respect to substrate binding and positive catalytic cooperativity. Whether the SMP catalyze a slow unisite rate of ATP synthesis or no net ATP synthesis takes place when only a single catalytic site on F_1 is occupied by substrate is an interesting mechanistic question yet to be investigated.

ACKNOWLEDGMENTS: The studies described above were supported by United States Public Health Service Grant DK08126.

REFERENCES

Berden, J.A., Hartog, A.F., & Edel, C.M. (1991) Biochim. Biophys. Acta 1057, 151-156.

Bizouarn, T., de Kouchkovsky, Y., & Haraux, F. (1991) Biochemistry 30, 6847-6853.

Cross, R.L., Grubmeyer, C., & Penefsky, H.S. (1982) J. Biol. Chem. 257, 12101-12105.

Cunningham, D., & Cross, R.L. (1988) J. Biol. Chem. 263, 18850-18856.

Dupuis, A., Issartel, J.-P., Lunardi, J., Satre, M., & Vignais, P.V. (1985) Biochemistry 24, 728-733.

Emanuel, E.L., Carver, M.A., Solani, G.C., & Griffiths, D.E. (1984) Biochim. Biophys. Acta 766, 209-214.

Feldman, R.I., & Sigman, D.S. (1982) J. Biol. Chem. 257, 1676-1683.

From, A.H.L., Zimmer, S.D., Michurski, S.P., Mohanakrishnan, P., Ulstad, V.K., Thoma, W.J., & Ugurbil, K. (1990) Biochemistry 29, 3731-3743.

Fromme, P., & Gräber, P. (1990) Biochim. Biophys. Acta 1016, 29-42.

Grubmeyer, C., Cross, R.L., & Penefsky, H.S. (1982) J. Biol. Chem. 257, 12092-12100.

Hansen, M., & Smith, A.L. (1964) Biochim. Biophys. Acta 81, 214-222.

Hatefi, Y., & Lester, R.L. (1958) Biochim. Biophys. Acta 27, 83-88.

Hekman, C., Matsuno-Yagi, A., & Hatefi, Y. (1988) Biochemistry 27, 7559-7565.

Hekman, C., Tomich, J.M., & Hatefi, Y. (1991) J. Biol. Chem. 266, 13564-13571.

Hekman, C., & Hatefi, Y. (1991) Arch. Biochem. Biophys. 284, 90-97.

Junesch, U., & Gräber, P. (1985) Biochim. Biophys. Acta 809, 429-434.

Junge, W. (1987) Proc. Natl. Acad. Sci. U. S. A. 84, 7084-7088.

Kamp, F., Astumian, R.D., & Westerhoff, H.V. (1988) Proc. Natl. Acad. Sci. U. S. A. 85, 3792-3796.

Matsuno-Yagi, A., & Hatefi, Y. (1985) J. Biol. Chem. 260, 14424-14427.

Matsuno-Yagi, A., & Hatefi, Y. (1986) J. Biol. Chem. 261, 14031-14038.

Matsuno-Yagi, A., & Hatefi, Y. (1988) Biochemistry 27, 335-340.

Matsuno-Yagi, A., & Hatefi, Y. (1989) Biochemistry 28, 4367-

4374.

Matsuno-Yagi, A., & Hatefi, Y. (1990a) J. Biol. Chem. 265, 82-88.

Matsuno-Yagi, A., & Hatefi, Y. (1990b) J. Biol. Chem. 265, 20208-20313.

Milgrom, Y.M., & Murataliev, M.B. (1987) FEBS Lett. 222, 32-36.

Mitchell, P. (1966) Chemiosmotic Coupling in Oxidative and Photosynthetic Phosphorylation. Glynn Research, Bodmin, Cornwall, England.

Mitchell, P. (1976) J. Theor. Biol. 62, 327-367.

Muraguchi, M., Yoshihara, Y., Tunemitu, T., Tani, I., & Higuti, T. (1990) Biochem. Biophys. Res. Commun. 168, 226-231.

Penin, F., Archinard, P., Moradi-Ameli, M., & Godinot, C. (1985) Biochim. Biophys. Acta 810, 346-353.

Quick, W.P., & Mills, J.D. (1987) Biochim. Biophys. Acta 893, 197-207.

Quick, W.P., & Mills, J.D. (1988) Biochim. Biophys. Acta 932, 232-239.

Sakamoto, J., & Tonomura, Y. (1983) J. Biochem. 93, 1601-1614.

Schäfer, G. (1982) FEBS Lett. 139, 271-275.

Slater, E.C. (1987) Eur. J. Biochem. 166, 489-504.

Stroop, S.D., & Boyer, P.D. (1987) Biochemistry 26, 1479-1484.

Van der Bend, R.L., Duetz, W., Colen, A.M.A.F., van Dam, K., & Berden, J.A. (1985) Arch. Biochem. Biophys. 241, 461-471.

Vinkler, C. (1981) Biochem. Biophys. Res. Commun. 99, 1095-1100.

Walker, J.E., Lutter, R., Dupuis, A., & Runswick, M.J. (1991) Biochemistry 30, 5369-5378.

Westerhoff, H.V., Melandri, B.A., Venturoli, G., Azzone, G.F., & Kell, D.B. (1984) Biochim. Biophys. Acta 768, 257-292.

Wong, S.-Y., Matsuno-Yagi, A., & Hatefi, Y. (1984) Biochemistry 23, 5004-5009.

Yoshida, M. (1983) Biochem. Biophys. Res. Commun. 114, 907-912.

Zoratti, M., Favaron, M., Pietrobon, D., & Azzone, G.F. (1986) Biochemistry 25, 760-767.

Adenine Nucleotides in Cellular Energy Transfer and Signal Transduction
S. Papa, A. Azzi & J.M. Tager (eds)
© 1992 Birkhäuser Verlag, Basel/Switzerland

NUMBER, LOCALISATION AND FUNCTION OF THE NON-CATALYTIC ADENINE NUCLEOTIDE BINDING SITES OF MITOCHONDRIAL F_1-ATPase

Jan A. Berden, Cees M. Edel and Aloysius F. Hartog

E.C. Slater Institute for Biochemical Research, University of Amsterdam, Plantage Muidergracht 12, 1018 TV Amsterdam, The Netherlands

SUMMARY: Upon dissociation of bovine-heart F_1 in 3M LiCl one of the two non-exchangeable nucleotides is lost and one remains bound to the residual $\alpha_3\gamma\delta\epsilon$ moiety. Reconstitution of this $\alpha_3\gamma\delta\epsilon$ moiety with ß-subunits to intact F_1 requires the presence of an adenine nucleotide triphosphate in the anti-configuration. The site at which the second "non-exchangeable" nucleotide is bound is a potentially catalytic ß-site.
Additionally two different exchangeable non-catalytic sites are identified. The high-affinity site is quite specific for adenine nucleotide diphosphates in the anti-configuration and binding of an adenosine diphosphate to this site in the presence of Mg^{2+} causes partial inhibition of the ATPase activity. The low-affinity non-catalytic site has no preference for nucleotides in the anti-configuration. Binding of a nucleotide to this site does not affect the Vmax of ATP hydrolysis, but destroys the negative cooperativity of ATP hydrolysis. Covalent modification of either of the two exchangeable non-catalytic sites with 2-nitreno-ADP results in labeling of Tyr-368 of a ß-subunit, while labeling with 8-nitreno-ADP results in equal labeling of α- and ß-subunits.
In total, therefore, two α/ß-pairs contain a catalytic and a regulatory site each, while the sites on the third pair contain the two tightly bound "non-exchangeable" adenine nucleotides.

INTRODUCTION

For the understanding of the basic catalytic mechanism of ATP hydrolysis by isolated F_1 it is necessary that the 6 adenine nucleotide binding sites (1,2) are defined and characterized. The presence of 3 ß-subunits and the demonstration that the catalytic

48

sites are located on ß-subunits have prompted the idea that the enzyme contains three catalytic sites (3). The multi- phasic kinetics (4,5) seemed to support this idea, but at the other hand it was clear that at some nucleotide binding sites the nucleotides do not exchange during catalysis (6,7) and that also one or two regulatory sites have to be present (8,9). In a recent paper (10) we have given arguments for our conclusion that only two of the three ß-sites are really catalytic and that the third one is in fact occupied with one of the two non-exchangeable nucleotides. This conclusion implies a basic asymmetry of the enzyme, one α/ß pair being not involved in catalysis and containing two non-exchangeable (structural) nucleotides, in agreement with the asymmetry detected with electron microscopy (11). The other two α/ß pairs, then, contain a catalytic and a non-catalytic (regulatory) site each and interact with each other. A schematic model is presented in Fig. 1. In the present paper we will show that F_1 contains one non-exchangeable nucleotide binding site with properties that fit very well with localisation at a potentially catalytic ß-site and we also identify two exchangeable non-catalytic interface nucleotide binding sites of which one is responsible for the observed negative cooperativity of ATP hydrolysis (12,13).

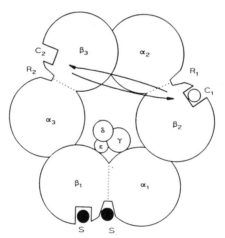

Fig. 1. Model for isolated F_1-ATPase. S, structural adenine nucleotide binding site; C, catalytic site; R, regulatory site.

MATERIALS AND METHODS

The F_1 preparations used in the present study have been isolated according to the method of Knowles and Penefsky (14) and stored in liquid nitrogen in the presence of 4 mM ATP and 2 mM EDTA. After removal of free and loosely bound nucleotides by ammonium sulphate precipitation and repeated column centrifugation the enzyme contained three tightly bound nucleotides.

The dissociation-reconstitution procedure will be described in ref. 15, the methods for labeling of the enzyme with nitreno-adenine nucleotides are reported in ref. 16 and the methods used for the synthesis of azido-adenine nucleotide analogs in refs 15, 16 and 17. The synthesis of $[^3H]NAP_3$-2-azido-ADP will be described elsewhere (in preparation).

The details of the FPLC and HPLC procedures, as well as the determination of the radioactivity can be found in ref. 15. The ATPase activity is determined using an ATP-regenerating system as described previously (17).

RESULTS AND DISCUSSION

To test our hypothesis about the mechanism of ATP hydrolysis by isolated F_1 (10) we wanted to investigate three implications: (i) one of the non-exchangeable nucleotides is bound at a potentially catalytic ß-site; (ii) the enzyme contains two exchangeable non-catalytic sites; (iii) one of the exchangeable non-catalytic sites is responsible for the negative cooperativity of ATP hydrolysis.

Specific labeling of a non-exchangeable nucleotide binding site.

Isolated F1 contains three tightly bound nucleotides of which one can be exchanged with medium-nucleotides (18). This binding site is directly involved in both uni- and multi-site catalysis of ATP hydrolysis. The other two bound nucleotides do not exchange. We used dissociation in the presence of LiCl to remove tightly bound nucleotides. After treatment of F_1 with 3 M LiCl for 90 or 180 s at pH 7.0 or 6,2, respectively, the activity is nearly zero, but reconstitution is possible after removal of the salt and addition of ATP (15,19,20). After dissociation and one column centrifugation step still one ADP/F_1 is bound to the protein. FPLC analysis of the LiCl-treated sample (Fig. 2), followed by TDAB-polyacrylamide gel electrophoresis (Fig. 3), showed that the enzyme was split into two fractions, an $\alpha_3\gamma\delta\epsilon$ moiety and monomeric ß-subunits. The $\alpha_3\gamma\delta\epsilon$ fraction contained bound ADP, although the nucleotide partially dissociated during passage over the FPLC column. The binding site of this ADP in intact F_1 is clearly an interface site.

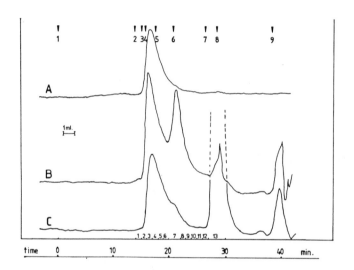

Fig.2. FPLC elution profile (absorbance at 280 nm) of F_1, eluted from a Superose 12 column at pH 6.2. The flow rate was 0.5 ml/min. A, isolated F_1; B, F_1 dissociated with 3M LiCl in the absence of added nucleotides; C, F_1 reconstituted in the presence of 250 μM ATP. The numbers 1-13 on the abscissa correspond with the collected and analysed fractions shown in Fig. 3.

Fig. 2C shows that the reconstituted enzyme is very similar to the native enzyme. Some non-reconstituted ß-subunits can be separated from the intact F_1. The content of tightly bound nucleotides and the ATPase activity are the same for native and reconstituted enzyme (Table I). Reconstitution requires (rapid) addition of ATP

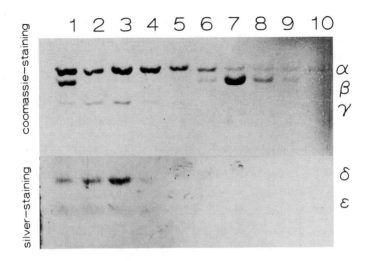

Fig.3. Polyacrylamide TDAB-gel electrophoresis (12% polyacryl-amide) of FPLC fractions of Fig. 2. Lane 1, purified F_1 (fraction 4 of Fig. 2 A); lanes 2-10, fractions 2-10 of Fig. 2 B.

Table I. Characterization of isolated and reconstituted F_1.

	Protein conc mg/ml	Spec. act. μmol/min/mg	Bound nucleotides mol/mol F_1		
			ADP	ATP	total
F_1 in buffer A	2,2	149	1,70	1,22	2,9
F_1 in buffer A after FPLC (fig. 3 trace A)	0,24	174	2,08	1,02	3,1
F_1 in buffer A after dissociation/reconstitution and FPLC (fig. 3 trace C)	0,116	149	1,79	0,85	2,6

or a suitable analog after removal of LiCl and is more efficient when ATP is also present during the dissociation step.

When dissociation and reconstitution are performed in the presence of 2-azido-$[\alpha^{32}P]$ATP, the reconstituted enzyme contains about 2 tightly bound 2-azido-$[\alpha^{32}P]$ATP/F_1, of which one is bound at the tight catalytic site and one at the originally tight site, which became empty upon dissociation of the enzyme. The nucleotide at the tight catalytic site can be exchanged with ATP and the residually bound 2-azido-ATP can be partially bound covalently upon illumination. Table II shows that the covalent binding is relatively low, but this covalent binding of 2-nitreno-ATP does not result in inhibition of the enzyme.

Table II. Bound nucleotides (mol/mol F_1) after dissociation/ reconstitution with [^{14}C]ATP (A) or 2-N$_3$-[α^{32}P]-ATP (B) followed by ammoniumsulphate precipitation and two column centrifugation steps.

	Remaining non-exchangeable nucleotide	(A) [14C–ATP]		(B) 2N3–[α32P]–ATP						
		before turnover	after turnover	before turnover	UV–illum. covalent	% inh	after turnover	UV–illum. covalent	% inh	
pH 7.0	0.6	1.7	0.64	2.0	1.46	70	0.46	0.22	3	
	0.7	2.0	0.70							
		2.1	1.1							
pH 6.2	1.2	1.9	0.73	2.3	1.0	60	0.80	0.20	6	
		1.7	0.80	2.4	0.77	32	0.83	0.21	1	

Starting F₁ in buffer A contains 3 tightly bound nucleotides.

The procedures, involving protease digestion and HPLC analysis of peptides, intended to determine the site of attachment of the label, did not give clear results. Most label was recovered in the void volume of the hydrophobic column, suggestive for labeling of a small trypsine fragment. Both the Tyr-368- and the Tyr-345-containing peptides were hardly labeled (cf.ref. 21). It is clear, then, that the site of binding of the 2-azido-ATP is not conformationally identical with the catalytic sites or the exchangeable non-catalytic sites. The elution profile from the ion-exchange column showed that most label was converted into ADP. This result, together with the finding that only nucleotide triphosphates (in the anti-configuration) induce reconstitution, supports the proposal that the involved site is a potentially catalytic site, a ß-site.

Modification of the high-affinity non-catalytic site.
Cross and coworkers (18) have identified a non-catalytic site with

a high affinity for ADP and analogues in the anti-configuration when Mg^{2+} is present. We have studied this site using 2-azido-ADP, NAP_3-ADP and NAP_3-2-azido-ADP. Incubation of F_1, containing three tightly bound nucleotides, with ADP or one of the mentioned analogues in the concentration range of 10-100 μM in the presence

Table III. Inhibition of ATPase activity after incubation in the dark with ADP or ADP-analogues in a Mg^{2+} containing medium. The incubation medium consisted of 50 mM Tris-HCl, pH 7.5, 150 mM sucrose, 0.2 mM EDTA and 4 mM $MgCl_2$. At the used concentrations of nucleotides, the inhibition was maximal.

ligand	INH. (%)
100 μM ADP	42
100 μM $2N_3$ADP	26
2.5 mM $8N_3$ADP	3
100 μM NAP_3-ADP	35
100 μM NAP_3-$2N_3$ADP	45

of Mg^{2+}, the enzyme activity is partially inhibited (see Table III). After a column centrifugation step the inhibition is still present and using radioactive ligands it could be established that about 2 moles of ligand remain bound per mole of enzyme. After addition of ATP one mole is exchanged and the other remains bound. With time the ligand slowly dissociates and the enzyme activity increases. When isolated enzyme is first incubated with Mg^{2+} before the loosely bound nucleotides are removed, the final enzyme preparation contains about 4 moles of adenine nucleotides per mole in stead of three. Addition of ADP or one of the mentioned analogues does not induce inhibition any more and after column centrifugation and ATP addition no added nucleotide diphosphate remains bound. It is clear, then, that the site of binding of the inhibitory ligand is the site which binds ADP in the presence of Mg^{2+}. This ADP is not exchanged at a time scale of a few minutes. The affinity of 8-azido-ADP for this site is much lower, since after a column centrifugation step the 8-azido-ADP does not remain bound and in an activity assay no inhibition is found. After illumination in the presence of 2-azido-$[\alpha^{32}P]$ADP we analysed the site of covalent attachment. After proteolysis with trypsin the labeled peptide fragments were isolated and sequenced. When only the non-catalytic site was labeled (in the presence of ATP) the labeled fragment was R21 of the ß-subunit, most label being attached to Tyr-368 and some to His-367. When the illumination of the enzyme was performed in the absence of added ATP, also the R20 peptide was labeled, the label being attached to Tyr-345 and also Ile-344. These results, then, fit nicely with the reported data (21). The inhibition of the ATPase activity by nucleotide diphosphates bound at this non-catalytic site does not change significantly upon covalent binding of the ligand. Non-covalent binding causes a inhibition of 26-45%, the highest inhibition being induced by NAP$_3$-2-azido-ADP and the lowest by 2-azido-ADP. The inhibition induced by covalent binding is the same for all analogues, 42-45%. Kinetic analysis showed that only the Vmax is affected, not the (apparent) Km values. On the effect of ATP bound at this site we have speculated (10) that it enhances the dissociation of ADP from the catalytic site.

Table IV shows that after incubation of F_1 with 100 μM 2-azido-ADP (+ Mg^{2+}), followed by illumination in the presence of ATP, more than 1 mole per mole of F_1 is bound. Apparently, an additional non-catalytic site is available for 2-azido-ADP, exhibiting a lower affinity. The inhibition of ATPase activity is not increased above

Table IV. Incubation of F_1-ATPase with 100μM 2N3-$[\alpha^{32}P]$ADP in Mg^{2+} containing medium. A: without removal of excess label. B: after removal of exess label via a column centrifugation step. For incubation medium and activity measurements, see table 3. After illumination (2 and 3) the label was covalently bound, after dark incubation (1) the label was not covalently bound.

	A.		B.	
	INH. (%)	mole lab. / mole F$_1$	INH. (%)	mole lab. / mole F$_1$
1. DARK	28.5 ± 1.7		17.1 ± 0.85	(3.4 ± 0.3)
2. ILL.	95.2 ± 1.4	3.85 ± 0.15	90.7 ± 3.6	2.15 ± 0.05
3. ILL. + T.O.	41.3 ± 1.7	1.5 ± 0.1	42.7 ± 3.9	0.85 ± 0.05

42%, indicating that binding of 2-azido-ADP to this second exchangeable non-catalytic site does not inhibit the enzyme.

Modification of the low-affinity non-catalytic site.

In earlier studies in our laboratory we had found that with 8-azido-ATP in the presence of EDTA 2 sites could be covalently occupied, one catalytic and one non-catalytic (22). It was concluded that occupation of the non-catalytic site does not inhibit the enzyme since the inhibition (linearly related with the occupation of the catalytic site) was not influenced by the additional occupation of the non-catalytic site (at higher concentrations of ligand). We now found that in the presence of EDTA also 2 moles of 8-nitreno-ADP per mole of enzyme can be covalently bound, the ratio of the labeling of ß- and α-subunits being 3 and the inhibition of enzyme activity being linear with bound ligand. When the illumination was performed in the presence of added ATP to remove 8-azido-ADP from the catalytic site, the residual covalent binding resulted in an equal distribution of the label over the α- and ß-subunits without inhibition of the enzyme activity. In the presence of Mg^{2+} 3 moles per mole of F1 could be covalently bound (Figs 4 and 5). The ratio of ß- and α-labeling was 2, indicative for the binding of one catalytic site (labeling of a ß-subunit) and 2 non-catalytic sites (labeling of 1 α- and 1 ß-subunit). When first one non-catalytic site was labeled with 8-azido-ADP in the presence of ATP and afterwards NAP_3-2-azido-ADP was added in the presence of Mg^{2+}, the inhibitory effect of the NAP_3-2-azido-ADP on the ATPase activity was still present. The inhibitory binding site of the NAP_3-2-azido-ADP is therefore different from the binding site of 8-azido-ADP. Using radioactive NAP_3-2-azido-ADP we could also show that the presence of the 8-nitreno-ADP at its binding site did not prevent the covalent binding of the NAP_3-2-azido-analogue upon illumination. The conclusion may be drawn that in isolated F_1, containing 3 tightly bound adenine nucleotides, of which 2 are not exchangeable, still 2 nucleotides can be bound to non-catalytic sites. Finally: The negative cooperativity of ATP hydrolysis by F_1 has been interpreted as due to binding of substrate to a third catalytic site (3). An alternative explanation has also been forwarded (12,13): the affinity of the catalytic sites for the substrate depends on the

56

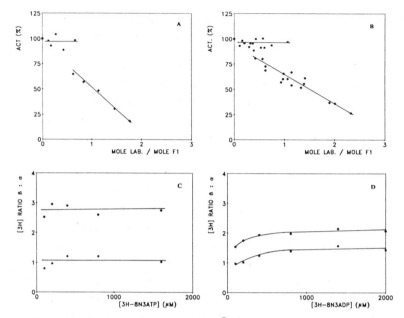

Fig. 4. Labeling of F_1-ATPase with $[^3H]$8-azido-ATP (**A** and **C**), or $[^3H]$8-azido-ADP (**B** and **D**), in EDTA containing medium (50 mM Tris-HCL pH 7.5, 150 mM sucrose and 4mM EDTA), after UV illumination with turn-over (■), or without turn-over (♦) . **A** and **B** : F_1-ATPase activity related to amount of covalently bound 8-nitreno-ATP (**A**) and 8-nitreno-ADP (**B**). **C** and **D**: distribution of label over α- and β- subunits for 8-nitreno-ATP (**C**) and 8-nitreno-ADP (**D**). The subunits were separated by TDAB gel electroforesis, coomassie stained bands were cut out of the gel, extracted and counted for radioactivity.

ligand bound to a non-catalytic site.

Since we now have succeeded in binding 8-nitreno-ADP or 8-nitreno-ATP to a rapidly exchangeable non-catalytic site, the model could be tested. We incubated the enzyme with a high concentration of 8-azido-ATP or 8-azido-ADP, added ATP to remove the azido-nucleotide from the catalytic site and illuminated the sample for 20 min. The whole procedure was repeated and the final amount of covalently bound ligand was between 0.7 and 0.8 mole per mole of enzyme. The ATPase activity of the modified enzyme, measured under Vmax conditions, was slightly decreased relative to the control. This slight inhibition is possibly due to some covalent binding of the nitreno-compound to a catalytic site. The kinetic analysis is presented in Fig. 5 as Lineweaver-Burke

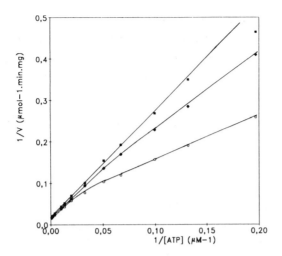

Fig. 5. Kinetics of ATP hydrolysis, as Lineweaver-Burke plot, by F_1 covalently modified with 8-nitreno-ATP (■), or 8-nitreno-ADP (♦), using UV illumination during turn-over. Control F_1 (o) was illuminated in the abscence of label.

plots. The covalent binding of either nitreno-ADP or nitreno-ATP changes the negative cooperativity drastically and even at low concentrations of substrate a high Km value is found, in agreement with the predictions of the model. Both nitreno-ATP and nitreno-ADP at the non-catalytic site induce the low-affinity form of the catalytic sites. Our data also nicely fit with the observations reported by Jault et al. (23) with mutant yeast enzyme in which the replacement of Gln-173 of the α-subunit by leucine results in the disappearance of the negative cooperativity. The mutant enzyme binds ATP at the regulatory site with a lower affinity than the wild-type enzyme does. The catalytic sites of the enzyme, therefore, stay in the high-affinity state at increasing concentrations of ATP.

CONCLUSION: Isolated F_1 contains two non-exchangeable nucleotides of which one is bound at an interface site and the other at the potentially catalytic ß-site of the same α/ß-pair. The two other α/ß-pairs contain a catalytic and a non-catalytic site each. One of the non-catalytic sites binds adenosine diphosphates in the anti-configuration with high affinity, especially in the presence of Mg^{2+} and this binding lowers the catalytic activity of the enzyme. The

58

other non-catalytic site shows a lower affinity for nucleotides and is not specific for the anti-configuration. Binding of an adenine nucleotide at this site lowers the affinity of the catalytic sites for ATP and binding of ATP at this site causes the negative cooperativity of ATP hydrolysis by F_1.

Acknowledgements: This work was supported in part by grants from the Netherlands Organization for the Advancement of Scientific Research (N.W.O.) under the auspices of the Netherlands Foundation for Chemical Research S.O.N.).

REFERENCES

1. Wagenvoord, R.J., van der Kraan, I. and Kemp, A. (1979) Biochim. Biophys. Acta 548, 85-95
2. Cross, R.L. and Nalin, C.M. (1982) J. Biol. Chem. 257, 2874-2881
3. Senior, A.E. and Wise, J.G. (1983) J. Membr. Biol. 73, 105-124
4. Gresser, M.J., Myers, J.A. and Boyer, P.D. (1982) J. Biol. Chem. 257, 12030-12038
5. Roveri, O.A. and Calcaterra, N.B. (1985) FEBS Lett. 192, 123-127
6. Rosing, J., Harris, D.A., Slater, E.C. and Kemp, A. (1975) J. Supramol. Struct. 3, 284-296
7. Garret, N.E. and Penefsky, H.S. (1975) J. Supramol. Struct. 3, 469-478
8. Lardy, H.A., Schuster, S.M. and Ebel, R.E. (1975) J. Supramol. Struct. 3, 214-221
9. Ebel, R.E. and Lardy, H.A. (1975) J., Biol. Chem. 250, 191-196
10. Berden, J.A., Hartog, A.F. and Edel, C.M. (1991) Biochim. Biophys. Acta 1057, 151-156
11. Boekema, E.J., Berden, J.A. and van Heel, M.G. (1986) Biochim. Biophys. Acta 851, 353-360
12. Recktenwald, D. and Hess, B. (1977) FEBS Lett. 76, 25-28
13. Stutterheim, E., Henneke, A.M.C. and Berden, J.A. (1980) Biochim. Biophys. Acta 592, 415-430
14. Knowles, A.L. and Penefsky, H.S. (1972) J. Biol. Chem. 247, 6617-6623
15. Hartog, A.F., Edel, C.M., Lubbers, F.B. and Berden, J.A., submitted for publication
16. van Dongen, M.B.M., de Geus, J.P., Korver, T., Hartog, A.F. and Berden, J.A. (1986) Biochim. Biophys. Acta 850, 359-368
17. Sloothaak, J.B., Berden, J.A., Herweijer, M.A. and Kemp, A. (1985) Biochim. Biophys. Acta 809, 27-38
18. Kironde, F.A.S. and Cross, R.L. (1987) J. Biol. Chem. 262, 3488-3495
19. Wang, J.H. (1985) J.Biol. Chem. 260, 1374-1377
20. Nieboer, P., Hartog, A.F. and Berden, J.A. (1987) Biochim. Biophys. Acta 894, 277-283
21. Cross, R.L., Cunningham, P., Miller, C.G., Xue, Z., Zhou, J.-M. and Boyer, P.D. (1987) Proc. Natl. Acad. Sci. USA 84, 5715-5719
22. van Dongen, M.B.M. and Berden, J.A. (1986) Biochim. Biophys. Acta 850, 121-130
23. Jault, J.M., Di Pietro, A., Falson, P. and Gautheron, D.C. (1991) J. Biol. Chem. 266, 8073-8078

Adenine Nucleotides in Cellular Energy Transfer and Signal Transduction
S. Papa, A. Azzi & J.M. Tager (eds)
© 1992 Birkhäuser Verlag, Basel/Switzerland

FLUOROALUMINUM AND FLUOROBERYLLIUM COMPLEXES AS PROBES OF THE
CATALYTIC SITES OF MITOCHONDRIAL F1-ATPase

Joël Lunardi, Alain Dupuis, Jérôme Garin, Jean-Paul Issartel,
Laurent Michel, André Peinnequin and Pierre Vignais

Laboratoire de Biochimie (CNRS UA 1130)/DBMS, Centre D'Etudes
Nucléaires, 85X, 38041 Grenoble Cedex, France and Laboratoire de
Biochimie, Faculté de Médecine de Grenoble.

SUMMARY: The mechanism by which fluoride and aluminum or
beryllium in combination with ADP inhibit mitochondrial and
bacterial F_1-type ATPases was investigated. Inhibition required
the presence of Al^{3+} or Be^{2+}, F^-, Mg^{++} and ADP, and developed
more rapidly with anions which activate F_1-ATPase activity. ADP-
fluorometal complexes bound quasi-irreversibly to isolated F_1.
However, the inhibition of the ATPase activity of the membrane
bound-F_1 caused by AlF_4^- and ADP was released upon energization
of the membrane. The effect of uncouplers suggested a direct
role of the proton motive force. Direct measurements of ADP,
magnesium, aluminum, beryllium and fluoride indicated that two
types of fluorometal complexes bound to the inhibitory sites:
$ADP_1 Mg_1 Al_1 F_4$ and $ADP_1 Mg_1 Be_1 F_x (H_2O)_{4-x}$ with $1<x<3$.
Fluoroaluminate and fluoroberyllate mimicked phosphate and
formed abortive complexes with ADP in the catalytic sites of F_1.
Each mol of fully inhibited enzyme retained 2 mol of inhibitory
complexes. Stoichiometries obtained with GDP and photolabeling
experiments conducted with radiolabeled 2-N_3-ADP showed
unambiguously that nucleotide-fluorometal complexes interacted
with catalytic sites. Binding data are discussed in terms of a
stochastic model in which the three cooperative sites of F_1-
ATPase function in interactive pairs.

The ubiquitous enzyme Fo-F1 ATPase [(H$^+$-transporting ATPase)(EC.3.6.1.34)] plays a central role in energy transduction in mitochondria, chloroplasts, and many bacteria by catalyzing the synthesis of ATP from ADP and inorganic phosphate (Pi) [for review, see Senior (1988)]. F1 consists of five different subunits associated in a stoichiometry of a3b3gde (Penefsky, 1979) and is catalytically active as an ATPase, GTPase or ITPase (Schuster et al, 1975). A total of six nucleotide-binding sites located in the three a and the three ß subunits of mitochondrial, bacterial and chloroplast F1 have been demonstrated (Dunn and Futai, 1980; Cross and Nalin, 1982; Lunardi and Vignais, 1982; Perlin et al, 1984; Issartel et al, 1986). Three of these sites rapidly exchange nucleotides, exhibit broad specificity, act in a cooperative manner and are considered as catalytic sites (Cross and Nalin, 1982, Gresser et al, 1982). The other three sites contain adenine nucleotides that are not readily exchangeable with nucleotides from the medium (Kironde and Cross, 1986). Photo or chemically reactive analogs of nucleotides have been widely used to characterize the different sites (for review, see Vignais and Lunardi, 1985). However, the precise assignment and function of the different nucleotide binding sites during the catalytic events are still a matter of debate. Several models for the mechanism of ATP hydrolysis and synthesis have been proposed: alternative site mechanism (Boyer, 1989), dual-site mechanism (Lübben et al, 1984; Berden et al, 1991, Issartel et al, 1991a) or one site mechanism (Wu et al, 1989). Interactions of fluorometal-ADP complexes, a new class of F1-ATPase inhibitors (Lunardi et al, 1988), with nucleotide binding sites present on mitochondrial and bacterial F1 will be discussed with respect to the nature of the inhibitory process and its relation to the catalytic mechanism of F1.

A number of enzymes involved in nucleotide binding or phosphate group transfer have proved to be sensitive to millimolar concentrations of fluoride anions. Sternweis and Gilman (1982) reported that micromolar concentrations of aluminum or beryllium are required for fluoride-dependent activation of adenylate cyclase to occur. Bigay et al (1987) proposed that the fluorometal complexes AlF_4^- or BeF_3 combine with GDP in the nucleotide binding site of transducin. The GDP-AlF_3 complex was postulated to mimic GTP, and promote activation of transducin, as does GTP (for review, see Chabre, 1990). On the other hand, analysis of the mechanism of aluminum toxicity in some physiological processes has led various workers to postulate the formation of rather stable complexes between Al^{3+} and high-energy phosphate components like nucleoside di- and triphosphates (Macdonald et al, 1987). Likewise, beryllium alone might be responsible for the inhibition of different enzymes (Robinson et al, 1986). Sutton et al (1987) reported an inhibition of proton-translocating ATPases in oral bacteria by fluoride. These authors suggested that the effect of fluoride was mediated through a modification of proton conductivity resulting from a direct interaction of fluoride with the enzyme at the level of the Fo sector.

Postulating that fluorometal complexes might form an abortive product with ADP in the catalytic site of F_1-ATPases,we investigated the effect of aluminum, beryllium and fluoride on ATPase activity.

INHIBITION OF H^+-TRANSPORTING ATPase BY TIGHTLY BOUND NUCLEOSIDE DIPHOSPHATE -FLUOROMETAL COMPLEXES

ATPase activities of mitochondrial F_1 and **E. coli** F_1 were strongly inhibited when enzymes were incubated in the presence of ADP, $AlCl_3$ or $BeCl_2$, and high concentrations of NaF. As indicated in table I, addition of both $AlCl_3$ and NaF is required for enzyme inhibition. Using nucleotide-depleted F_1, no significant inhibition of ATPase activity was detected in the presence of $AlCl_3$ and NaF, unless a further addition of ADP was done. These results indicated that inhibition of MF_1 or ECF_1 could not occur unless ADP, metals and fluoride anions were all present in the medium. Other nucleoside diphosphates, which are recognized by the catalytic sites of MF_1, like GDP or IDP, can be substituted to ADP in the inhibitory process. Inhibition was found to develop in a time dependent manner. This indicated that binding of inhibitory components to the enzyme was a slow process or, more likely, that inhibition of the enzyme required a transition from an active conformation to an inactive one. Kinetics of inhibition are affected by the nature of the anions present in the assay medium. Anions known as activators of the ATPase activity like sulfate or sulfite (Ebel and Lardy, 1975) promoted inhibition (Issartel et al, 1991a). Interestingly, inhibition required Mg^{++} to develop (Issartel et al, 1991a). This indicated that inhibition induced in the presence of aluminum or beryllium did not result from a displacement of magnesium by these metals as it was earlier suggested (Martin, 1986).

Table I. Effect of $AlCl_3$ and NaF on ATPase activity of F_1

	ADP (μM)	$AlCl_3$ (μM)	NaF (mM)	Activity *
F_1	60	–	–	60
	60	30	–	59
	60	–	5	58
	60	30	5	5
Depleted F_1	–	–	–	77
	–	30	–	76
	–	–	5	74
	–	30	5	71
	60	30	5	3

* ATPase specific activities measured at 30°C were expressed in μmol of ATP hydrolyzed per min and per mg.

A major feature of the inhibition developed in the presence of ADP, $AlCl_3$ or BeF_2 and NaF was the quasi irreversibility of the process. When inhibited F_1 was incubated in the presence of EDTA or citrate, two well known chelators of Al^{3+}, or in a medium containing a large excess of ADP or ATP in the presence of either EDTA or MgCl$_2$, the enzyme remained inhibited. Similarly, glycerol treatment used to deplete F$_1$ from its tightly bound nucleotides resulted only in a very slight recovery of the activity (Issartel et al, 1991a). The ATPase activity of the membrane bound F$_1$ in submitochondrial particles was also inhibited after incubation with ADP, AlCl$_3$ and NaF. However, in contrast with soluble MF$_1$, the inhibition of ATPase activity in coupled submitochondrial particles could be reversed upon addition of succinate. This result and the fact that uncouplers like FCCP or inhibitors of oxidative-phosphorylation like antimycin prevented the reversal of ATPase inhibition by succinate suggested that the proton motive force generated by respiration was essential for displacement of the ADP-fluoroaluminate complex from its binding site (Lunardi et al, 1988).

NATURE OF THE FLUOROALUMINATE AND FLUOROBERYLLATE SPECIES RESPONSIBLE FOR F$_1$ INHIBITION

Al^{3+} and Be^{2+} are known to combine with different numbers of fluoride anions, depending on fluoride concentration, to generate stable complexes (Goldstein, 1964)(figure 1). Therefore the biological effects obtained in the presence of millimolar and micromolar concentrations of fluoride and aluminum or beryllium respectively, might be due to fluorometal complexes. In aqueous solution, fluoroberyllate complexes are tetracoordinated and AlF$_4^-$ is currently assumed to be the active tetracoordinated species for fluoroaluminates although it has been reported that AlF$_4^-$ in solution might occur as the hexacoordinated species (H$_2$O)$_2$AlF$_4^-$ (Nelson and Martin, 1991). Whereas fluoroaluminate complexes might adopt various geometries, all fluoroberyllate complexes are tetracoordinated and isomorphous to phosphate group (Corbridge, 1974). From these data it has been postulated that AlF$_4^-$ and fluoroberyllate

64

complexes could act as phosphate analogs. The nucleoside-
fluorometal complexes formed in the nucleotide binding sites of

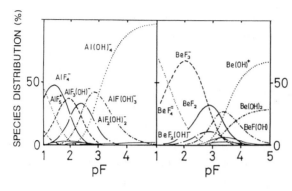

Figure 1. Distribution
of fluorometal species
as a function of
fluoride concentration

various enzymes would be recognized as nucleoside triphosphate
analogs (Paris and Pouysségur, 1987; Carlier et al, 1988,
Missiaen et al, 1988; Lunardi et al, 1988). According to this
hypothesis the binding of one molecule of NDP and one molecule
of fluorometal in the nucleotide binding site would facilitate a
reversible interaction between the ß phosphate of the bound NDP
and the fluorometal. In agreement with this proposal would be
the fact that an extensive multinuclear NMR study led to the
conclusion that ternary complexes: ADP-BeF and ADP-BeF$_2$ are
generated spontaneously in solution (Issartel et al, 1991b).
However, in presence of millimolar concentrations of Mg^{2+}, the
ADP-BeF$_x$ species present in solution are displaced to form ADP-
Mg and fluoroberyllate complexes. Therefore, under the
conditions used in most biochemical experiments, ADP-Mg and BeF$_x$
might bind to the nucleotide binding site in an independant
manner. Although NDP-fluorometal complexes might act by
mimicking NTP ligands at the nucleotide binding site of enzymes,
up to now this hypothesis had not yet received experimental
support (Martin, 1988; Jackson, 1988). Taking advantage of the
virtually irreversible nature of the inhibition of F$_1$-ATPase by
fluoroaluminate or fluoroberyllate, Dupuis et al (1989) measured
the ratio of ADP, beryllium/aluminum and fluoride entrapped in

F_1 as a function of the observed inhibition. Direct determination of [^{3}H]ADP, beryllium and fluoride tightly bound to the inhibited F_1-ATPase revealed a stoichiometry of bound species of 1 mol [^{3}H]ADP, 1 mol beryllium and 1 or 3 mol fluoride depending on the initial fluoride concentration. Atomic absorption analysis of the inhibitory complex bound to F_1 showed that inhibition of F_1-ATPase developed in parallel with the tight binding of aluminum and magnesium to F_1. A ratio of F^- to Al^{3+} of 4 was found in these conditions. These results indicated that two types of complexes with tetracoordinated aluminum or beryllium could occur in the inhibited state of F_1: $ADP_1Mg_1Al_1F_4$, and $ADP_1Mg_1Be_1F_x(H_2O)_{4-x}$ with $1<x<3$. As presented in figure 2, the fluorometal complexes might mimic phosphate and form an abortive complex with ADP at the catalytic site(s) of F_1.

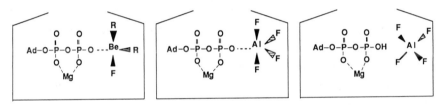

Figure 2. Schematic representation of the nucleotide fluorometal complexes bound to inhibited F_1-ATPase (R= OH or F).

IDENTIFICATION OF THE NUCLEOTIDE BINDING SITES OF F_1 OCCUPIED BY THE ADP-FLUOROMETAL INHIBITORY COMPLEXES

Full inhibition of ATPase activity was linearly related to the tight binding of 2 mol of [^{3}H]ADP per mol F_1, whatever the metal used i.e., Al^{3+} or Be^{2+}. As the amount of bound beryllium or aluminum ions was equal to that of bound [^{3}H]ADP (Dupuis et al, 1989), one could assume that complete inhibition of F_1 required the binding of 2 mol beryllium or aluminum and 2 mol [^{3}H]ADP per mol of F_1.

As previously stated, MF_1 contains at least six nucleotide binding sites which belong to two different classes: catalytic and non catalytic. Therefore, it was of interest to identify which class of nucleotide binding sites was involved in the inhibitory process induced in the presence of fluorometals. In the presence of [^{3}H]GDP, a nucleotide which is essentially recognized by catalytic sites only (Cunningham and Cross, 1988), a linear relationship between F_1 inhibition and the binding of [^{3}H] GDP, similar to that obtained with ADP, was observed. In the absence of fluorometals, mapping studies of the catalytic sites present on the ß subunits have been performed with radiolabeled 2-N$_3$-ADP, a highly specific photoreactive analog of ADP. The bound radioactivity was localized on few amino acids residues adjacent to tyrosine 345 of ß-F$_1$ (Garin et al, 1986). In the presence of AlCl$_3$ and NaF, 2-N$_3$-ADP was able to induce inhibition of F$_1$-ATPase by forming 2-N$_3$-ADP-fluorometal complexes. As described for ADP, full inactivation of F$_1$ was attained upon binding of 2 mol of 2-N$_3$-[8-^{3}H]ADP per mol F$_1$ in the dark. When F$_1$ inhibited under these conditions was phototoirradiated the bound radioactivity was restricted to two residues: Tyr 345 and Pro 346 of ß-F$_1$ (Garin, 1989) .Altogether these data suggest that the two ADP-fluorometal complexes whose binding is required for complete inhibition were entrapped at two identical catalytic sites (Issartel et al, 1991b).

FLUOROALUMINUM AND FLUOROBERYLLIUM NUCLEOSIDE DIPHOSPHATE COMPLEXES AS PROBES OF THE F$_1$-ATPase MECHANISM

A consensus favors the view that all three catalytic sites in F$_1$ function and exhibit strong cooperativity. One important point which is still debated however, is whether the three sites function in a sequential order (alternative site model) or whether they function in pairs, in a random fashion (stochastic model). The determination of the minimum number of catalytic sites which must be derivatized in order to inhibit multisite catalysis might discriminate between a sequential or random order for a three site model (Cross, 1988). Modification of a single catalytic site per F_1 should be sufficient to inhibit the whole catalytic cycle if the order is sequential. In contrast, if the process is random, modification of only one catalytic site will leave the remaining two sites with the ability to catalyze ATP hydrolysis by an alternative bi-site mechanism. Unfortunately, results found in the literature concerning the binding stoichiometry of F_1 with respect to chemical modifiers and photolabels are far from homogenous. The amount of covalently bound reagents resulting in full inactivation of F_1-ATPase ranged between 1 and 3 (for reviews, see Vignais and Lunardi, 1985; Senior, 1988).

A complicating factor in the evaluation of the binding data was introduced by the fact that the precise localisation of the binding of chemical modifiers or photolabels was not always clearly determined. Furthermore, one has to keep in mind that covalent binding of the ligands might disturb interactions between the different sites. In this context, ADP-fluorometal complexes have the advantage of binding very tightly, although non covalently, to clearly identified catalytic sites. Considering the binding stoichiometry which lead to 50% inhibition of F_1, i.e., 1 mol of ADP-fluorometal complex per mol F_1, it is clear that this stoichiometry is not readily explained by the sequential order mechanism. To fit with this model, one could have to postulate that full inhibition was promoted by the binding of one ADP to one catalytic site and that binding of the second ADP to the second catalytic site resulted from a concomitantly induced very fast change of conformation of this site. Although we cannot rule it out totally, the observed kinetic data do not favour such an interpretation. A more direct explanation of the results obtained with ADP-fluorometals is provided by the stochastic model which is based on the cooperative interaction of pair of catalytic subunits at random. In this mechanism, for a saturating concentration of substrate, three possible pairs of catalytic subunits operate in a cooperative manner. As illustrated in figure 3, the catalytic pathway of ATP hydrolysis might consist of two fast steps and a slow one. The first two steps (thick arrows) correspond to multisite catalysis (Grubmeyer et al, 1982; Cross et al, 1982). The hydrolysis of the third bound molecule of ATP occurs by a slow unisite catalytic process, unless refilling of other subunits with ATP initiates another multisite catalytic cycle. Under steady state conditions, one multisite catalytic cycle results in the fast hydrolysis of two ATP per F_1. According to this scheme, binding of one mol of fluorometal to one subunit is accompanied by a 50% inhibition and only one pair of subunits remain able to perform fast hydrolysis since refilling of the third subunit is hampered by the presence of bound ADP-fluorometal at the catalytic site. The binding of a second mole of inhibitor to one of the two remaining subunits would lead to a nearly complete inhibition of F_1 since the enzyme with only one unmodified catalytic site would be unable to perform fast hydrolysis of ATP.

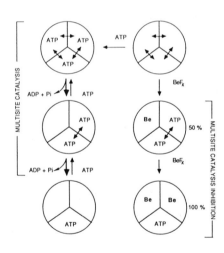

Figure 3. Cooperativity between pairs of catalytic subunits according to a stochastic model. Double arrows indicate interactions between catalytic subunit. In the bottom row, the single remaining catalytic site filled with ATP is postulated to carry out unisite catalysis. On the right-hand side, sequential binding of 2 mol of fluoroberyllates per mol of F_1 is represented. Multisite catalysis of the F_1 molecules tagged by fluoroberyllate is inhibited by 50 and 100% for a stoichiometry of 1 and 2 mol of bound fluoroberyllate per mol of F_1, respectively. "Be" represents inhibitory complexes containing ADP, Mg, Be and F.

Such a stochastic model has been proposed earlier by Lübben et al (1984) to account for the results of photolabeling of F_1 by azidonaphtoyl-ADP. In agreement with the hypothesis of a stochastic model, Miwa et al (1989) reported experiments in which F_1 from a thermophilic bacterium, reconstituted with one mutated ß subunit incorporated in a $\alpha_3\beta_3\gamma$ complex exhibited 46% of the activity of the wild-type complex. On the other hand, the one to one stoichiometry observed for inhibition of F_1-ATPases with ligands such as nitrobenzofurazan, phenylglyoxal, azidonitrophenylphosphate etc., , fits well with the alternative turnstile mechanism. Yet, the stochastic model might explain these data, provided that one postulates that the binding of one of these ligands to one ß subunit of F_1 is able to induce an asymmetric arrangement of the three subunits. In this situation, strong interaction between one of the two remaining ß subunits and the small γ, δ and ϵ subunits might prevent its participation in the catalytic cycle.

REFERENCES

Berden, J.A., Hartog, A.F., and Edel, C.M. (1991) Biochim. Biophys. Acta 1057, 151-156.

Bigay, J., Deterre, P., Pfister, C., and Chabre, M. (1987) EMBO J., 6, 2907-2913.

Boyer, P.D. (1989) FASEB J. 3, 2164-2178.

Carlier,M.F., Didry, D., Melki, R., Chabre, M., and Pantaloni, D. (1988) Biochemistry, 27, 3555-3559.

Chabre, M (1990) Trends Biochem. Sci. 15, 6-10.

Corbridge, D.E.C. (1974) in: The Structural Chemistry of Phosphorus, Elsevier, Amsterdam.

Cross, R.L. (1988) J. Bioenerg. Biomembr. 20, 395-405.

Cross, R.L., and Nalin, C.M. (1982) J. Biol. Chem. 257,2874-2881.

Cross, R.L., Grubmeyer, C., and Penefsky, H.S. (1982) J. Biol. Chem. 257, 120101-12105.

Cunningham, D., and Cross, R.L. (1988) J. Biol. Chem. 263, 18850-18856.

Dunn, S.D., and Futai, M. (1980) J. Biol. Chem. 255, 113-118.

Dupuis, A., Issartel, J-P., and Vignais, P.V. (1989) FEBS Lett. 255, 47-52.

Ebel, R.E., and Lardy, H.A. (1975) J. Biol. Chem. 250, 191-196.

Garin, J. (1989) Thesis, Université J. Fourier, Grenoble, France

Garin, J., Boulay, F., Issartel, J-P., Lunardi, J., and Vignais, P.V. (1986) Biochemistry, 25, 4431-4437.

Goldstein, G. (1964) Anal. Chem. 36, 243-244.

Gresser, M.J., Myers, J.A., and Boyer, P.D. (1982) J.Biol.Chem.

257, 12101-1205.

Grubmeyer, C., Cross, R.L., and Penefsky, H.S. (1982) J. Biol. Chem. 257, 12092-12100.

Issartel, J-P., and Vignais, P.V. (1984) Biochemistry 23, 6591-6595.

Issartel, J-P., Lunardi, J., and Vignais, P.V. (1986) J. Biol. Chem. 261, 895-901.

Issartel, J-P., Dupuis, A., Lunardi, J., and Vignais, P.V. (1991a) Biochemistry 30,4726-4733.

Issartel, J-P., Dupuis, A., Morat, C., and Girardet, J-L. (1991b) Eur. Biophys. J. 20, 115-126.

Jackson, G.E. (1988) Inorg. Chem. Acta 151, 273-276.

Kironde, F.A.S., and Cross, R.L. (1986), J. Biol. Chem. 261, 12544-12549.

Lübben, M., Lücken, U., Weber, J., and Schäfer, G. (1984) Eur. J. Biochem. 143, 483-490.

Lunardi, J., and Vignais, P.V. (1982) Biochim. Biophys. Acta 682, 124-134.

Lunardi, J., Dupuis, A., Garin, J., Issartel, J-P., Michel, L., Chabre, M., and Vignais, P.V. (1988) Proc. Natl. Acad. Sci. USA 85, 8958-8962.

Macdonald, T.L., Humphreys, W.G., and Martin, R.B. (1987) Science 236, 183-186.

Martin, R.B. (1986) Clin. Chem. (Winston-Salem, N.C.) 32, 1797-1806.

Martin, R.B. (1988) Biochem. Biophys. Res. Commun.(1988) 155, 1194-1200.

Missiaen, L., Wuytack, F., De Smedt, H., Vrolix, M., and Casteels, R. (1988) Biochem. J. 253, 827-833.

Miwa, K., Ohtsubo, M., Denda, K., Hisabori, T, Date, T., and Yoshida, M. (1989) J. Biochem. 106, 679-683.

Nelson, D.J., and Martin, R.B. (1991) J. Inorg. Biochem. 43, 37-43.

Paris, S., and Pouysségur, J. (1987) J. Biol. Chem. 262,1970-1976.

Penefsky, H.S. (1979) Adv. Enzymol. Relat. Areas Mol. Biol; 49, 223-280.

Perlin, D.S., Latchney, R.L., Wise, J.G., and Senior, A.E. (1984) Biochemistry, 23, 4998-5003.

Robinson, J.D., Davis, R.L., and Steinberg, M.(1986) J. Bioenerg. Biomemb. 18, 521-531.

Senior, A.E.(1988) Physiol. Rev. 68, 177-231.

Schuster, S.M., Ebel, R.E. and Lardy, H.A. (1975) J. Biol. Chem. 250, 7848-7853.

Sternweis, P.C., and Gilman, A.G. (1982) Proc. Natl. Acad. Sci. USA 79, 4888-4891.

Sutton, S.V.W., Bender, G.R., and Marquis, R.E. (1987) Infect. Immun. 55, 2597-2603.

Vignais, P.V., and Lunardi, J. (1985) Ann. Rev. Biochem. 54, 977-1014.

Wu, J.C., Lin, J., Chuan, H., and Wang, J.H. (1989) Biochemistry 28, 8905-8911.

Adenine Nucleotides in Cellular Energy Transfer and Signal Transduction
S. Papa, A. Azzi & J.M. Tager (eds)
© 1992 Birkhäuser Verlag, Basel/Switzerland

REGULATION OF ATP SYNTHESIS IN RAT HEART CELLS

David A. Harris and Anibh M. Das

Department of Biochemistry, South Parks Rd., Oxford OX1 3QU, U.K.

SUMMARY: As the work rate of the heart increases, its rate of mitochondrial ATP synthesis must increase. Quantitative studies show that a passive response of the ATP synthase to variations in [ADP], [P_i] or $\Delta\mu_{H+}$ cannot account for the changes in rate observed in vivo. Using cultured cardiomyocytes as an experimental system, we have shown a direct regulation of the mitochondrial ATP synthase to occur within cells, such that the capacity of this enzyme for turnover increases with increased work rate. This activation is triggered by intramitochondrial Ca^{2+}, but its precise molecular mechanism is unknown. This regulation is defective in pathological states of chronic heart overload, such as hypertension.

INTRODUCTION

Under normal physiological conditions, the heart is fully aerobic, with over 90% of its ATP made by oxidative phosphorylation in the mitochondria. Furthermore, the amount of ATP within the cell at any instant is sufficient to power beating for only 10-15 s. When variations in work load (e.g. in exercise) – up to 10-fold in the rat – are also considered, it is clear that ATP synthesis in heart mitochondria must be precisely regulated to balance utilisation in the cytoplasm. Various mechanisms have been proposed to account for this control, and these are summarised below.

'Respiratory control'

ATP synthesis rates in vivo are commonly believed to be substrate limited. In other words, ATP utilisation is assumed to cause increased levels of ADP (and P_i), which serve as substrates for ATP synthesis, whose rate thus increases. This type of model derives from early studies of 'respiratory control' and 'coupling' in intact mitochondria.

In fact, this idea can be expanded into three distinct regulatory mechanisms. These are:

(i) the thermodynamic model - in which the synthase maintains ATP and ADP at equilibrium with $\Delta\mu_{H+}$ (Erecinska & Wilson, 1982). Increases in the ADP/ATP ratio thus lead to an increased net flux of ADP phosphorylation.

(ii) the translocator model - in which the ATP synthase is limited in its turnover rate by the rate of ADP supply through the adenine nucleotide translocator (Groen et al., 1982) which, in turn, is limited by the supply of ADP to the outside of the mitochondrion.

(iii) the kinetic model - in which synthase turnover increases with extramitochondrial [ADP] in a simple, Michaelian manner.

The first two models appear not to apply to the heart. Firstly, the ATP synthase does not appear to act at equilibrium in heart cells (Kingsley-Hickman et al., 1986). Secondly, translocator activity is high relative to synthase activity (i.e. not limiting for ATP synthesis) in heart mitochondria (Doussiere et al., 1984). It is possible, however, that model (iii) operates in heart, i.e. the simple substrate dependence of the ATP synthase on cellular [ADP] can modulate ATP synthesis rates in vivo.

In the normal heart, however, the importance of this mechanism appear to be minimal. Measurements indicate that free cellular ADP concentrations are maintained within a narrow range under physiological conditions (Balaban, 1986; Unitt et al., 1989), and that any variations are too small to account for the observed changes in the rate of oxidative phosphorylation. In other words, quantitively model (iii) can account for little of the variation in ATP synthase rate seen under normal physiological conditions.

Dehydrogenase activation

An alternative model for regulation of oxidative phosphorylation has been proposed by McCormack and coworkers (for a review, see McCormack et al., 1990). In this model, regulation is exerted on the mitochondrial dehydrogenases, which are stimulated as work rate increases. Thus intramitochondrial [NADH] rises, oxidation rates rise (via an increasing saturation of NADH dehydrogenase) and consequently $\Delta\mu_{H+}$ and the ATP synthesis rate rise. In this model, the messenger that signals cytoplasmic work load is intramitochondrial $[Ca^{2+}]$ which, reflecting cytoplasmic $[Ca^{2+}]$, switches on pyruvate, α-oxoglutarate and isocitrate dehydrogenases.

There is considerable evidence that mitochondrial dehydrogenases are activated/inactivated in vivo by this mechanism. However, again this model cannot quantitatively explain the increases in ATP synthesis rate observed in vivo - where $\Delta\mu_{H+}$ appears to fall rather that to rise under conditions of increased ATP synthesis (Moreno-Sanchez et al., 1990; LaNoue & Bang Wan, 1991). It must be concluded, therefore, that the ATP synthase itself is directly regulated within the cell.

METHODS

Calcium tolerant cardiomyocytes were prepared from male Wistar rats (250-300g) except for experiments with hypertensive animals, when SHR (experimental) and Wistar-Kyoto (control) animals were used. Myocytes were cultured using the rapid attachment method of Piper et al. (1982). ATP synthase capacity was estimated from the oligomycin-sensitive ATPase activity, at saturating ATP concentrations, in sonicated extracts of these cultures (Das and Harris, 1990a). Cellular ATP was measured on neutralized perchloric acid extracts of cultured cells, using firefly luciferase, as previously (Das and Harris 1990c).

Error bars indicate S.E.M. (n>7).

RESULTS

ATP synthase regulation in the normal heart

To test the hypothesis that the ATP synthase is directly regulated, we developed a method for monitoring ATP synthase activity in heart cells. This method utilises cultured ventricular myocytes from adult rat heart, and measures ATP synthase capacity, the potential of this enzyme for turnover in vivo (Das & Harris, 1990a). We have shown that, as the workload on a heart is acutely increased, the capacity of its mitochondrial ATP synthase for turnover is increased. This occurs via an increase in V_{max} of the assembly of synthase enzymes i.e. by a true change in the kinetic parameters of this enzyme. The enzyme is thus actively regulated in vivo – it does not simply respond passively to substrate levels as implied in previous regulatory models (above).

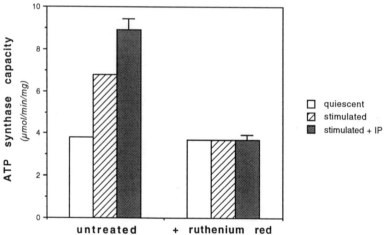

Figure 1: Variations in ATP synthase capacity of cardiomyocytes with increased work load. Cultured cardiomyocytes (10^5 cells per plate) were incubated in 25mM HEPES, 110mM NaCl, 10mM glucose, 2.6mM KCl, 1.2mM KH_2PO_4, 1.2mM $MgSO_4$, 1mM isoascorbate, 1mM $CaCl_2$, pH 7.4 (NaOH) (BUFFER I) under 100% O_2 at 37°C. Where indicated, isoprenaline (10µM) and/or ruthenium red (3µg/ml) were added. After 3 min. preincubation, the cells were stimulated electrically (10Hz, 20V/cm), where indicated, for 2 min. and their ATP synthase capacity measured as described.

Fig. 1 shows the extent of this regulation. In quiescent myocytes, the capacity of the ATP synthase for turnover is about 3.8 µmol/min/mg cell protein. Electrical stimulation increased this value by some 50%, and stimulation in the presence of positive inotropic agents (which increase the work rate of these cells) approximately doubled the capacity. This switch is rapid – the change was complete within 30s of its initiation – and reversible when the stimulus was removed (Das & Harris, 1989), as is expected for a regulatory mechanism operative in vivo. NADH oxidase activity in cell sonicates, measured as a control for mitochondrial recovery (in these and the following experiments) does not change under these conditions (data not shown).

Fig. 1 also shows that the trigger for this switch is intramitochondrial $[Ca^{2+}]$; the increase in capacity was blocked by ruthenium red, which prevents Ca^{2+} entry into the mitochondria from the cytoplasm. (It was also prevented by verapamil, which blocks Ca^{2+} entry into the cytoplasm from the medium (Das & Harris, 1990a) It appears, therefore, that the mitochondrial ATP synthase is activated in the cell in response to the rise in cytoplasmic $[Ca^{2+}]$ caused by increased contractile activity and/or positive inotropes.

This phenomenon can be reproduced in isolated mitochondria from rat heart (Fig 2). Note that in heart mitochondria, the 'resting' level of ATP synthase capacity was around 10 µmol/min/mg protein, consistent with the value of 3.8 µmol/min /mg cell protein (above) and the estimate of 40% heart cell protein as mitochondrial (Idell-Wenger et al., 1978).

This experimental system allows the investigation of the $[Ca^{2+}]$ sensitivity of the regulatory mechanism. Varying free $[Ca^{2+}]$ between 10^{-9} and 10^{-2}M showed that the transition from low capacity to high activity synthase occurs in the range 0.1-1 µM Ca^{2+}, believed to be the physiological range. This is the range over which the dehydrogenases of the tricarboxylic acid cycle have been shown to be activated. Fig. 2 also shows that Ca^{2+} entry into the mitochondria is necessary for ATP synthase activation, since both Ca^{2+} entry and synthase activation are blocked by ruthenium red.

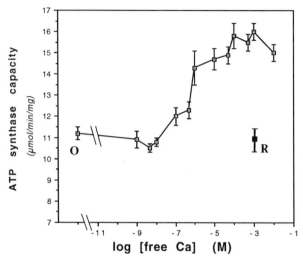

Figure 2: Response of the ATP synthase of rat heart mitochondria to external Ca^{2+}. Rat heart mitochondria (respiratory control index >15) were added to isotonic Ca^{2+} buffers at 37°C at the indicated concentrations to a final concentration of 1mg protein/ml. After 2 min., the mitochondria were broken by sonication and their ATP synthase capacity measured as above. O = value at zero added Ca^{2+}. R = value after preincubating mitochondria with ruthenium red (5nmol/mg).

Thus it appears that rises in intramitochondrial [Ca^{2+}], in response to rises in cytoplasmic [Ca^{2+}], lead in vivo to the coordinate activation of the dehydrogenases and the ATP synthase. This would allow increases in oxidation rate and ATP synthesis rates with minimal changes in [ADP] or $\Delta\mu_{H+}$ - just as is observed in the perfused heart (Balaban, 1986).

ATP synthase regulation in anoxia and ischaemia

In the absence of oxygen, the mitochondrial ATP synthase will not synthesize ATP. This must be made in the cytoplasm by substrate level phosphorylation. Under these conditions, indeed, the synthase is potentially capable of hydrolysing ATP since, when the mitochondrial membrane is de-energised, it can operate in reverse.

We investigated ATP synthase regulation in cultured rat cardiomyocytes exposed to anoxia. (Das & Harris, 1990b). Fig. 3

shows that, under anoxia, the ATP synthase capacity fell (rapidly) to about half the value observed in aerobic, quiescent cells. (Similar results were obtained if mitochondrial electron transfer was blocked by cyanide, or uncoupled with CCCP - data not shown.) In other words, under conditions where the synthase cannot function in the forward direction, its activity is switched off. This may limit loss of ATP from the cells by mitochondrial ATP hydrolysis under these conditions - indeed, cellular ATP is virtually unaltered during periods of anoxia of up to 30 min. This switch too is reversible; normal ATP synthase capacity was rapidly restored on reoxygenation (Fig. 3). Similar results have been observed in dog heart muscle made anoxic by sealing inside plastic bags (Rouslin, 1983).

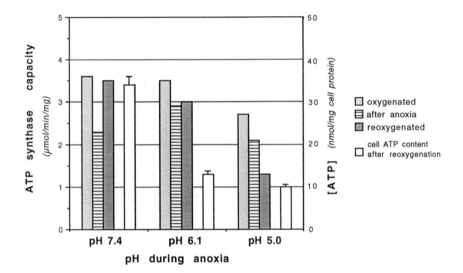

Figure 3: Effects of anoxia and pH on ATP synthase regulation Cultured cardiomyocytes were incubated in BUFFER I, adjusted to the pH values indicated by addition of lactic acid, under 100% O_2. O_2 was then replaced by N_2, and the cells incubated for 15 min (anoxia). Finally, the pH of the buffer was adjusted to pH 7.4 where necessary, and the supply of O_2 restored (reoxygenation). The ATP synthase capacity of cultured cardiomyocytes was measured after each stage, and the cellular ATP content was measured after reoxygenation (see Methods). The initial ATP content of the cells was 38 nmol/mg protein.

In vivo, cardiac anoxia is normally accompanied by restriction of cardiac blood flow, leading to the accumulation of lactate and a drop in local pH. These conditions – anoxia + low pH – together constitute ischaemia, and it is this condition (rather than anoxia alone) which leads to the irreversible damage of cardiac muscle in heart infarction. In fact, it is the period of the following reperfusion, when oxygen and normal pH are restored, when ischaemic damage occurs in vivo.

We mimicked this situation by lowering the pH in our anoxic cell cultures with lactic acid. Fig. 3 shows that if the pH was lowered to 6.1 or below during anoxia, the ATP synthase did not recover when the original conditions were restored, and a large drop in cellular ATP levels occurred. Thus irreversible loss of function in ischaemia is associated with irreversible loss of ATP synthase regulation. Such effects may be related to abnormal movements in cell Ca^{2+} during ischaemia and reperfusion (the 'calcium paradox') but further work is necessary to identify causal relationships in this system.

ATP synthase regulation in chronic work overload

In certain pathological conditions, such as hypertension, the heart is required to sustain high work rates continuously for periods of weeks or months. Under these conditions, chronic changes in the heart occur; the ventricles enlarge (hypertrophy) and the likelihood of heart failure is increased.

We produced chronic work overload in rat heart by two procedures – genetically, using spontaneously hypertensive (SHR) rats (pressure overload), and pharmacologically, by administering thyroxine over a period of 7 days (volume overload). These very different procedures had different effects on mitochondrial protein content; SHR cells have a normal complement of the ATP synthase protein while in thyroxine treated rats, this protein is raised by about 25% (Das & Harris, 1990c, 1991).

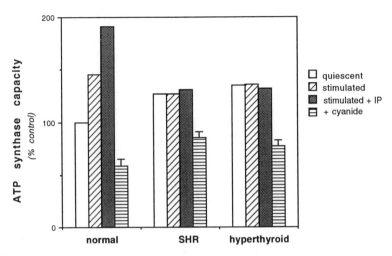

Figure 4: Regulation of the ATP synthase of cardiomyocytes after chronic heart overload. Cardiomyocytes were prepared from control rats (see Methods), SHR rats and rats treated with thyroxine (350μg/kg body wt per day for 7 days). Their ATP synthase capacity was measured in quiescent cells, after electrical stimulation, and after isoprenaline treatment as described in Fig. 1. Sodium cyanide, where indicated, was present at 1mM.

Cardiomyocytes from these rats were cultured, and showed normal ATP content and and contractility. In both cases, hypertrophy of the heart was reflected in the myocytes, which were 30-50% longer than cells from normal hearts. Investigation of their ATP synthase capacities (Fig. 4), however, showed clearly that regulation of this enzyme in these cells was abnormal. First, quiescent cells from both SHR and thyroxine-treated rats had an ATP synthase capacity about 30% higher than cells from normal animals. Secondly, up-regulation of the synthase in response to an increased work load was absent in the treated rats (despite there being a large potential excess capacity of synthase in the mitochondria). However, down-regulation of the synthase in response to cyanide (or anoxia) was normal – a decrease of about 40% from the capacity in the respective quiescent cells was observed in all cases (Fig. 4).

This loss in regulation was accompanied by an inability to maintain cellular ATP levels. In normal cells, ATP levels remained within narrow limits (38-40 nmol/mg cell protein) irrespective of electrical or hormonal stimulation (Das & Harris, 1990a). In cells from hypertensive or hyperthyroid animals, in contrast, electrical or hormonal stimulation led to a drop in ATP levels of 20-30% within 2 min of initiating the stimulus (Das & Harris, 1991, Harris & Das 1992). This drop was transient, at least in the case of SHR cells, suggesting that ATP synthesis can be speeded up by means other than by synthase regulation (see above). However, this work suggests that regulation of the synthase is necessary for maintaining a constant energy balance. It may be that irreversible damage occurs if this ATP deficit in the abnormal cells becomes too great, but this point is not investigated in the present work.

DISCUSSION

Regulation of the ATP synthase in vivo

From a quantitative assessment of previous models for regulation of mitochondrial ATP synthesis in heart, it was predicted that a direct regulation of the synthase enzyme should occur. The above work indicates that this enzyme is directly regulated within heart cells, and that this regulation is a response to the mitochondrial import of Ca^{2+} ions from the cytoplasm (Fig. 1,2).

This work also shows how pathological situations affect ATP synthase regulation. Under ischaemic conditions (Fig. 3), reversibility of down-regulation is lost, and the cells suffers irreversible loss of ATP. In contrast, chronic heart overload leads to loss of up-regulation of the synthase (Fig. 4). This condition leads to only transient imbalances in cellular ATP levels in our cell preparations - but even such transient effect may be a contributory factor in heart failure in the long term.

Molecular mechanism of ATP synthase regulation

The mechanism for ATP synthase regulation in vivo is unknown. However, the above results limit the type of mechanisms that are possible. Low to high activity transitions (above) are too fast to involve de novo protein synthesis. Regulation cannot involve allosteric interaction of a small molecule regulator (e.g.Ca^{2+} ions) with F_1F_o, as the activated state persists through cell breakage and dilution/removal of Ca^{2+} in the assay medium. It is also unlikely that regulation involves protein kinase activity, since work over many years has been unable to show phosphorylation of the ATP synthase complex.

Down-regulation of the ATP synthase, in anoxia or ischaemia, has been shown in dog heart, using immunochemical methods, to be due to the ATPase inhibitor protein, IF_1, found normally in mitochondria (Rouslin and Pullman, 1987). This protein binds tightly to the ATP synthase (inhibiting both synthesis and hydrolysis) and is displaced from its inhibitory site with increasing membrane potential (Power et al., 1983). There is some doubt as to whether enough of this protein is present in rat heart mitochondria to inhibit all ATP synthase molecules (Rouslin, 1987), but it is likely that, even here, IF_1 may be the down-regulator.

The mechanism of up-regulation is less clear. Although it is tempting to assign activation of the ATP synthase to the reversal of inactivation i.e. to implicate release of IF_1 in the process, there is as yet no evidence to support this view. On the contrary, neither IF_1 nor F_1F_o appear to respond to Ca^{2+} - above implicated as the likely secondary messenger in synthase activation - in isolated membrane preparations. Further, as can be seen from Fig. 4, up regulation of the ATP synthase can be lost in pathological states such as hypertension without loss of down-regulation, perhaps suggesting two separate mechanisms. One possibility for the role of regulator is a second mitochondrial protein, the Calcium Binding Inhibitor (CaBI) isolated by Yamada and Huzel (1988), which inhibits ATP hydrolysis by the uncoupled ATP synthase. The inhibitory effect of CaBI is abolished around $1\mu M$ added Ca^{2+}.

However, detailed studies of the nature of this protein and its interaction with F_1F_0 remain to be undertaken.

ACKNOWLEDGMENTS: We thank the British Heart Foundation (DAH,AMD) and the Wellcome Trust (DAH) for support.

REFERENCES

Balaban, R.S. (1986)   Science 232, 1121-1124

Das, A.M. and Harris, D.A. (1989)   FEBS Lett. 256, 97-100

Das, A.M. and Harris, D.A. (1990a)   Cardiovasc. Res. 24, 411-417

Das, A.M. and Harris, D.A. (1990b)   Biochem. J. 266, 355-361

Das, A.M. and Harris, D.A. (1990c)   Amer. J. Physiol. 259, H1264-H1269

Das, A.M. and Harris, D.A. (1991)   Biochim. Biophys. Acta. 1096, 284-290

Doussiere, J. Ligeti, E., Brandolin, G. and Vignais, P.V. (1984) Biochim. Biophys. Acta 766, 492-500

Erecinska, M. and Wilson, D.F. (1982) J. Memb. Biol. 70, 1-14

Groen, A.K., Wanders, R.J.A., Westerhoff, H.V., van de Meer, R. and Tager, J.M. (1982)   J. Biol. Chem. 257, 2754-2757

Harris, D.A. and Das, A.M. (1992)   Biochem. J. in press

Idell-Wenger, J.A., Grotyohann, L.W. and Neely, J.R. (1978) J. Biol. Chem. 253, 4310-4318

Kingsley-Hickman, P. Sato, E.Y., Andreone, P.A., St. Cyr, J.A., Michurski, S., Foker, J.E., From, H.A.L., Petein, M. and Ugurbil, K. (1986) FEBS Lett. 198, 159-163

LaNoue, K.F. and Bang Wan (1991) J. Mol. Cell. Cardiol. 23 (Suppl. III) 36

McCormack, J.G., Halestrap, A.P. and Denton, R.M. (1990) Physiol. Rev. 70, 391-425

Moreno-Sanchez, M., Hogue, B.A. and Hansford, R.G. (1990) Biochem. J. 268, 421-428

Piper, H.M., Probst, I., Schwartz, J.F. and Spiekermann, P.G. (1982)   J. Mol. Cell. Cardiol. 14, 397-412

Power, J., Cross, R.L. and Harris, D.A. (1983) Biochim. Biophys. Acta 724, 128-141

Rouslin, W. (1983) J. Biol. Chem. 258, 9657-9661

Rouslin, W. (1987) Amer. J. Physiol 252, H622-H627

Rouslin, W. and Pullman, M.E. (1987)   J. Mol. Cell. Cardiol. 19, 661-668

Unitt, J.F., McCormack, J.G., Reid, D., MacLachlan, L.K. and England, P.J. (1989) Biochem. J. 262, 293-301

Yamada, E.W. and Huzel, N.J. (1988) J. Biol. Chem. 263, 11498-503

UNCOUPLING ACTION OF FATTY ACIDS ON HEART MUSCLE MITOCHONDRIA AND SUBMITOCHONDRIAL PARTICLES IS MEDIATED BY THE ATP/ADP ANTIPORTER

Vladimir P. Skulachev[1], Vera I. Dedukhova[1], Elena N. Mokhova[1], Anatoly A. Starkov[1], Eduardo Arigoni-Martelli[2], Valentina A. Bobyleva[3]

[1]Department of Bioenergetics, A.N. Belozersky Institute of Physico-Chemical Biology, Moscow State University, Moscow 119899, USSR; [2]Sigma-Tau, Pomezia, Roma, Italy; [3]Institute of General Pathology, Modena, Italy

Abbreviations: $\Delta\Psi$, Transmembrane electric potential difference; Atr, atractylate; BSA, bovine serum albumin; CAtr, carboxyatractylate; DNP, 2,4-dinitrophenol; EDTA, ethylenediaminetetraacetic acid; EGTA, ethylene glycol-bis (ß-aminoethyl ether) - N,N,N',N'-tetracetic acid; FCCP, p-trifluoromethoxycarbonylcyanide phenylhydrazone; MOPS, morpholinopropane sulphonate; PCB$^-$, phenyldicarbaundecaborane; SDS, sodium dodecyl sulfate; TPP$^+$, tetraphenyl phosphonium.

SUMMARY: Effect of the ATP/ADP-antiporter inhibitors on the palmitate-induced uncoupling is studied in heart muscle mitochondria and the inside-out submitochondrial particles. In both systems palmitate is found to decrease the respiration-generated membrane potential. In mitochondria this effect is specifically abolished by carboxyatractylate (CAtr), non penetrating inhibitor of antiporter. In the particles CAtr does not abolish the palmitate-induced potential decrease. At the same time, bongkrekic acid, penetrating inhibitor of the antiporter, suppresses palmitate effect on the potential both in mitochondria and particles. Palmitoyl-CoA which is known to inhibit the antiporter in mitochondria as well as in particles decreases the palmitate uncoupling efficiency in both systems.
Palmytoyl carnitine does not substitute for either palmitate or palmitoyl-CoA. The data are in agreement with hypothesis that the ATP/ADP-antiporter is involved in the action of free fatty acids as natural uncouplers of oxidative phosphorylation.

INTRODUCTION

Fatty acids are known to differ from classical protonophorous uncouplers in that they penetrate through phospholipid bilayer only in its protonated form. As to their anions, they are non-penetrants for the bilayer (Walter and Gutknecht, 1984; Andreyev et al. 1989). This is why they fail to uncouple in cytochrome oxidase proteoliposomes (Andreyev et al., 1989). To explain the mechanism of fatty acid uncoupling in mitochondria we suggested that there are protein(s) in the inner mitochondrial membrane that facilitate transport of fatty acid anionic species (Andreyev et al., 1989; 1987 a; 1987 b; Skulachev 1988, 1991).

In brown fat which is specialized in heat production by means of the fatty acid-mediated uncoupling, the role of fatty acid anion porter was postulated to be performed by the "uncoupling protein" thermogenin (Skulachev, 1991; Garlid, 1990).

Recently this suggestion seems to be confirmed by Jezek and Garlid (Garlid, 1990; Jezek and Garlid, 1990) who found that thermogenin transports various monovalent anions which can substitute for fatty acids in activating the thermogenin-linked H^+-conducting activity. Both these properties increase with the increase in the lenght of the hydrocarbon chain of the anion.

Thermogenin is absent from tissues other than brown fat, but nevertheless fatty acids can uncouple oxidation an phosphorylation in these tissues (for review, see Skulachev

1988; 1991). We assumed (Andreyev et al., 1989; Skulachev 1988; 1991) that is such cases, the role of the fatty acid anion porter is performed by the ATP/ADP antiporter, a protein which is very similar to thermogenin in the amino acid sequence, domain structure and some other properties (Aquila et al. 1985; 1987; Bouilland et al. 1986). It was suggested that the adenine nucleotide anion-translocating machinery of the antiporter hydrophobic anions with no ATP^{4-} (ADP^{-3})-specific gate involved (Andreyev et al., 1989; Skulachev 1988; 1991). In agreement with the above hypothesis, the following observations were done in our group: (1) The ATP/ADP antiporter inhibitors (CAtr, Atr, bongkrekic acid, pyridoxal 5-phosphate) and its substrate (ADP anion) specifically inhibit the fatty acid-induced uncoupling in the liver and skeletal muscle mitochondria, and are without effect on the FCCP-induced uncoupling (Andreyev et al., 1989; 1987 a; b); (2) CAtr-sensitive uncoupling is inherent not only in fatty acids but also in other hydrophobic monovalent anions such as dodecyl sulfate and cholate (Brustovetsky et al., 1990 a); (3) The burst in heat production during arousal of the hibernating ground squirrels is accompanied (i) by the appearance of the CAtr-sensitive uncoupling in heart; skeletal muscle and liver mitochondria and (ii) by an increase in the free fatty acid concentration in tissues and in isolated mitochondria (Brustovetsky et al., 1990 b, 1991).

The observation (1) was recently confirmed and extended by Schonfeld (1990) who showed that the uncoupling activity of a fatty acid varies in different tissues being proportional to the ATP/ADP antiporter content (heart > kidney > liver).

In this paper, we compared the uncoupling effect of fatty acids in heart muscle mitochondria and inside-out submitochondrial particles. It was found that the uncoupling efficiencies of palmitate in these two systems are similar but in the particles, the non-penetrating ATP/ADP antiporter inhibitor, CAtr, fails to inhibit the uncoupling. On the other hand, bongkrekic acid which is penetrating inhibitor is effective in both mitochondria and particles.

MATERIALS AND METHODS

Mithochondria were isolated from the rabbit or rat heart muscle. Cooled muscles, previously separated from fat and tendons, were minced and were passed through stainless steel press with holes about 1 mm diameter.

Tissue was then homogenized for 3 min, using a Teflon pestle in a glass (Pirex) homogenizer, the tissue: isolation medium ratio being 1:8.

After the first centrifugation (10 min, 600 g), the supernatant was decanted and filtered through 4 layer of gauze. The mixture obtained was centrifuged (10 min, 12000 g), the sediment was suspended in 1 ml of the isolation medium (250 mM sucrose, 10 mM MOPS, 1 mM EDTA, pH 7.4) supplemented with BSA (3 mg/ml), then the medium without BSA was added. The final mitochondrial precipitate (10 min, 12000 g) was suspended in the isolation medium with BSA. The mitochondrial suspension (70-90 mg/ml) was stored on ice.

The inside-out AS-submitochondrial particles were prepared according to Kotlar and Vinogradov (1990).

In the majority of the experiments, the incubation medium contained 250 mM sucrose, 10 mM MOPS, and 0.5 mM EGTA, pH 7.4. Glutamate (5 mM) and malate (5 mM) or 5 mM succinate were used as oxidation subtrates. In the latter case, the medium contained 1 μM rotenone.

Oxygen consumption was recorded with a Clark-type oxygen electrode and LP-7E polarograph, the incubation mixture being stirred by a glass-covered magnetic stirrer bar. The concentration of mitochondrial protein was 1 mg/ml (glutamate and malate) or 0.4 mg/ml (succinate). The temperature was 26°C.

The synthetic penetrating ions (TPP$^+$ and PCB$^-$) were used as $\Delta\Psi$ probes (Grinius et al., 1970; Bakeeva et al., 1970). The penetrating ion concentration in solution was measured with TPP$^+$ and PCB$^-$-sensitive electrodes (Kamo et al., 1979). Oligomycin, MOPS, palmitic acid, palmitoyl-CoA, CAtr, Atr and fatty acid-free BSA were from Sigma; EDTA, EGTA, ADP, GDP, rotenone, palmitoyl-L-carnitine and DNP from Serva ; malate from Fluka; and FCCP from Boehringer.

RESULTS

Intact mitochondria

As seen in Fig. 1, curves a-c, CAtr was strongly inhibitory for respiration of rat heart mitochondria, stimulated by palmitate in the presence of oligomycin. The inhibition was released by DNP. ADP (curve c), Atr (curve d, palmitoyl CoA (curve h) and bongkrekic acid (not shown) were also inhibitory but their effects were not so strong as that of CAtr. Moreover, CAtr added after Atr failed to inhibit the palmitate-stimulated respiration as strong as without Atr (curve d). The CAtr-sensitive stimulation of respiration could also be inducted by SDS but at concentration higher than that of palmitate (curves e and f). Palmitoyl carnitine could not substitute for neither palmitate or palmitoyl CoA (curve i and j).

Inhibiting concentrations of Atr and Catr for palmitate-stimulated respiration proved of the same order of magnitude as those for State 3 respiration (Fig. 2).

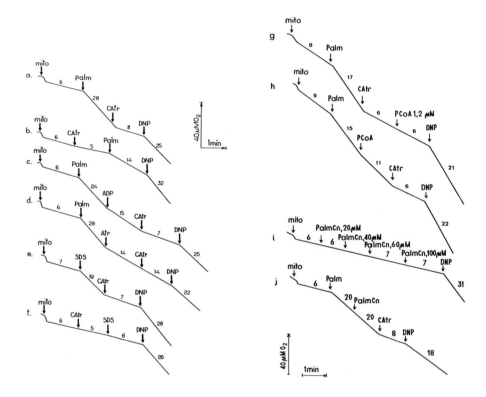

Fig.1 Effect of the ATP/ADP antiporter inhibitors and ADP on the respiration of heart mitochondria. Incubation mixture: 0.25 M sucrose, 10 mM MOPS, 0.5 mM EGTA, 5 mM glutamate, 5 mM malate, oligomycin (1 μg/ml), BSA (0.2 mg protein/ml), pH 7.4, temperature 27°C. Additions: mito, a-d, i, j, rat heart mitochondria (1 mg/ml); e, f, rabbit heart mitochondria (1 mg protein/ml); g, h, rat heart mitochondria (2.2 mg/ml); Palm, 30 μM palmitic acid; PCoA, 1.2 μM palmitoyl CoA; PalmCn, palmitoylcarnitine; 2 μM CAtr, 40 μM ADP; 90 μM SDS; 5 μM (a-d) and 8 μM (e-j) DNP. In g, h, incubation mixture was supplemented with 0.5 mM MgCl$_2$.

90

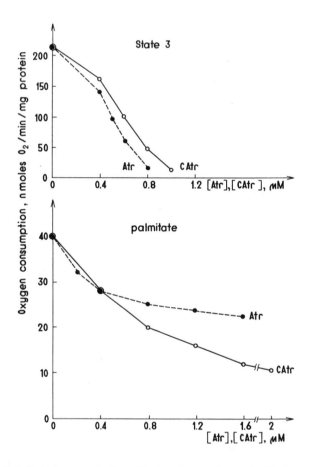

Fig.2 Inhibition of the State 3 and palmitate-stimulated
      respiration of rabbit heart mitochondria by CAtr and
      Atr.
Incubation mixture as in Fig. 1, but oligomycin was omitted.
The incubation mixture was supplemented with 0.7 mM $MgCl_2$, 1
mM KH_2PO_4, 10 mM glucose; hexokinase (3U/ml), 100 μM ADP;
and mitochondria (0.6 mg protein/ml) (upper figure) or with
0.5 mM $MgCl_2$, 40 μM palmitic acid and mitochondria (1 mg
protein/ml) (lower figure).

Stimulation of the mitochondrial respiration by palmitate was accompanied by a $\Delta\Psi$ decrease which was completely of partially reversed by CAtr or bongkrekic acid, respectively (Fig, 3). ADP was also foud to increase the palmitate-lowered $\Delta\Psi$ but not as strongly as CAtr (Fig, 4A). The $\Delta\Psi$ decrease by low concentration of DPN was partially reversed by ADP and CAtr. On the other hand, ADP and CAtr were almost without effect when FCCP was used as uncoupler.

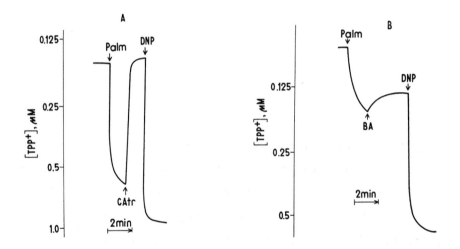

Fig. 3 Effects of CAtr and bongkrekic acid (BA) on palmitate induced $\Delta\Psi$ decrease in mitochondria.
Incubation mixture: 0.25 M sucrose, 10 mM MOPS, 0.5 mM EGTA; 5 mM succinate; 1 µM rotenone, oligomycin (2 µg/ml), BSA (0.4 mg/ml), 1.5 µM TPF$^+$, pH 7.4, Temperature 26°C. Additions: mito, rat heart mitochondria (0.8 mg protein/ml), 1 µM CAtr, 20 µM DNP, 5 µM BA, 40 µM (A) and 20 µM (B) palmitic acid.

92

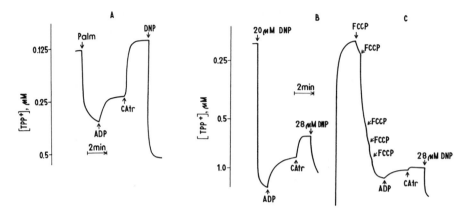

Fig. 4   Sensitivity of effects of different   uncouplers   to
        ADP and CAtr.
Experimental   conditions as in Fig. 1. Additions: rat   heart
mitochondria (A,   0.6 mg and B, 0.8 mg protein/ml),   20   µM
ADP,   1   µM   CAtr,   20   µM   DNP   (A),   120   nM   FCCP   (final
concentration), 1.5 µM TPP.

Inside-out submitochondrial particles

        Inside-out beef heart submitochondrial particles   were
also found to respond to the palmitate addition by   lowering
of ΔΨ .   However, in this case CAtr did   not   suppress   the
palmitate   effect.   On   the   other   hand,   bongkrekic   acid
reversed   the   palmitate effect.   Again, as in mitochondria,
subsequent   addition   of an artificial   protonophore   (FCCP)
caused   a   ΔΨ   decrease (Fig. 5). Palmitoyl   CoA   proved   to
decrease   the   palmitate effect in   both   mitochondria   and
submitochondrial particles (not shown).

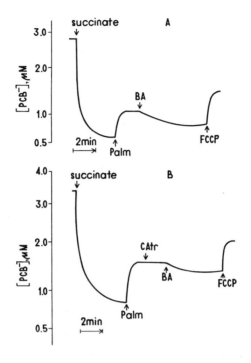

Fig. 5 Effects of bongkrekic acid, CAtr and FCCP on palmitate-induced decrease of ΔΨ in submitochondrial particles.

Incubation mixture: 0.25 M sucrose, 10 mM MOPS, 0.5 mM EGTA, submitochondrial particles (0.45 mg protein/ml), 1 μM rotenone, oligomycin (4 μg/ml), 15 μM PCB⁻, pH 7.3, Temperature 26°C. Additions: 5 mM succinate; 1 μM CAtr; BA, 8 μM bongkrekic acid; palm, 30 μM palmitic acid; 1 μM FCCP.

DISCUSSION

The data presented above clearly indicate that the membrane sidedness is critical for CAtr inhibition of the palmitate-induced uncoupling in heart muscle mitochondria and their inside-out particles. CAtr effectively abolishes the uncoupling in the mitochondria but not in the particles. On the other hand, some inhibiting effects of bongkrekic acid were seen in both mitochondria and particles.

Just these relationships could be predicted assuming that it is the ATP/ADP antiporter that mediates the uncoupling effect of free fatty acids. It is know that the CAtr and bongkrekate binding sites of the antiporter are localized on outer and inner surfaces of the inner mitochondrial membrane, respectively, whereas sites for palmitoyl-CoA are on both membrane surfaces. It has also been found that CAtr cannot penetrate through the membrane whereas bongkrekic acid can (for review see Vignais et al., 1985).

Thus comparison of heart mitochondria and particles gives one more piece of evidence that the ATP/ADP antiporter is somehow involved in the fatty acid uncoupling. Apparently the antiporter also facilitates uncoupling by SDS and low concentrations of DNP (Figs. 1 and 4). Such an effect may be explained by translocation of anions of these compounds by the antiporter, resembling in this respect thermogenin.

The results described on heart mitochondria confirmed our previous observations made on liver and skeletal muscle mitochondria (Andreyev et al., 1989; 1987 a, b), i.e. among specific inhibitors of the antiporter, CAtr is the most potent agent for suppressing the fatty acid-induced uncoupling, whereas ADP, Atr, palmitoyl CoA and bongkrekic acid are less effective. Atr added before CAtr prevents, at least for some time, complete inhibition of the palmitate-stimulated respiration by CAtr (see also Andreyev et al., 1989), It should be emphasized that the Atr concentration used was quite sufficient to completely inhibit State 3 respiration (Fig. 2). A probable explanation of this fact is that all the tested inhibitors combine with the interface-localized, substrate-specific antiporter gates rather than with the non specific anion-translocating mechanism inlaid in the membrane core. Apparently, modification of gate decelerates the translocator mechanism and, as a result, inhibits the fatty acid anion translocation rate. Such a deceleration may vary depending on the kinf of inhibitor.

ACKNOWLEDGEMENTS

The authors are very grateful to Prof. A.D. Vinograd for providing as with a preparation of tightly-coupled submitochondrial particles, and Sigma-Tau for financial support.

REFERENCES

Andreyev, A.Yu., Bondareva, T.O., Dedukhova, V.I., Mokhova, E.N., Skulachev, V.P. and Volkov, N.I. (1987a) FEBS Lett. 226, 265-269.
Andreyev, A.Yu., Bondareva, T.O., Dedukhova, V.I., Mokhova, E.N., Skulachev, V.P., Tsofina, L.M., Volkov, N.I. and Vygodina, T.V. (1989) Eur. J. Biochem. 182, 585-592.
Andreyev, A.Yu., Volkov, N.I., Mokhova, E.N. and Skulachev, V.P. (1987b) Biol. Membrany 4, 474-478 (Russ.).
Aquila, H., Link, T.A. and Klingenberg, M. (1985) EMBO J. 4, 2369-2376.
Aquila, H., Link, T.A. and Klingenberg, M. (1987) FEBS Lett. 212, 1-9.
Bakeeva, L.E., Grinius, L.L. Jasaitis, A.A., Kuliene V.V., Levitsky, D.O., Liberman, E.A., Severina, I.I. and Skulachev, V.P. (1970) Biochim. Biophys. Acta 216, 13-21.
Bouilland, F., Weissenbach, J. and Ricquier, D. (1986), J. Biol. Chem. 261, 1487-1490.
Brustovetsky, N.N., Dedukhova, V.I., Yegorova, M.N., Mokhova, E.N. and Skulachev, V.P. (1990a) FEBS Lett. 272, 187-189.
Brustovetsky, N.N., Amerkhanov, Z.G., Yegorova, M.N., Mokhova, E.N. and Skulachev, V.P. (1990b) FEBS Lett. 272, 190-192.
Brustovetsky, N.N., Amerkhanov, Z.G., Yegorova, M.N., Mokhova, E.N. and Skulachev, V.P. (1991) Biokhimiya 56, 947-953 (Russ).

Garlid, K.D. (1990) Biochim. Biophys. Acta 1018, 151-154.

Jezek, P. and Garlid, K.D. (1990) J. Biol. Chem. 265, 19303-19311.

Grinius, L.L., Jasaitis, A.A., Kadziauskas, J.P., Liberman, E.A., Skulachev, V.P., Topali, V.P. and Vladimirova, M.A. (1970) Biochim. Biophys. Acta 216, 1-12.

Kamo, N., Muratsuga, M., Hongoh, R. and Kobatake, Y. (1979) J. Membr. Biol. 49, 105-121.

Kotlylar, A.B. and Vinogradov, A.D. (1990) Biochim. Biophys. Acta 1019, 151-158.

Schonfeld, P. (1990) FEBS Lett. 264, 246-248.

Skulachev, V.P. (1988) Membrane Bioenergetics, Springer Verlag, Berlin.

Skulachev, V.P. (1991) FEBS Lett. (accepted).

Vignais, P.V., Block, M.R., Boulay, F., Brandolin, G., Lauquin, G.J.M. (1985) In: Structure and properties of cell membranes (G. Bengha, Ed.) CRC Press, v. 11, pp. 139-179.

Walter, A. and Gutknecht, J. (1984) J. Membr. Biol. 77, 255-264.

MITOCHONDRIAL FUNCTION AND BIOGENESIS IN CULTURED MAMMALIAN
CELLS WITHOUT FUNCTIONAL RESPIRATORY CHAINS

Coby Van den Bogert[1], Nicole H. Herzberg[2], Johannes N.
Spelbrink[1], Henk L. Dekker[1], Brenda H. Groen[1] and Piet A.
Bolhuis[2]

[1]E.C. Slater Institute for Biochemical Research; and [2]Department
of Neurology, University of Amsterdam, Meibergdreef 15, 1105 AZ
Amsterdam, The Netherlands

SUMMARY: Long-term impairment of mitochondrial gene expression
is possible in cultured mammalian cells grown under specific
conditions. This allowed us to study the expression and the
function of nuclearly coded mitochondrial proteins in cells that
contain no mitochondrial gene products and that are therefore
depleted of functional oxidative phosphorylation systems. These
cells appeared to synthesise nuclearly coded mitochondrial
proteins at near-normal rates. The proteins are imported into
the mitochondria, implying that a mitochondrial membrane
potential can still be generated. We conclude that mitochondrial
gene expression does not influence the expression of nuclear
mitochondrial genes, at least not under the culture conditions
used.

INTRODUCTION

Mitochondria contain a number of copies of small DNA molecules
(mtDNA). In mammalian cells mtDNA encodes only 13 polypeptides
and a full set of mitochondrial tRNA's and rRNA's. The majority
of enzymes and other compounds involved in the transcription and
translation of the mitochondrial polypeptide genes (a process
which takes place inside the mitochondria) are thus of nuclear
genetic origin. Mitochondrial gene expression therefore depends
largely on the expression of nuclear genes and leads to the
formation of 13 polypeptides (Nelson, 1987). All the polypeptides resulting from mitochondrial gene expression are

subunits of enzymes involved in oxidative phosphorylation: 7 are
part of complex I (NADH-dehydrogenase), 1 of complex III (bc_1
complex), 3 of complex IV (cytochrome c oxidase) and 2 of
complex V (ATP synthase). The many other subunits of these
complexes as well as those of complex II (succinate dehydrogenase) are nuclear gene products (Chomyn & Attardi, 1987). The
formation of these complexes is dependent on the nuclear gene
products involved in the import of the nuclearly coded subunits
and in the assembly of mitochondrially and nuclearly coded
subunits.

The oxidative phosphorylation system requires reduction
equivalents for its function. These are generated in cultured
cells mainly in the Krebs-cycle (Donelly & Schleffler, 1976).
Moreover, this cycle produces several metabolic intermediates
which can be used as precursors in anabolic pathways inside or
outside the mitochondria. Import systems for mitochondrial
enzymes involved in intermediary metabolism and transport
systems for substrates and products are required also.

Both biogenesis and function of mitochondria therefore
depend on complex interactions between nuclear and mitochondrial gene expression. In turn, this implies that
mitochondrial dysfunction might result from an impaired
expression of nuclear genes and/or mitochondrial genes. In the
past few years, many research groups have shown that mutations
of mtDNA are related to the occurence of mitochondrial diseases
in man (Lander & Lodish, 1990). Frequently, the mutations result
in a decreased expression of all the mitochondrial polypeptide
genes, for instance when deletions involve one or more tRNA's or
if point mutations in tRNA's are present. In several cases both
mutated mtDNA and normal mtDNA are present, the ratio between
both populations most likely being important for the severity of
the clinical symptoms.

Effects of an impaired expression of mtDNA can be studied in
cultured human cells as model systems. Mitochondrial gene
expression can be blocked in these cells either by inhibition of
mitochondrial protein synthesis by for instance tetracyclines or

by inhibition of replication and transcription of mtDNA by ethidium bromide. In the former case, the concentration of mitochondrially coded polypeptides will be halved at every cell division since their synthesis is blocked. In the latter case the result will be the same, but this is now achieved by reducing the concentration of mitochondrial genes by 50 % at every division. We analysed the effects of a decreasing concentration of mitochondrial gene products on cellular function, and studied the effects of impaired mtDNA expression on the expression of nuclear genes for mitochondrial proteins, using transformed and primary human cell lines as model systems.

MATERIALS AND METHODS

Primary human fibroblasts, Molt4 cells (a human leukemic cell line) and other human tumor cell lines were cultured as described before, in either DME or RPMI as culture medium (Van den Bogert et al., 1991). Primary human myoblasts were cultured in DME supplemented with 20 % FCS and 2 % chicken embryo extract. Determination of the activity of mitochondrial enzymes, immunoprecipitation, electrophoresis, blotting followed by immunodetection and immunofluorescence in stitu of these enzymes have also been described before (Van den Bogert et al., 1988; Van den Bogert et al., 1991). The ATP and ADP contents of the cells were measured by an FPLC method after extraction of the cells with ice-cold methanol. Mitochondrial ATP production was measured in cells permeabilized with digitonin (Granger & Lehninger, 1982; Kroon et al., 1985).

RESULTS AND DISCUSSION

Many authors have shown that proliferation of mammalian cells cultured in media such as RPMI 1640 is arrested after 1-2 cell divisions if mitochondrial gene expression is inhibited continuously. This suggests that reduction below 50 % of the normal value of the activity of 4 of the 5 enzyme complexes involved in oxidative phosphorylation leads to a shortage of ATP-generating capacity. The observation that increasing ADP but decreasing ATP levels precede proliferation arrest by the tetracyclines in, for instance, Molt4 cells (Van den Bogert et al., 1988) support this assumption. An increased rate of glycolytic ATP synthesis is apparently not able to compensate the reduced capacity for oxidative phosphorylation under these conditions.

We tested the possibility that proliferation arrest might be avoided or postponed if the cells were cultured under conditions where their ATP demand could be less. This condition appeared to be met by culturing Molt4 cells in DME medium, a medium which contains substantially higher concentrations of amino acids and vitamins than RPMI. Accumulation of ADP or decreasing ATP levels were not observed during inhibition of mitochondrial protein synthesis in Molt4 cells cultured in DME and the cells were able to perform about 5 divisions. After 3-4 divisions in the presence of inhibitors, mitochondrial ATP production was below the detection limit, as was the activity of enzymes which are partly coded by the mitochondrial genome (Table I). If the activity of the respiratory chain is that much affected, it can be expected that NADH accumulates to an extent that the activity of the Krebs-cycle is severely impaired. This might explain that addition of pyruvate to DME and raising the CO_2 tension to 10 % appeared to prevent proliferation arrest of cells without a functional oxidative phosphorylation system. Pyruvate (and CO_2) most likely serve as alternate sources for indispensable metabolic precursors under these conditions, as has been described before for mutant Chinese Hamster Ovary cells (Whitfield, 1985). The effects of inhibition of mitochondrial protein synthesis on proliferation of Molt4 cells are summarised in Table I. They lead to the conclusion that Molt4 cells can grow without functional mitochondria by adapting their growth rate (and size) provided that they are cultured in a rich medium which contains metabolic precursors which normally result from mitochondrial intermediary metabolism.

This conclusion is, however, not specific for Molt4 cells or for inhibition of mitochondrial protein synthesis. Several other cell lines, including primary human fibroblasts and myoblasts, showed a comparable, medium type related response to inhibition of mitochondrial gene expression, achieved either by inhibition of translation, transcription or replication of mtDNA.

Table I: Effects of inhibition of mitochondrial protein synthesis on Molt4 cells cultured in different media.

concentration of mt-gene products(1)		mt-ATP synthesis (1)	ATP level (2)		proliferation	
			RPMI	DME	RPMI	DME
100	(0)	100	52	53	normal	normal
81	(0.3)	nd	43	50	retarded	normal
60	(0.6)	nd	31	51	retarded	normal
50	(1.0)	71	26	50	retarded	normal
41	(1.3)	65	8	49	arrested	normal
31	(1.7)	nd	–	48	arrested	normal
23	(2.1)	26	–	51	arrested	normal
17	(2.5)	nd	–	47	arrested	normal
13	(3.0)	8	–	48	arrested	retarded
6	(4.0)	–		35		retarded
–	(>5)	–		36		retarded

The cells were cultured in DME or RPMI, supplemented with 10 % FCS, 1 mM pyruvate and 15 μg doxycycline/ ml at a CO_2 tension of 10 %. Mitochondrial ATP synthesis was only measured in cells cultured in DME. (1): % of control value; (2): nmol/10^6 cells; nd: not determined; –: not detectable; values between brackets: number of cell divisions.

This allowed us to obtain cells depleted of mtDNA and/or mitochondrial gene products, or, in other words since oxidative phosphorylation is excluded under these conditions, allowed us to obtain cells without functional mitochondria. These cells can be used as model systems for human cells in which oxidative phosphorylation is impaired because of, for instance, mtDNA mutations (see Introduction). One of the questions we addressed is whether or not nuclear genes for mitochondrial proteins are still expressed and, if so, where these proteins are localised. Measurements of the activity of mitochondrial matrix enzymes such as citrate synthase revealed no differences in the specific activity between control cells and cells without functional mitochondria. Immunofluorescence studies showed, moreover, the same intracellular distribution of these enzymes. This was also found for mitochondrial membrane proteins of exclusive nuclear genetic origin like the adenine nucleotide translocator. Surprisingly, also nuclear subunits of mitochondrial enzymes of dual genetic origin such as subunit IV of cytochrome c oxidase and subunits of the F_1 part of ATP synthase could be demonstrated in this way.

Metabolic labelling showed that the rate of synthesis of nuclearly coded cytochrome c oxidase subunits was hardly affected, whereas quantitative immunoblotting revealed that their cellular concentration was reduced about three-fold. We and others (Kuzela et al., 1988; Hayashi et al., 1990; Van den Bogert et al., 1991) have shown before that the concentration of nuclearly coded cytochrome c oxidase subunits decreases in line with the disappearance of the mitochondrially coded subunits during the initial stages of inhibition of mitochondrial gene expression, without significant effects on their rate of synthesis. It might be, therefore, that unassembled nuclearly coded subunits are degraded slower when inhibition of mtDNA expression has lasted longer.

Analysis of the mobility of nuclear subunits revealed that they are mainly present in the mature form, suggesting that import and processing of these polypeptides is not affected. In turn, this implies that mitochondrial membranes are present over which a potential can be generated, a prerequisite for the import of most mitochondrial proteins. The results of staining with fluorescent dyes which accumulate specifically in the mitochondria because of the mitochondrial membrane potential support this: cells without functional respiratory chains also showed staining of mitochondria like organelles (Fig. 1). The stained structures appeared roundish and swollen, instead of the

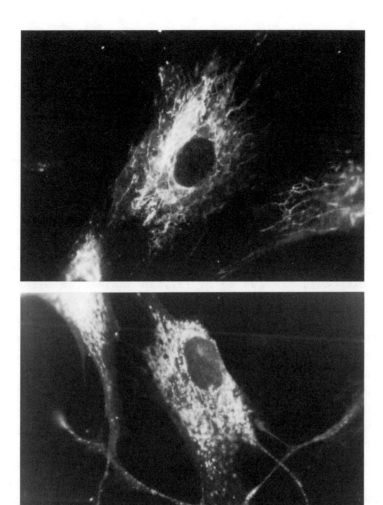

<u>Fig 1:</u> Morphology of mitochondria in normal myoblasts and in myoblasts depleted of mtDNA.
Upper panel: normal human myoblasts; lower panel: human myoblasts depleted of mtDNA by prolonged treatment with ethidium bromide.
The cells were stained without fixation by adding DIO to the cultures. Next the cells were inspected by fluorescence microscopy. Identical pictures were obtained in immuno-fluorescent studies of methanol-fixed cells labelled with antibodies against nuclearly coded mitochondrial proteins.

thread-like structures in control cells. This might be due to a reduced protein content of the mitochondrial inner membrane, which normally consists of proteins for about 80 % . Interestingly, a similar aberrant morphology of mitochondria is frequently observed in cell lines from patients with mitochondrial diseases (Ruiters et al., 1991).

One may wonder how a mitochondrial membrane potential can be generated in the absence of respiration and oxidative phosphorylation. For rho° yeast cells, it has been demonstrated that import of ATP into the mitochondria is essential for survival of these mutants which do not contain mtDNA (Hapala, 1989). In preliminary experiments we, likewise observed that inhibition of the adenine nucleotide carrier impaired the proliferation of Molt4 cells without functional respiratory chains. Imported ATP can be hydrolysed inside the mitochondria by the F_1 ATPase, but proton pumping by the F_O part of the ATPase is highly unlikely, since this would require the presence of mitochondrially coded subunits. Possibly, pumps for other ions are involved in the generation of a mitochondrial membrane potential in cells without a functional oxidative phos-phorylation system.

From the present study we conclude that several types of human cells can be cultured in the absence of mitochondrial gene products provided that the appropriate medium is used. Unexpectedly, no significant influence of mitochondrial gene expression on the expression of nuclearly coded mitochondrial proteins was found in these cultures. Possibly, mitochondrial gene expression is regulated entirely by chromosomal gene products, without any interaction in the opposite direction. Alternatively, a feed-back mechanism does exist, but is not operational in cells which are cultured under conditions where mitochondrial function is not essential.

Acknowledgements: Part of this study was made possible by a fellowship of Coby Van den Bogert of the Royal Netherlands Academy of Arts and Sciences and a grant of The Prinses Beatrix Fonds. Paul Arends is thanked for his help in preparing the manuscript.

REFERENCES

Chomyn, A. and Attardi, G. (1987) In: Current Topics in Bioenergetics (C.P. Lee, Ed), Vol. 15, Academic Press, San Diego, pp. 295-331

Donelly, M. and Schleffler, I.E. (1976) J. Cell Physiol. 89, 39-52

Granger, D.L. and Lehninger, A.L. (1982) J. Cell Biol. 95, 527-535

Hapala, I. (1989) Biochem. Biophys. Res. Commun. 159, 612-617

Hayashi, Y-I., Tanaka, M., Sato, W., Ozawa, T., Yonekawa, H. and Otha, S. (1990) Biochem. Biophys. Res. Commun. 167, 216-222

Kroon, A.M., Holtrop, M., Fries, H., Melis, T. and Van den Bogert, C. (1985) Biochem. Biophys. Res. Commun. 128, 1190-1195

Kuzela, S., Luciaková, K., Kolarov, J. and Nelson, B.D. (1989) In: Molecular Basis of Membrane-Associated Diseases (A. Azzi, Z. Drahota and S. Papa, Eds), Springer Verlag, Berlin, pp. 359-367

Lander, E.S. and Lodish, H. (1990) Cell 61, 925-926

Nelson, B.D. (1987) In: Current Topics in Bioenergetics (C.P. Lee, Ed), Vol. 15, Academic Press, San Diego, pp. 221-272

Ruiters, M.H.J., Van Sponsen, E.A., Skjeldal, O.H., Stromme, P., Scholte, H.R., Pzyrembel, H., Smit, G.P.A., Ruitenbeek, W. and Agsteribbe, E. (1991) J. Inher. Metab. Dis., 14, 45-48

Van den Bogert, C., Muus, P., Haanen, C., Pennings, A., Melis, T.E. and Kroon, A.M. (1988) Exp. Cell Res. 178, 143-153

Van den Bogert, C., Pennings, A., Dekker, H.L., Boezeman, J.B.M., Luciaková, K. and Sinjorgo, K.C.M. (1991) Biochim. Biophys. Acta 1097, 87-94

Whitfield, C.D. (1985) In: Molecular Cell Genetics (M.M. Gottesmann, Ed), John Wiley and Sons, New York, pp. 545-588

REGULATION BY ADENINE NUCLEOTIDES, CYCLIC NUCLEOTIDES AND PROTEIN
SYNTHESIS OF THE MATURATION OF RAT LIVER MITOCHONDRIAL FUNCTIONS

François Cretin[1], Jane Comte[2], Alain Dorier[3] and Catherine Godinot[4]

Laboratoire de Biologie et Technologie des Membranes du CNRS,
Université Claude Bernard de LYON I, 69622 VILLEURBANNE - FRANCE

SUMMARY: After birth, oxidative phosphorylation capacity of rat
liver mitochondria increases within 2-3 hours. (1) One of the most
important factors limiting the rate of oxidative phosphorylation,
at birth is the rate of ATPase-ATPsynthase. After mitochondrial
maturation, the main step controlling succinate oxidation coupled
to ATP synthesis becomes the supply of reducing equivalents.
(2) This maturation which is accompanied by an increase in the
mitochondrial adenine nucleotide contents (Aprille and Asimakis,
1980, Arch. Biochem. Biophys. 201, 564-575) can be inhibited by
glucose injection at birth. Glucose injection also prevented the
adenine nucleotide increase and slightly decreased the content of
liver cAMP and cGMP. Dibutyryl-cAMP prevented the glucose effect
while a rise of cGMP induced by nitroprusside or 8-Br-cGMP
inhibited mitochondrial maturation, as did glucose.
(3) Glucose did not significantly modify the rate of
[^{35}S]methionine incorporation into mitochondrial proteins. However,
the normal increase in the rate of ATP synthesis could be
prevented by in vivo injection of either cycloheximide or
thiamphenicol, specific inhibitors of nuclear and mitochondrial
protein synthesis, respectively. (4) In conclusion,the improvement
of the ATPase-ATPsynthase efficiency observed shortly after birth
involves both a direct regulation of the enzyme due to the
increase of the adenine nucleotide pool, this increase being under
hormonal control, and a new synthesis of ATPase-ATPsynthase
subunits encoded by nuclear and mitochondrial DNA.

---

[1]Some of the results presented here are part of the thesis of
F.Cretin who was supported by the MRT; present address: Ecole
Normale Supérieure de Lyon, 46, Allée d'Italie, 69634, Lyon.
[2]Present address UPR 412, UCB-LYON I, 69622 Villeurbanne. [3]From
the "Institut Universitaire de Technologie" UCB-LYON I, 69622
Villeurbanne. [4]To whom correspondence should be addressed.

INTRODUCTION

ATP production in the fetal liver is mainly glycolytic and its main source of energy comes from maternal glucose. At the end of gestation, glycolytic flux increases and glycogen accumulates in newborn rat liver (Greengard, 1970). This stock will be exhausted within 12 hr of extrauterine life (Girard et al, 1981). Mitochondria purified from newborn rat liver exhibit a rate of ATP synthesis coupled to substrate oxidation which is lower than that of the adult rat (Hallman, 1971)) A rapid increase of the rate of ATP synthesis occurs during the 2-3 hr following birth (Pollak, 1975; Aprille and Asimakis, 1980); this is due to an increase in the state 3 rate of substrate oxidation and not to leaky or uncoupled mitochondria that would result in an increased state 4 (see Aprille, 1986 for review).

In this paper, we will show that an activation of the ATPase-ATP synthase activity occurs during the maturation of the bioenergetic functions of the mitochondria shortly after birth. This activation can be related to the increase of intramitochondrial adenine nucleotides and/or be due the synthesis of new proteins. In addition, variations of the level of cAMP and cGMP in the cell can also trigger the activation or the inhibition of this maturation.

MATERIALS AND METHODS

Animals: Adult or dated pregnant Wistar rats were obtained at the "Institut Universitaire de Technologie" of Lyon and treated as described by Meister et al (1983) and Comte et al (1986).

Methods: Previously reported techniques were used for: mitochondria preparation (Aprille and Asimakis, 1980); determination of oxygen consumption rate and RCR (respiratory control ratio) (Baggetto et al, 1984); control strength analysis (Groen et al, 1982); ATP (Lamprecht and Stein, 1965), ADP (Adam, 1965), cAMP and cGMP (Comte et al, 1986) estimations; immunoprecipitation of F_1-ATPase (Cretin et al, 1991) and *in vivo* incorporation of [^{35}S]methionine into mitochondrial proteins (Hochberg et al, 1972).

RESULTS AND DISCUSSION

Steps involved in the limitation of oxidative phosphorylation at birth: The increase in the rate of oxidative phosphorylation which results in a full development of bioenergetic functions *in vivo* can also be obtained *in vitro* by incubation of mitochondria from newborn rats with 0.5 mM ATP (Pollak, 1975). With adult mitochondria, ATP had no effect on substrate oxidation. To determine which reactions are the most important in limiting the rate of succinate oxidation at birth in the presence or absence of ATP, the effects of limiting amounts of inhibitors specific for various reactions involved in oxidative phosphorylation were studied (Baggetto et al, 1984). The data were analyzed as described by Groen et al (1982) to quantify the amount of control or "control strength" exerted by different steps of oxidative phosphorylation on the rate of succinate oxidation by newborn mitochondria. A similar study was performed with adult rat liver mitochondria under the same conditions. The data are summarized in Table I.

Carboxyatractylate which inhibits nucleotide translocation (Luciani et al, 1971) contributed to control the rate of succinate oxidation by 15% in the absence of ATP in the newborn and by 29% in the adult. Preincubation with ATP of newborn rat liver mitochondria suppressed the control strength exerted by the rate of nucleotide translocation. This can be explained by an activation of adenine nucleotide translocase due to the increased mitochondrial adenine nucleotide pool size through the carboxyatractylate-insensitive transporter which exchange ATP-Mg with matrix Pi (Austin and Aprille, 1984).

The rate of electron transfer between cytochrome b and cytochrome c1 and the rate of cytochrome oxidase did not control the rate of succinate oxidation in newborn rats while cytochrome oxidase participated for 5% control in adult.

Table I: Control strength exerted by different steps of oxidative phosphorylation in adult and newborn rat liver mitochondria.

Step	Inhibitor	Control strength		
		Newborn		Adult
		− ATP	+ ATP	− ATP
Nucl. translocation	Carboxyatractylate	0.15	0	0.29
Succinate transport	Butylmalonate	0.12	0.51	0.08
Succinate oxidation	Malonate	0.08	0.44	0.15
Cyt. b − cyt. c_1	Antimycin A	0	0	0
Cytochrome oxidase	KCN	0	0	0.05
ATP synthase	Oligomycin	0.58	0.13	0

Control strength was determined as described by Baggetto et al (1984). The values are the average of 3 determinations except in the case of the cytochrome b − cytochrome c1 and of cytochrome oxidase for the newborn rat liver, for which only 2 experiments were performed

When the supply of reducing equivalents in the case of the newborn was inhibited by limiting either the rate of succinate transport by n-butylmalonate (inhibitor of the dicarboxylic anions transport in mitochondria: Johnson and Chappel, 1973) or by malonate, competitive inhibitor of succinate oxidation, the control strength could be estimated in the absence of ATP to 12% and 8%, respectively. This control strength increased up to 51% and 44%, in the presence of ATP. In the adult, the control strength induced by succinate transport and succinate oxidation was of 8 and 15%, respectively. ATP had no significant effect on this control strength, in the adult.

When limiting concentrations of oligomycin were added to specifically inhibit the ATPase-ATPsynthase complex there was no significant effect on the rate of oxygen uptake in the adult. No significant control was therefore exerted by the ATPase-ATP synthase in the adult. On the contrary, a control strength of 58% could be estimated for the newborn in the absence of ATP. After addition of ATP, this control strength was decreased to 13%.

The inhibitions observed with mitochondria prepared from 2 hr-old rats in the presence of increasing concentrations of oligomycin (control strength, 0.16) or of *n*-butylmalonate (control strength, 0.48) was similar to those observed with mitochondria from newborn rats treated with ATP.

In conclusion, when the rate of succinate oxidation in newborn rat liver mitochondria, was increased, *in vivo*, within 2 hr, or *in vitro* by preincubation with ATP , this rate depended heavily on the supply of reducing equivalents while, in the absence of ATP, the ATPase-ATPsynthase was the most important rate controlling factor in the newborn.

What event increases the ATPase-ATPsynthase activity after birth? Is it a synthesis of new proteins or an activation?

<u>Synthesis of enzymes involved in oxidative phosphorylation during the perinatal period:</u> Data related to succinate dehydrogenase and cytochrome oxidase development obtained in several laboratories have been summarized by Aprille (1986). Succinate dehydrogenase increases slightly but regularly from 4 days before birth until 2 days after birth; afterwards, a doubling of activity occurs between 2 and 8 days after birth. Cytochrome oxidase activity decreases by more than 50% between 4 days and 1 day before birth and then increases relatively slowly until birth and more rapidly until 10 days after birth.

An increase in the specific activities of succinate dehydrogenase, cytochrome oxidase and F1-ATPase was observed 1 or 2 hr after birth by Valcarce et al (1988). The increase observed for succinate dehydrogenase (52%) was higher than for F1-ATPase (30%). Therefore these results are not likely to explain why, when measuring control strength (see Table I), the ATPase became less limiting while the succinate oxidation became more limiting. The same authors reported a doubling of the amount of ß subunit titrated with a polyclonal antibody in homogenates or mitochondria from rat liver between birth and 1 hr after birth.In a second paper, Izquierdo et al, (1990), have measured the *in vivo* incorporation of [^{35}S]methionine, into the F1-ATPase ß subunit of 1 hr-old rats after immunoprecipitation of F1 with an anti-F1 antibody. The ß subunit was labeled while no incorporation could

Table II: Effect of inhibition of cytoplasmic protein synthesis by cycloheximide on the mitochondrial oxidative phosphorylation capacity in the two hours following birth.

Rat treatment	Control	Cycloheximide	% of Control
N	7	7	
RCR	3.33 ± 0.25	2.35 ± 0.18**	77.8
ADP/O	1.48 ± 0.08	1.33 ± 0.08*	89.9

Rats (3-5) were injected intraperitoneally with cycloheximide (250 µg) dissolved in 50 µl of 0.9% NaCl at birth and 10 min later. Control rats (3-5) received only NaCl. They were maintained for 2 hr in a humidified crib at 37°C. The rate of succinate oxidation was tested for each type of mitochondria in the presence of 10 mM succinate, 10 mM Tris-KCl, 225 mM sucrose, 5 mM $MgCl_2$, 1 mM EDTA, 10 mM potassium phosphate, before (state 4) and after (state 3) addition of 0.1 mM ADP. N is the number of experiments. RCR (Respiratory Control Ratio) is the ratio between the rate of oxygen consumption before and after addition of ADP. ADP/O is the ratio between the number of µmoles of ADP added to initiate state 3 to the number of µatoms of oxygen used up during state 3. Results are presented as average of N experiments ± SEM. The means were compared using the paired t test; **$P < 0.01$; *$P < 0.05$.

be detected in the α subunit immunoprecipitated with the same antibody. This suggests that the surge in the synthesis in ß subunit apparent 1 hr after birth was not integrated into the assembled ATPase complex and, eventually, that the antibody immunoprecipitated preferentially free newly synthesized ß subunits.

Table II shows that, when rats were injected with cycloheximide, which inhibits cytoplasmic protein synthesis, the RCR was decreased (the state 3 rate of succinate oxidation was decreased while state 4 was not modified, not shown):the RCR of rats injected with cycloheximide was 77.8% of that of the control rats ($P<0.01$). The ADP/O ratio was also decreased but only by 10% ($P<0.05$).

Table III shows that, after injection at birth of thiamphenicol which inhibits mitochondrial protein synthesis, the

Table III: Effect of inhibition of mitochondrial protein synthesis by thiamphenicol on the mitochondrial oxidative phosphorylation capacity in the two hours following birth.

Rat treatment	Control	Thiamphenicol	% of Control
N	4	4	
RCR	2.59 ± 0.15	1.98 ± 0 .18***	76.4
ADP/O	1.47 ± 0.27	1.23 ± 0.07	84.0

Conditions as described in Table II except that the rats were injected with thiamphenicol instead of cycloheximide .***P< 0.001.

state 3 rate of succinate oxidation was also decreased. The RCR of thiamphenicol treated rats was 76.4% that of the controls. These results are in agreement with those of Valcarce et al (1988) and indicate that both cytoplasmic and mitochondrial protein synthesis are required for postnatal maturation of bioenergetic functions.

As shown previously, glucose injection at birth, inhibits mitochondrial maturation (Meister et al, 1983). In addition the presence of glucose decreases the rate of Leucine incorporation into proteins of hepatocytes isolated from newborn rat liver (Gautheron et al, 1988). The rate of incorporation of [35S]methionine into mitochondrial proteins was compared in rats injected at birth with 0.9% NaCl in the presence or absence (controls) of glucose. No significant difference between control rats and glucose-treated rats could be observed either on the overall rate of incorporation of [35S]methionine into mitochondrial proteins or in the incorporation of [35S]methionine into the subunits of the ATPase-ATPsynthase in mitochondria isolated from 2 hr-old rats. Indeed, no change was detected in ATPase-ATPsynthase labeled in vivo, immunoprecipitated from a Triton-X100 extract of mitochondria prepared from control or glucose treated rats, separated by SDS-gel electrophoresis and analyzed by autoradiography (data not shown). This means that the effect of glucose is not correlated with the synthesis of new mitochondrial proteins or new ATPase-ATPsynthase subunits. Another (or other) factor(s) should be involved.

Evolution of mitochondrial adenine nucleotide concentration during the perinatal period: As previously described by Aprille and Asimakis (1980), the mitochondrial adenine nucleotides increase during the 2 hr following birth. In our experiments (Table IV), the adenine nucleotide contents increased from 2.35 nmol/mg protein at birth to 5.67 nmol/mg protein 2 hr after birth. Glucose injection diminished this increase (3.72 nmol/mg protein). The effect of glucose injection on the improvement of RCR could be mediated by prevention of the mitochondrial adenine nucleotide augmentation normally observed after 2 hr of extrauterine life.

Table IV: Influence of glucose injection at birth on the amount of nucleotides present in rat liver mitochondria.

Rat age	Injection at birth	Adenine nucleotides (ATP+ADP+AMP) (nmol/mg protein)	N	RCR	N
Newborn	–	2.35 ± 0.07^a	3	1.9 ± 0.11	12
2 hr-old	NaCl	$5.67 \pm 0.43^{a,b}$	4	2.5 ± 0.3	4
2 hr-old	Glucose	3.72 ± 0.31^b	4	1.9 ± 0.34	4
Adult	–	8.86 ± 0.11^a	3	6.3 ± 0.2	3

Liver mitochondria were prepared from rats immediately after birth (newborn rats), from 2 hr-old rats or from adult rats. The 2 hr-old rats were injected intraperitoneally immediately after birth either with 50 μl of 0.9 % NaCl or with 25 mg of glucose. [a]To compare newborn rats, 2 hr-old rats treated with NaCl and adult rats, normal t-test was used: $P<0.001$, in all cases. [b]The means of NaCl- and glucose-treated rats from the same litter were compared by the paired t-test: $P<0.005$.

Cyclic nucleotides level regulates mitochondrial maturation: As mentioned above, a glucose injection to newborn rats immediately after birth delays the increase in the rate of ATP synthesis coupled with succinate oxidation normally occuring during the first postnatal hr (Meister et al, 1983). ATP production in the fetal liver is mainly glycolytic, glucose being supplied by the mother blood. Birth induces a decrease of insulin/glucagon molar ratio from 10 before birth to 1 after birth (Girard et al, 1973).

Glucose injection at birth is known to increase insulinemia (Girard et al, 1976; Martin et al, 1981) and thus should oppose to the decline of insulin/glucagon ratio.

The levels of cAMP and cGMP have been titrated in liver of rats injected at birth with 50 µl of 0.9% NaCl in the presence or absence of glucose (Comte et al, 1986; Meister et al, 1988). The results are summarized in Fig 1. In control rats, a slight increase of cAMP and of cGMP was observed 10 min after birth. The level of cAMP decreased until 45 min and then increased 2 hr after birth. There was a slight decrease in cGMP between 10 min and 2 hr after birth. For glucose-treated rats, there was a decrease of cAMP from 10 to 45 min after birth, followed by an increase. However, this increase was more limited than for controls. The cGMP level significantly decreased 45 min and 2 hr after birth. An injection at birth of an alkylxanthine, Pentoxifylline, a drug known to increase the cellular cAMP level by inhibiting cAMP-phosphodiesterases (Mauduit et al, 1984) prevented the transient decrease observed 20 and 45 min after

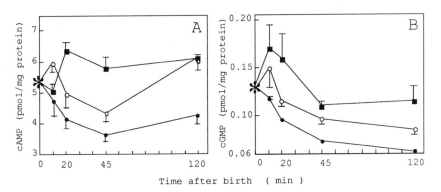

Time after birth (min)

Fig. 1: Influence of glucose or Pentoxifylline injection at birth on the cAMP and cGMP levels in newborn rat liver during the 2 hr following birth. Rats were injected at birth with 50 µl of 0.9% NaCl (O) containing 25 mg of glucose (●) or the therapeutic dose of 10mg/kg of Pentoxifylline (■). Rats were killed at the time indicated, the livers were removed and frozen in liquid nitrogen within 10 s after animal death. cAMP and cGMP were estimated as previously (Meister et al, 1988).

116

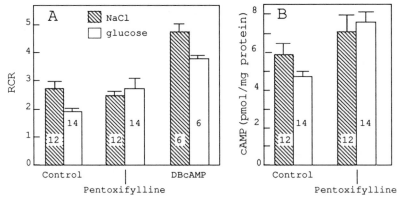

Fig. 2: Improvement of mitochondrial maturation by increase of hepatic cAMP. Rats were injected at birth with NaCl, glucose or Pentoxifylline as in Fig 1 or with DBcAMP (7 µmol/kg). RCR was measured as in Table II and cAMP as in Fig 1 in 2 hr-old rats. The figures in the blocks represent the number of experiments; The vertical bars represent the standard error of the mean.

birth. The apparent increase in cGMP level was not significant by two-way analysis of variance and pairwise comparison of Pentoxifylline-treated and control rats.

Fig 2 shows that, when Pentoxifylline was injected at birth simultaneously with glucose, the inhibitory effect of glucose on RCR was prevented. In addition, injection at birth of dibutyryl-cAMP (DBcAMP) did not change the state 4 rate of succinate oxidation, increased the state 3 and, hence the RCR of mitochondria isolated 2 hr after birth. In conclusion, an increase of hepatic cAMP was correlated with the improvement of mitochondrial bioenergetic functions, which occurs within the 2 hr following birth. In addition, an increase of cAMP could also release the inhibition of maturation for glucose injected rats.

Hepatic cGMP level changes after birth or after Pentoxifylline treatment were barely significant and the glucose-induced decrease in cGMP was not very important, which suggested little influence of cGMP on mitochondrial maturation. However, Fig 3 shows that an injection at birth of nitroprusside, known to stimulate guanylate cyclase and to increase intracellular cGMP (Ignarro and Fadowitz, 1985) drastically increased cGMP, 10 min after birth (which decreased, thereafter, not shown) and prevented

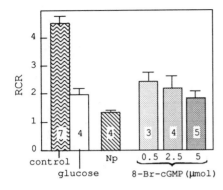

Fig 3: Inhibition of mitochondrial maturation by effectors increasing hepatic cGMP. Rats were injected at birth with 50 μl 0.9% NaCl containing or not either 25 mg glucose or 10 μg nitroprusside (Np) or the indicated concentration of 8-Br-cGMP in μmol/Kg. Other experimental conditions are described in Table II. The figures in the blocks indicate the number of experiments and the bars indicate the standard error of the mean.

the increase of RCR in the 2 hr following birth. Similarly, increasing concentrations of 8-Br-cGMP decreased the RCR.

In conclusion, an increase in hepatic cAMP induced by Pentoxifylline in glucose-treated rats or by DBcAMP in normal or glucose-treated rats could improve the bioenergetic functions of newborn rat liver. Conversely, an increase in cGMP induced by nitroprusside or 8-Br-cGMP inhibited this maturation process.

CONCLUSION

The increase in the efficiency of the ATPase-ATPsynthase, inducing an increase in the rate of substrate oxidation coupled to ATP synthesis and observed shortly after birth could be correlated to two main factors: (1) an increase of mitochondrial adenine nucleotides might be responsible for a direct activation of the enzyme by saturating catalytic or regulatory nucleotide binding sites; and (2) a new synthesis of ATPase-ATPsynthase subunits that could increase the total amount of active enzyme. However, none of these factors is sufficient in itself. In addition, an hormonal regulation of this maturation process can be exerted by cyclic nucleotides which could act on adenine nucleotides accumulation.

Acknowledgements: This work was supported by the "Association Française des Myopathies", the "Région Rhône-Alpes" and the "Centre National de la Recherche Scientifique" Thanks are due to C. Van Herrewege for the art work, D. Gouttay for expert animal care and to Dr. C. Lopez-Mediavilla for stimulating discussions.

118

REFERENCES

Adam, J. (1965) in Methods of Enzymatic Analysis (2nd ed)
    Bergmeyer, H.U., ed., Academic Press, New York pp. 620-625
Aprille, J.R. (1986) In: Mitochondrial Physiology and Pathology
    (G. Fisher, ed) USA, pp 66-99
Aprille, J.R. and Asimakis, G.K. (1980) Arch. Biochem. Biophys.
    201, 564-575
Austin, J. and Aprille, J. R. (1984) J. Biol. Chem.259, 154-160
Baggetto, L.G., Gautheron, D.C.and Godinot, C. (1984) Arch.
    Biochem. Biophys.232, 670-678
Comte, J., Meister, R.,Baggetto, L.G.,Godinot, C.and Gautheron,
    D.C. (1986) Biochem. Pharmacol. 35, 2411-2416
Cretin, F., Baggetto, L., Denoroy, L. and Godinot, C. (1991)
    Biochim. Biophys. Acta, 1058, 141-146
Gautheron, D.C., Comte, J. Meister, R. and Godinot, C. (1988)
    Pathol. Biol. 36, 1056-1059
Girard, J.R., Cuendet, G.S.,Marliss, E.B., Kevran, A., Rieutort,
    M. and Assan, R. (1973) J. Clin. Invest, 52, 3190-3200
Girard, J.R., Guillet, I., Marty,J., Assan, R. and Marliss, E.B.
    (1976) Diabetologia, 12, 327-337
Girard, J.R., Pegorier, J.P., Le Turque, A. and Ferré, P.(1981) In
    Physiological and Biomedical basis for perinatal Medicine"
    (Mousset-Couchard and Minkowski,eds) Karger, Basel, pp. 90-96
Greengard, O.,(1970) in "Biomedical Action of Hormones"
    (Litwack, G., ed.) Academic Press, New York, pp. 53-87
Groen, A.K., Wanders, R.J.A., Westerhoff, H.V., Van der Meer, R.
    and Tager, J.M. (1982) J. Biol. Chem. 257, 2754-2757
Hallman, M. (1971) Biochim. Biophys. Acta, 253, 360-372
Hochberg,A., Zahlten, R., Stratman, F.W. and Lardy, H.A. (1972)
    Biochemistry, 11, 3143-3153
Ignarro, L.J. and Fadowitz, P.J. (1985) Annu. Rev. Pharmacol.
    Toxicol. 25, 171-191
Izquierdo, J.M., Luis, A.M. and Cuezva, J.M. (1990) J. Biol.
    Chem. 265, 9090-9097
Johnson, R.N. and Chappell, J.B. (1973) Biochem. J. 134, 769-774
Lamprecht, W. and Stein, P; (1965) in Methods of Enzymatic
    Analysis (2nd ed.) Bergmeyer, H.U., ed., Academic Press, New
    York pp. 967-975
Luciani, S., Martini, N., Santi, R.(1971) Life Sci, 10, 961-968
Martin, A., Caldes, T., Benito, M. and Medina, J.M. (1981)
    Biochim. Biophys. Acta, 672, 262-267
Mauduit, P., Herman, G. and Rossignol, B. (1984) Am. J. Physiol.
    246, C37
Meister, R., Comte, J.,Baggetto, L.G.,Godinot, C.and Gautheron,
    D.C.(1983) Biochim. Biophys. Acta, 722, 36-42
Meister, R., Comte, J.,Baggetto, L.G.,Godinot, C.and Gautheron,
    D.C.(1988) Biochim. Biophys. Acta, 936, 67-73
Pollak, J. K. (1975) Biochem. J.150, 477-488
Valcarce, C., Navarette, R.M., Encabo, P., Loeches, E., Satrùste-
    gui, J. and Cuezva, P. (1988) J. Biol. Chem. 263, 7767-775

Adenine Nucleotides in Cellular Energy Transfer and Signal Transduction
S. Papa, A. Azzi & J.M. Tager (eds)
© 1992 Birkhäuser Verlag, Basel/Switzerland

INHIBITION OF THE MITOCHONDRIAL ATPase BY POLYBORATE ANIONS
DESIGNED FOR NEUTRON CAPTURE THERAPY

Zdenek Drahota, Vladislav Mares, Hana Rauchova and Martin Kalous

Institute of Physiology, Czechoslovak Academy of Sciences,
CS-142 20 Prague, Czechoslovakia

SUMMARY: Mercapto undecahydro dodecaboron designed for neutron capture therapy and several new carboranes showed inhibitory effects on mitochondrial ATPase activity. For mercapto and chloroderivates of carboranes 50 % inhibition of ATPase activity was found at 100-200 uM concentration. For dodecaboranes 50 % inhibition was found at 10-times higher concentration.

INTRODUCTION

Neutron capture therapy (NCT) is based on the interaction of Boron-10 atoms with relatively harmless neutrons followed by release of alpha-particles with much higher energy. The energy of particles released is absorbed in biological material at a very short distance and cells in close vicinity of boron molecules are lethally damaged (see Hatanaka, Sweet 1975). For NCT of brain tumors undecahydro dodecaboron was designed (Hatanaka, Sweet 1975, Hatanaka 1986).

Pharmacological experimental studies of the acute toxicity of mercapto undecahydro dodecaboron revealed LD-50 as 308mg/kg. Toxic effect of the drug can be explained by its interaction with metabolic pathways using pyridoxal phosphate as cofactor (Kliegel

1980), however more research is required to understand molecular mechanisms of undesirable side effects of polyboron compounds used for NCT. A new group of chloro and mercapto derivates of carboranes was recently synthesized (Plešek et al, 1984, Burian et al. 1989). They appeared to be several times more toxic and thus not suitable for NCT, but they can be used as tool for evaluation of various mechanisms through which polyboron anions may affect various areas of cell metabolism.

In experiments presented in this paper we have concentrated our attention on the evaluation of toxic effects of both dodecaboranes and new compounds, carboranes. We concentrated our attention or reactions of cell energy metabolism, especially on mitochondrial ATPase, the key enzyme in cell bioenergetics.

RESULTS AND DISCUSSION

On membrane preparations of frozen-thawed liver mitochondria we have tested the effect of two dodecaboranes and two dicabononaboranes (mono and dichlor derivates) on activity of mitochondrial ATPase. Data presented in Fig. 1 show that both groups of polyboranes inhibit the enzyme activity in the concentration range of mM that was found critical for inhibition of cell growth in experiments with isolated cells (Mares et al. 1992). In agreement with previous biological toxicity tests the inhibitory effect of dicarbo nonaboranes was more pronounced than that of dodecaboranes. The inhibitory effect of mercapto dicarbo nonaboron was quite similar to that of chloro compounds.

In agreement with higher toxicity in in vivo experiments, in the case of dicardo nonaboranes the 50 % inhibition of the mitochondrial ATPase activity occurred at about 10-times lower concentration than that required for 50 % inhibition by dodecaboranes.

The inhibitory effect of polyboranes was quite similar when frozen-thawed or intact coupled and uncoupled mitochondria were used. In intact mitochondria the inhibition of the ATPase activity was accompanied by swelling. When both processes are

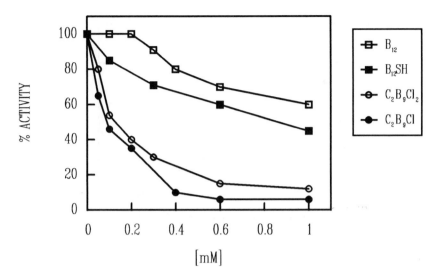

Fig. 1: Inhibition of the mitochondrial ATPase by various polyboron anions. ATPase activity was measured as release of inorganic phosphate in the medium: 100mM KCl, tris-HCl, 3mM Mg, 5mM ATP, pH-7.4. Frozen-thawed rat liver mitochondria were used.

compared (see Fig. 2), it is evident that they occur at the similar concentration range of polyboron anions. However, they are not directly related, because the same inhibitory effect of polyboron anions can be observed also in experiments with disrupted mitochondria (see Fig. 1).

The inhibitory effect of dicarbo nonaboranes was not quite specific for the ATPase. In similar concentration range as ATPase also other mitochondrial respiratory-chain enzymes were inhibited. Our data indicate that glycerolphosphate dehydrogenase is inhibited at lower concentrations and succinate dehydrogenase at higher concentrations than ATPase activity.

122

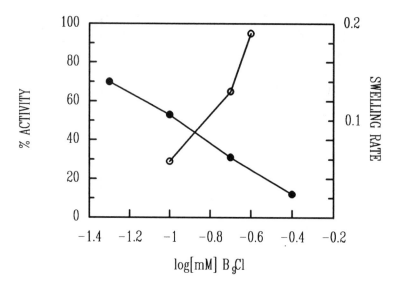

Fig. 2: Inhibition of mitochondrial ATPase (full circles) and increase in swelling rate (empty circles) at various concentrations of monochlor dicarbo nonarboron. Swelling rate was expressed as change of optical density at 520nm per min. ATPase activity was determined as in Fig. 1. Intact rat liver mitochondria were used.

When the fluorescence anisotropy was determined in membranes incubated with DPHT at various concentrations of dicarbo nonaboron, of change in membrane microviscosity was found at the concentration range in which the activity of ATPase was inhibited.

Our data on swelling induced by polyboron anions are in agreement with previous findings indicating that boron anions (e.g. tetraphenyl boron) as lipophylic compounds can easily penetrate through biological membranes and modify membrane permeability properties (Utsuma, Packer 1967). Inhibitory effects of boron anions on membrane translocators and membrane bound enzymes were also described. These inhibitory effects were explained as binding of lipophylic boron anion to the hydrophobic part of the membrane and formation of the high negative charge on the membrane surface that acts to repeal metabolite anions (Meisner 1973). In our case, however, negative surface charge can not fully explain the inhibition of mitochondrial ATPase. But it is quite evident that the inhibitory effect is realized through the binding of lipophylic polyboron anion on the hydrophobic part of the mitochondrial membrane, because we could not detect any inhibitory effect of polyboron on solubilized enzyme deprived of the membrane sector.

Polyboranes including carboranes were already used as membrane potential probes.In liposimes (Jasaitis et al. 1972), in submitochondrial particles (Grinius et al. 1970), as well as in intact mitochondria (Bakeva et al. 1970),it was demonstrated that energization or discharge state is accompanied by movements of polyboron anions across the membrane according to membrane potential change. These experiments were done with micromolar concentrations of polyboron which in our experimental conditions showed no effect on enzyme activity.

Toxic effects of dodecaboranes were explained by their interaction with metabolic pathways using pyridoxal phosphate as cofactor (Kliegel 1980). Our data showed that inhibition of metabolic reactions of cell energy transformation may be also involved.We may thus conclude that previous data in literature as well as our observations indicate that polyboron anions have at different concentrations various effects on biological membranes connected with the modification of membrane permeability properties and with the modification of functional activity of various membrane bound enzymes. All these effects should be considered in further development of therapeutic procedures using polyborates for neutron capture therapy.

124

REFERENCES

Bakeeva, L.E., Grinius, L.L., Jasaitis, A.A., Kuliene, V.V., Levitski, D.O., Lieberman, E.A., Severina, I.I., and Skulachev, V.P. (1970) Biochim.Biophys.Acta 216, 13-21

Baudysova, M., Mares, V., Kvitek, J., Cervena, J., Vacik, J., Hnatowicz, V., and Kunz, Z. (1988) Physiol.Bohemoslov. 37, 541-550

Burian, J., Janku, I., Kvitek, J., Mares, V., Prouza, Z., Spurny, F., Sourek, K., Stibr, B., and Strouf, O.(1989) In: Clinical Aspects of Neutron Capture Therapy, Plenum Publ. Corb., New York, pp. 39-48

Grinius, L.L., Jasaitis, A.A., Kadziankas, Yu.P., Lieberman, E.A., Skulachev, V.P., Topali, V.P., Tsofina, L.M. , and Vladimirova, M.A. (1970) Biochim.Biophys Acta 216, 1-12

Hatanaka, H., (1986) In: Boron-Neutron Capture Therapy for Tumors (H.Hatanaka, Ed.), Niigata Jap.Nishimura Co., pp. 349-480

Hatanaka, H., and Sweet, W.H. (1975) In: Slow Neutron Capture Therapy for Maligant Tumors, Adv.Biochemical Dosimetry, Proc. IAEA, Viena, pp. 147-178

Jasaitis, A.A., Nemecek, I.B., Severina, I.I., Skulachev, V.P., and Smirnova, S.M. (1972) Biochem.Biophys.Acta 275, 485-490

Kliegel, W. Bor in Biologie, Medizin und Pharmazie, Springer Verlag, Berlin 1980

Mares, V., Baudysova, M., Vitek, J., Hnatowicz, V., Cervena, J., Vacik, J. and Folbergrova, J.(1992) J.Pharmacol.Exp.Therap., in press

Plesek, J., Stibr, B., Drdakova, E. (1985) The methed of production of dodecahydro-closo-dodecaborate (2-) metals. Czechoslovak Author's Certificate and Patent No.288254

Plesek, J., Stibr, B., Drdakova, E., Jelinek, T. The method of production of tertiary amoniumborans Czechoslovak Author's Certificate and Patent No 242064, 1985.

Plesek, J., Jelinek, T., Drdakova, E., Hermanek, S., Stibr, B. (1984) A convenient preparation of $1-CB_{11}H^-_{12}$ and its C-amino derivation. Collect.Czech.Chem,Commun. 49, 1559.

Utsuma, K., and Packer, L. (1967) Arch. Biochem. Biophys. 122, 509-515

ACKNOWLEDGEMENT

The study was partly supported by the Czechoslovak Academy Research Grant No. 71104

Adenine Nucleotides in Cellular Energy Transfer and Signal Transduction
S. Papa, A. Azzi & J.M. Tager (eds)
© 1992 Birkhäuser Verlag, Basel/Switzerland

^{31}P MR SPECTROSCOPIC STUDIES OF METABOLIC LESIONS IN THE LIVER

Ruth M. Dixon

MRC Biochemical and Clinical Magnetic Resonance Unit, John
Radcliffe Hospital, Oxford, OX3 9DU, U.K.

SUMMARY: Energy metabolism in liver and muscle can be studied non-invasively by ^{31}P magnetic resonance spectroscopy. In muscle, the rate of ATP synthesis by oxidative phoshorylation appears to be controlled by free ADP (which can be measured from the creatine kinase equilibrium). In the liver, control of energy metabolism is not well understood, and free ADP levels cannot normally be measured. We have studied patients with a number of metabolic defects and specific metabolic loads, and shown that even when ATP levels fall, adenine nucleotides are lost from the cell in preference to allowing ADP levels to rise.

INTRODUCTION

^{31}P MRS enables us to study the biochemistry of tissues non-invasively, and to measure metabolite ratios (and in some cases concentrations), intracellular pH, and, sometimes, rates of reactions. Clinical MRS not only provides information that can assist the diagnosis or assessment of patients, but can also help us to understand underlying biochemical changes or adaptations to disease. Study of specific lesions can also throw light on the control of normal metabolism.

MUSCLE

Phosphorus MRS allows us to quantify some of the important compounds concerned with energy metabolism, namely phosphocreatine (PCr), adenosine 5'-triphosphate (ATP), inorganic phosphate (P_i), and H$^+$ (Figure 1). Most of the work in this area has concerned muscle metabolism (Taylor et al., 1983, Radda et al., 1988), so the insights gained in muscle will be briefly described, in order to compare them with studies of liver metabolism, where the control of energy metabolism is less well understood.

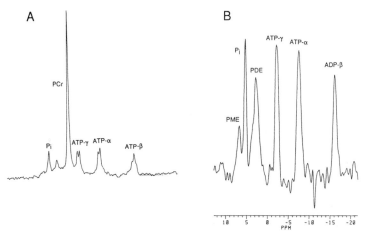

Figure 1. A. ^{31}P MR spectrum from normal muscle. B. ^{31}P MR spectrum from normal liver.

One can follow dynamic changes during exercise. In a normal subject, the PCr decreases and the P_i increases with exercise, while the P_i peak broadens and shifts, demonstrating an acidification in the tissue. At the end of exercise, the metabolites recover to their normal levels. ATP levels are generally constant throughout exercise. One can pose the question; what controls ATP synthesis during the sudden metabolic demand for energy and what causes it to slow down at the end of exercise, so as not to waste substrates?

If the demand for ATP increases, ATP is rapidly replenished by the creatine kinase equilibrium. This maintains ATP levels for a few minutes, while other sources (glycolysis, glycogenolysis, and oxidative phosphorylation) are activated.

The creatine kinase equilibrium also buffers the ATP/ADP ratio, and ADP (adenosine 5'-diphosphate) rises very little compared to the total turnover of ATP. Both P_i and ADP can increase oxygen consumption in the mitochondrion (Chance et al. 1986), but P_i is unlikely to be rate limiting at concentrations above 1 mM. The concentration of cytosolic ADP can be calculated from the creatine kinase equilibrium, which shows that free ADP increases from about 6 μM in normal muscle at rest to about 45 μM during exercise (Taylor et al., 1983), around the Km of 20-30 μM for mitochondrial O_2 consumption (Veech et al., 1979), thus ADP is a suitable candidate for controlling ATP synthesis.

The calculated free ADP can be compared with the rate of ATP synthesis after exercise, which is estimated from the initial rate of PCr recovery at the end of exercise (Taylor, 1989). A correlation between ADP at the end of exercise and PCr resynthesis rate can be seen in a wide range of subjects. Ageing subjects show on average a higher ADP in response to exercise but increase their rate of PCr recovery as well. The same is found in patients with heart failure. This correlation breaks down in subjects who cannot increase their rate of oxidative phosphorylation to meet demand. Patients with mitochondrial myopathies increase their ADP levels above 150 μM but the rate of PCr resynthesis does not increase (Arnold et al., 1985).

It is interesting that most of the ADP in the cell is "NMR invisible", that is, extracts of tissue suggest that ADP is present in the cell at millimolar concentrations, so should be visible in the spectrum. It appears that magnetic resonance spectroscopy detects only the cytosolic metabolites, thus the micromolar levels of ADP in the cytosol are below the levels of detection. It is postulated that the unseen ADP is bound to the contractile proteins, or sequestered in the mitochondria, and that the signals from bound molecules are broadened beyond detection.   In the same way, some of the cellular P_i may be undetectable (Taylor et al, 1983).

Normal subjects can exercise to such an extent that they deplete their muscles of ATP (Taylor et al., 1986).   On recovery, ATP does not reach its original level.   During the course of this exaustive exercise, ADP does not increase more than in much lighter exercise, and PCr recovery is slow.   It appears that under these conditions, breakdown of adenine nucleotides occurs that results in non-phosphorylated intermediates being lost from the cell.   Presumably, AMP is increased via the adenylate kinase equilibrium.   Subsequently, AMP deaminase converts AMP to IMP, which is dephosphorylated and lost from the cell, resulting ultimately in an increase in serum urate.   It seems that ADP levels do not rise, even at the expense of losing adenine nucleotides from the cell.

LIVER

The liver spectrum (Figure 1B) shows a major difference from muscle; it lacks phosphocreatine, since creatine kinase is not expressed in the liver.   ADP concentrations therefore cannot be calculated in the same way as in muscle, brain, or heart, but energy metabolism can be investigated in other ways.   Other differences in the spectrum are the much higher levels of phosphomonoesters (PME) and phosphodiesters (PDE) in the liver. The phosphodiester peak comes largely from phospholipid bilayers in the cells (in the liver this is mostly endoplasmic reticulum (Murphy et al., 1989)), and the phosphomonoester peak contains phosphorylated sugars, and other phosphorylated metabolites.

We have studied a number of metabolic lesions by ^{31}P MRS of the liver, and have followed the responses to particular metabolic stresses.

Fructose intolerance: Fructose metabolism in the liver involves the phosphorylation of fructose, followed by cleavage of fructose 1-phosphate by aldolase B to glyceraldehyde and

dihydroxyacetone phosphate (Figure 2).   Aldolase B is lacking
in fructosaemia, which is an autosomal recessive disorder.

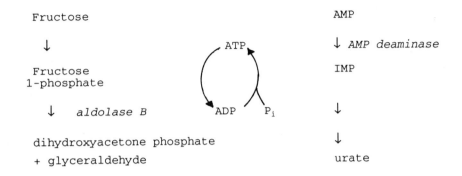

Figure 2. Hepatic fructose metabolism.

Fructose intolerant patients show an increase in fructose 1-
phosphate in the PME region of the liver spectrum following a
small amount of fructose (Oberhaensli et al., 1987).   An
unexpected finding in this study was that the asymptomatic
parents of fructosaemic children showed an abnormal response to
an oral fructose dose.   Six obligate heterozygotes were
compared with 9 controls.   The control subjects showed little
change in their ^{31}P MR spectra following 50 g fructose, whereas
the heterozygotes showed a rapid fall in P_i, while the PME,
containing fructose 1-phosphate, increased.

The ATP also sometimes falls; this suggests an explanation
for the high incidence of gout in fructosaemics and their
families.   Fructokinase uses ATP, which is replenished by
phosphorylation of ADP.   This uses inorganic phosphate, which
is replenished only very slowly from outside the liver (Sestoft
and Kristensen, 1979), so intracellular P_i levels fall, unless
the fructose 1-phosphate is rapidly metabolised.   AMP deaminase

is the rate limiting step in purine breakdown and is inhibited by P_i (Chapman and Atkinson, 1973). A decrease in P_i therefore allows AMP (increased via the adenylate kinase equilibrium) to be converted to IMP, and adenine nucleotides are lost from the cell. This eventually leads to an increase in serum urate, and gout.

These findings led to a recent study (Seegmiller et al., 1990), in which the question was posed: can a proportion of familial gout be attributed to heterozygosity for fructose intolerance?

We studied 12 unrelated gout patients, and found that 2 had abnormalities in fructose handling similar to those found in obligate heterozygotes. We then studied a number of relatives of these two patients, and found that the mother and 3 brothers of one of them, although free from gout, showed the same changes. The 6 responders showed greater increases in serum urate in response to the fructose load than did other subjects. Examples of spectra from two related subjects are shown in Figure 3. Subject LB shows essentially no changes following fructose (in common with other control subjects), while subject AB showed a rapid increase in PME, and decrease in P_i, followed by recovery, consistent with heterozygotes for fructosaemia. An earlier report (Stirpe et al, 1970) that restriction of fructose in the diet could benefit a proportion of sufferers from gout leads us to suggest that these patients could be heterozygotes for aldolase B deficiency.

Although oral fructose leads to minimal changes in normal subjects, intravenous fructose leads to much more dramatic changes, including a loss of ATP (Oberhaensli et al., 1986). This raises the question: is the ATP loss caused by the fall in P_i, relieving the inhibition of AMP deaminase, or does ADP increase, causing an increase in AMP via adenylate kinase?

This question has been addressed very elegantly by Koretsky and co-workers in Pittsburgh (Brosnan et al., 1991). Transgenic mice have been produced that express creatine kinase in their livers (Koretsky et al., 1990, Brosnan et al., 1990), and feeding with creatine results in metabolically active PCr being present. Intravenous fructose caused a drop in ATP in the normal mouse liver, while ATP in the transgenic mouse liver was protected. Feeding with various amount of creatine resulted in different amounts of hepatic PCr, and, at low PCr levels, ATP was not wholly preserved. ADP levels were calculated, and were found to rise by about the same amount (from about 60 mM to 150 mM), regardless of whether ATP was lost or not. Instead, the loss of ATP was related to the decrease in P_i, suggesting that AMP deaminase activity controlled the loss of ATP from the cell [14].

Glycogen storage disease, Type 1A: Another metabolic defect associated with hyperuricaemia and gout is glycogen storage disease type 1A (glucose 6-phosphatase deficiency). The fasting spectrum shows low P_i and high phosphomonoesters, which revert towards normal on feeding (Oberhaensli et al., 1988b). It is likely that glycolytic intermediates (particularly glucose 6-phosphate) accumulate during fasting, trapping P_i. The low P_i may increase purine breakdown via AMP deaminase. After a meal, the phosphomonoester decreases and the P_i increases, probably because glycogen synthesis is increased. An increase in P_i may result in an activation of phosphoribosylpyrophosphatase, the rate limiting enzyme in purine synthesis. Thus the variation in P_i during fasting-refeeding cycles may contribute to the increase in serum urate.

132

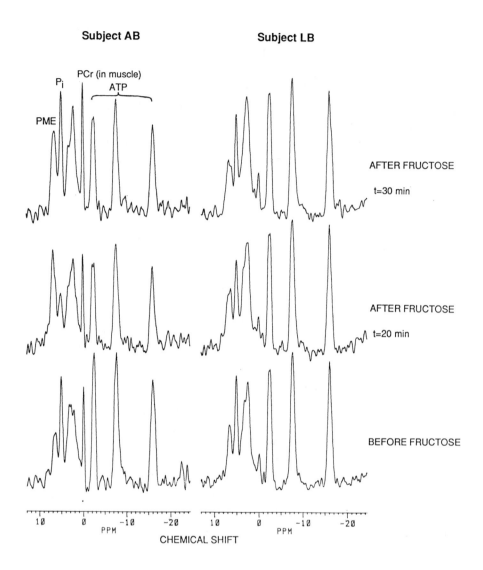

Figure 3. Spectra from two related subjects, showing differing responses to an oral fructose load (50 g).

Galactose intolerance: Galactosaemia, in contrast, is not associated with hyperuricaemia or gout.   The reasons for this are not clear, but the magnetic resonance findings may provide some insight.   An oral dose of 1 g galactose leads to an increase of about 100 % in the P_i region of the patient's liver spectrum, which could be due to galactose 1-phosphate, as this has a similar chemical shift to P_i.   With this dose of galactose, ATP is unchanged over 1h.   An animal study confirmed that the increase in the P_i region was indeed galactose 1-phosphate (Dixon et al., 1989, Radda et al., 1991).

The normal pathway for galactose metabolism shown in Fig. 4. Galactosaemia results from a deficiency in uridylyl transferase, preventing the conversion of galactose 1-phosphate to UDP-galactose.   An alternative pathway for galactose metabolism exists and is also shown in Fig. 4 (Isselbacher, 1957).   The P_i produced by hydrolysis of pyrophosphate may be enough to ensure that AMP deaminase is inhibited, and to conserve adenine nucleotide levels.

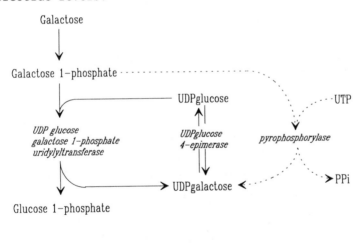

Figure 4. Pathways of galactose metabolism.

CONCLUSIONS

ADP appears to control oxidative phosphorylation in normal skeletal muscle (K_m 20-30 µM), although under some circumstances oxygen or subtrates become rate limiting. The creatine kinase equilibrium ensures that the ADP/ATP ratio is kept more or less constant, until PCr is exhausted. Under conditions of strenuous exercise that cause ATP depletion, ADP does not increase above 50 µM; instead AMP increases and adenine nucleotides are lost from the cell.

In the liver creatine kinase is not present, as ATP demands are generally much more constant than in muscle. ADP is probably an important regulator of oxidative metabolism in this tissue (Iles et al., 1985), as the concentration of hepatic free cytosolic ADP is low (about 50 µM) (Veech et al., 1979, Brosnan et al., 1990), but Pi may also be regulatory (Chance et al., 1991). When ATP demand increases, as when the liver is challenged by a metabolic load, ADP rises only moderately, and, as in muscle, adenine nucleotide depletion occurs. A low level of ADP is important in maintaining the free energy of hydrolysis of ATP, thus the liver appears to preserve low ADP levels even at the expense of nucleotide loss. Thus although the demands on muscle and liver are very different, both tissues may respond to an abnormal demand for ATP in a similar way.

These examples show that ^{31}P MRS is a powerful method for following biochemical changes in intact organs and cells. When it is possible to study alterations in specific enzyme levels (either in congenital diseases, or using transgenic technology), the control of energy metabolism can be investigated *in vivo*.

ACKNOWLEDGEMENTS: I gratefully acknowledge the many colleagues who were involved in the studies described, and in particular I thank Professor George K. Radda, Dr Peter Styles, Dr Doris Taylor, Dr Graham Kemp, and Dr Bheeshma Rajagopalan for helpful discussions during the preparation of this manuscript.

REFERENCES:

Arnold, D.L., Taylor, D.J., and Radda, G.K. (1985) *Ann. Neurol.*, **18**, 189-196.

Brosnan, M.J., Chen, L., van Dyke, T.A., and Koretsky, A.P. (1990) *J. Biol. Chem.*, **265**, 20849-20855.

Brosnan, M.J., Chen, L., Wheeler, C.E., van Dyke, T.A., and Koretsky, A.P. (1991) *Am. J Physiol.*, **260**, C1191-C1200.

Chance, B., Ikai, I., Tanaka, A., and Okuda, M. (1991) in Regulation of Hepatic Function, Alfred Benzon Symposium 30, Eds Grunnet, N., and Quistorff, B., Munksgaard, Copenhagen, 141-160.

Chance, B., Leigh, J.S., Kent, J., McCully, N.S., Clark, B.J., Maris, J.M., and Graham, T., (1986) *Proc. Natl. Acad. Sci. USA*, 9458-9462.

Chapman, A.G., Atkinson, D.E. (1973) *J. Biol. Chem.*, **248**, 8309-12.

Dixon, R.M., Kalderon, B., Rajagopalan, B., Angus, P.W., Oberhaensli, R.D., and Radda, G.K. (1989) *Abstracts of the Soc. Magn. Reson. Med. (Amsterdam)*, 69.

Iles, R.A., Stevens, A.N., Griffiths, J.R., and Morris, P.G. (1985) *Biochem. J.*, **229**, 141-151.

Isselbacher, K.J. (1957) *Science*, **126**, 652-654.

Koretsky, A.P., Brosnan, M.J., Chen, L., Chen, J., and van Dyke, T.A. (1990) *Proc. Natl. Acad. Sci. USA*, **87**, 3112-3116.

Murphy, E.J., Rajagopalan, B., Brindle, K.M., and Radda, G.K. (1989) *Magn. Reson. Med.*, **12**, 282-289.

Oberhaensli, R.D., Galloway, G.J., Taylor, D.J., Bore, P.J., Radda, G.K. (1986) *Br. J. Radiol.*, **59**, 695-699.

Oberhaensli, R.D., Rajagopalan, B., Taylor, D.J., Radda, G.K., Collins, J.E., Leonard, J.V., Schwartz, H., and Herschkowitz, N. (1987) *Lancet*, **ii**, 931-934.

Oberhaensli, R.D., Rajagopalan, B., Taylor, D.J., Radda, G.K., Collins, J.E., and Leonard, J.V. (1988) *Pediatric Res.*, **23**, 375-380.

Radda, G.K., Dixon, R.M., Angus, P.W., and Rajagopalan, B. (1991) in Regulation of Hepatic Function, Alfred Benzon Symposium 30, Eds Grunnet, N., and Quistorff, B., Munksgaard, Copenhagen, 433-446.

Radda, G.K., Rajagopalan, B., and Taylor, D.T. (1988) In Magnetic Resonance Annual, Raven Press, New York.

Seegmiller, J.E., Dixon, R.M., Kemp, G.J., Angus, P.W., McAlindon, T.E., Dieppe, P., Rajagopalan, B., and Radda, G.K. (1990) *Proc. Natl. Acad. Sci., USA*, **87**, 8326-8330.

136

Sestoft, L., and Kristensen, L.O. (1979) *Am. J. Physiol.*, **236**, C202-C210.

Stirpe, F., della Corte, E., Bonetti, E., Abbondanza, A., Abbati, A., and Stefano, F.D. (1970) *Lancet*, **ii**, 1310-1311.

Taylor, D.J., Bore, P.J., Styles, P., Gadian, D.G., and Radda, G.K. (1983) *Mol. Biol. Med.*, **1**, 77-94.

Taylor, D.J. (1989) in Anion Carriers of Mitochondrial Membranes, ed. Azzi, A., Springer Verlag, Berlin, 373-381.

Taylor, D.J., Styles, P., Matthew, P.M., Arnold, D.L., Gadian, D.G., Bore, P.J., and Radda, G.K. (1986) *Magn. Reson. Med.*, **3**, 44-54.

Veech, R.L., Lawson, J.R.W., Cornel, N.W., and Krebs, H.A. (1979) *J. Biol. Chem.*, **254**, 6538-6547.

Adenine Nucleotides in Cellular Energy Transfer and Signal Transduction
S. Papa, A. Azzi & J.M. Tager (eds)
© 1992 Birkhäuser Verlag, Basel/Switzerland

137

MITOCHONDRIAL MUTATIONS AND THE AGEING PROCESS

Anthony W. Linnane, Alessandra Baumer, Agapi Boubolas, Ryan Martinus, Ronald J. Maxwell, François Vaillant, Zhong-Xiong Wang, Chunfang Zhang and Phillip Nagley

Department of Biochemistry and Centre for Molecular Biology and Medicine, Monash University, Clayton, Victoria 3168, Australia

SUMMARY: We have proposed that the accumulation of random somatic gene mutations in the mitochondrial DNA (mtDNA) during life is an important contributor to the ageing process. The consequence of these mutations is the progressive loss of bioenergetic capacity to produce ATP. In support of this hypothesis, we have demonstrated the age-related occurrence of a 4977 bp mtDNA deletion in autopsy tissue materials of 11 human subjects using the polymerase chain reaction (PCR). Extensive PCR analyses of two adults showed the presence of multiple deletions. In a similar survey carried out on rats, the accumulation of an analogous 4834 bp mtDNA deletion was also shown to be age-related. The accumulation of mtDNA deletions with age, we suggest, is a general phenomenon and our ageing hypothesis may apply to all mammalian systems. As a model to examine the bioenergetic consequences of mitochondrial damage, we have considered the use of the human lymphoblastoid Namalwa cell line. We report that these cells can be grown aerobically or anaerobically without the energy contribution normally derived from mitochondrially produced ATP. The growth and viability of the Namalwa cell line grown under anaerobic conditions is dependent upon a redox sink (pyruvate/lactate and/or ferricyanide/ferrocyanide) for reoxidation of NADH. This dependency on pyruvate for growth can be used to demonstrate the effect of AZT upon the bioenergetic capacity of mitochondria.

INTRODUCTION

Ageing is a highly complex process involving the gradual decline of the physiological performance of individual tissues and complete organs. We have earlier proposed that somatic gene mutation and its accumulation in mitochondrial DNA (mtDNA) makes a significant contribution to the ageing process (Linnane et al., 1989; Linnane et al., 1991; Nagley et al., 1991). Human mtDNA is a 16569 bp double stranded circular molecule which encodes 13 protein subunits of the mitochondrial respiratory enzyme complexes as well as specific mitochondrial RNAs (two rRNAs and 22 tRNAs). The approximately 70 remaining protein subunits of these complexes are encoded by the nuclear genome, synthesized in the cytosol and imported into the mitochondria where they assemble with mtDNA-encoded subunits to form functional complexes. Distinguishing features of the mtDNA are: 1) replication of mtDNA is not related to the cell division cycle; 2) mtDNA is very tightly packed with little or no intervening sequences such that any mutation has the potential to affect either an essential gene or the regions containing the origins of replication; 3) mtDNA is not protected by a protein coat; 4) no efficient DNA repair system has been detected in mitochondria and 5) the mutation rate of mtDNA is 10-100 times higher than that of nuclear DNA. A contributing factor to such a high mutation rate is the exposure of mtDNA to high levels of oxygen free radicals which are known to be potent mutagens and are largely generated by mitochondrial oxygen metabolism (Harman, 1983; Hayakawa et al., 1991). In this article, we review our recent data in regard to the accumulation with age of large mtDNA deletions which support our hypothesis. The bioenergetics of mitochondrially impaired human cultured cells are examined in relation to the need for reoxidation of NADH by means of a redox sink. We also present evidence for an effect of AZT on mitochondrial energy metabolism.

AGE-RELATED 4,977 BP DELETION IN HUMAN MITOCHONDRIAL DNA

A 4,977 bp mtDNA deletion has been reported to occur frequently in patients affected by mitochondrial myopathies (Holt et al., 1988; Shoffner et al., 1989; Obermaier-Kusser et al., 1990) and Parkinson's disease (Ikebe et al., 1990). The commonality of this deletion in mitochondrial disease patients implies a possible "hot spot" in mtDNA (Schon et al., 1989) at two well characterised 13 bp direct repeats. In our initial study, we investigated the occurrence of this deletion in

Table I. Detection of an age-related 5 kb deletion in human mtDNA

SUBJECT	AGE	SEX	CAUSE OF DEATH	TISSUES ANALYSED	5 kb DELETION AFTER 30 CYCLES
1	80 minutes	male	congenital heart defect	heart, liver, mid-brain, psoas muscle	–
2	3 months	female	congenital heart defect	heart, liver, frontal lobe, substantia nigra, kidney, lung, adrenal gland, diaphragm, ovary, pancreas, spleen, tyroid	–
3	40 years	male	ruptured thoracic aorta	adrenal gland, frontal lobe, kidney, left ventricle, liver, lung, left atrium, pancreas, psoas muscle, right atrium, right ventricle, skeletal muscle	+
4	46 years	female	motor neuron disease	left ventricle, kidney, diaphragm, right ventricle, right atrium, liver	+
5	50 years	male	ruptured thoracic aorta	kidney, liver, lung, psoas muscle, skeletal muscle, spleen	+
6	69 years	female	primary carcinoma of lower bowel	heart, brain, liver, skeletal muscle, kidney	+
7	72 years	male	motor neuron disease	heart, liver, kidney, lung, skeletal muscle, adrenal gland	+
8	79 years	male	aortic aneurysm	substantia nigra, mid-brain	+
9	80 years	female	peptic ulcer	substantia nigra, mid-brain	+
10	82 years	female	bowel obstruction	substantia nigra, mid-brain	+
11	87 years	male	cardiac arrest	substantia nigra, mid-brain	+

Table II. Detection of multiple mitochondrial DNA deletions in a 69-year-old female subject

Deletion	Sequence*	Deletion observed**		
		Heart	Brain	Muscle
4977 bp	8470 · 13447 ACTACCACCTACCTCCCTCACCA (AAGCC.........CTTCAACCTCCCTCACCA) TTGGCAGCCT	+	+	+
5827 bp	7954 · · · · · · · · · · · · · · 13781 CCTACATACTICCCCC (ATTAT.........AACAATCCCCC) TCTACCTAAA	nd	nd	+
6335 bp	8470 · · · · · · · · · · · · · · · · 14805 ACTACCACCTACCTCCC (TCACC.........CATCGACCTCCC) CACCCCATCC	nd	nd	+
7436 bp	8637 · 16073 CCAAATATCTCATCAACAACCG (ACTAA.........TCACCCATCAACAACCG) CTATGTATTT	+	+	+
7635 bp	8433 · · · · · · · · · · · · · · · · 16068 CTATTCCTCATCACCCA (ACTAA.........TTGACTCACCCA) TCAACAACCG	nd	nd	+
7737 bp	7986 · · · · · · · · · · · · · · · · 15723 GGCGACCTGCGACTCCT (TGACG.........TTATTGACTCCT) AGCCGCAGAC	nd	nd	+
7856 bp	8027 · · · · · · · · · · · · 16071 CAGCTTCATGCCCATCG (TCCT.........ACTC) ACCCATCAACAACCGCT	nd	nd	+
8041 bp	8030 · · · · · · · · · · · · · 16071 GATTGAAGCCCCAT (TCGTA.........ACTCACCCAT) CAACAACCGC	+	+	nd
8044 bp	8030 · · · · · · · · · · · · · · · · · · 16071 GATTGAAGCCC (CCATTCGTA.........ACTCACCCA) TCAACAACCGC	nd	nd	+
5756 bp	7769 · · · · · · · · · · · · · · · · · 13525 TCAGGAAATAGAAAACCG (TCTGA.........TCATCGAAACCG) CAAACATA	nd	nd	+

*Direct repeat sequences are underlined and the numbers above each sequence indicate the position of the first nucleotide of each repeat. The numbering of nucleotides is according to that of Anderson et al. (1981). The breakpoints are indicated by parentheses and the nucleotides shown inside the parentheses (indicated by italics) were deleted.
**nd, not detected.

normal subjects of different ages (Table I). Tissues samples were derived from 11 human subjects ranging in age from 80 minutes to 87 years with no known mitochondrial disease. Total cellular DNA was extracted from tissue samples by standard procedures and was subjected to PCR amplification using PCR primers L7901 (nucleotide positions light strand 7901-7920) and H13631 (nucleotide positions heavy strand 13650-13631). The 4,977 bp deletion was found to occur in an age-related manner. This deletion was detected after 30 cycles of PCR in all adult tissues analysed. The detection of this deletion in the two infant tissues however required 60 amplification cycles indicating a markedly lower initial abundance (Linnane et al., 1990). DNA sequencing of the PCR amplified fragments covering the deletion breakpoints showed that the 4,977 bp deletion is the result of a recombination between two 13 bp direct repeats (ACCTCCCTCACCA) that occur at nucleotide positions 8470-8482 and 13447-13459. One copy of the 13 bp repeats is maintained in the subgenomic molecules, the other being part of the deleted 4,977 bp fragment.

MULTIPLE DELETIONS IN HUMAN MTDNA

Preliminary studies on the characterization of additional PCR-amplified DNA molecules representing other mtDNA deletions in an elderly and a middle aged individual are presented in Tables II and III. Table II shows an extensive analysis on three tissues (heart, brain and skeletal muscle) from a 69-year-old female (subject 6 in Table I). Ten different deletions were detected after 30 PCR cycles using several sets of primers. The deletions were flanked by direct repeats varying in size from 5 to 13 base pairs and the sizes of the deletions from 4,977 bp to 8044 bp. In this study not all deletions were detected in every tissue. It was shown that mtDNA deletions occur most commonly in skeletal muscle but nevertheless multiple deletions were detected in both brain and cardiac muscle. Furthermore, we found that the detection of a given deleted mtDNA molecule would depend on the choice of appropriate pairs of PCR primers (Zhang et al., submitted for publication). In a 46-year-old individual (subject 4 in Table I) multiple deletions have also been detected. A single pair of primers (L7901 and H13631), used in the PCR amplification of a kidney sample (see Table III) detected three deletions. In this case the subgenomic molecules were faintly visualised on agarose gels after 30 cycles of PCR and strongly after 60 cycles indicating the low level of occurrence. The deletions

Table III. Detection of multiple subgenomic mtDNA molecules in DNA isolated from a kidney sample from a 46 year-old female individual

Amplified fragment	Sequence/breakpoints*	Deletion size
0.14 kb	TGACAA(.......)AATTCT 8004 13614	5609 bp
0.25 kb	CTTCAA(.......)TACCTT 7941 13483	5542 bp
0.30 kb	GGCGAC(.......)ATAGGA 7981 13411	5429 bp

*The number indicates the position of the nucleotide above the last digit.

were not flanked by repeats. Extending this analysis to other tissues from this subject, different PCR products were generated from each tissue (data not shown). These results suggest that the extent and nature of the deletions may vary from one tissue to another, presumably dependent upon the different tissue metabolism and potential intra-tissue mutagens.

AGE-RELATED MTDNA DELETIONS IN RATS

Our proposal for an age related accumulation of mtDNA deletions is not restricted to human subjects. It would be predicted that these mutations would occur in an age related manner in other mammals and that the accumulation would be faster in mammals with shorter life spans presumably associated with higher metabolic rates.

The laboratory rat has a maximal life span of about 30 months (life potential may be longer) and as such at 9 months, would be expected to be equivalent to a 20 to 25-year-old human. This age group in normal human subjects has been shown to contain the 4977 bp deletion (Cortopassi and Arnheim, 1990). PCR analysis on total cellular DNA isolated from rats ranging in age from 8 to 140 weeks has shown an age related accumulation of a 4.8 kb deletion (Fig. 1). This deletion

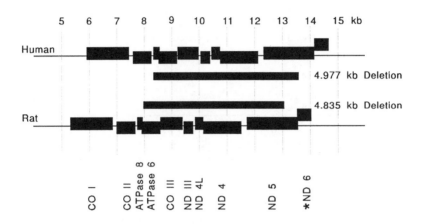

Fig. 1: Part of the linearised maps of the human and rat mtDNA showing the positions of the 5 kb human deletion and the 4.8 kb deletion in rat. The two maps are aligned by the nucleotide position numbers of Anderson et al. (1981) for human and Gadaleta et al. (1989) for rat.

occurs between two 16 bp direct repeats, situated at nucleotide positions 8103-8118 and 12937-12952. Young rats (8-12 weeks) did not show the deletion after 30 PCR amplification cycles. The 4.8 kb deletion was detected in rats aged 36 weeks and older, although in these preliminary studies, not in all tissues (Table IV). In a few cases there were indications of the occurrence of a second predicted 5.4 kb deletion occurring between two 13 bp direct repeats, situated at nucleotide positions 8046-8058 and 13462-13474 respectively. Detailed analyses of the rat tissues are continuing.

Table IV. Age related accumulation of deletions in mtDNA in different rat tissues.

AGE (WEEKS)	DELETION SIZE (Kb)			NUMBER OF RATS
	LIVER	HEART	MUSCLE	
8	nd	nd	nd	1/1
12	nd	nd	nd	5/5
36	4.8	nd	nd	1/1
52	4.8	nd	4.8	1/1
112	4.8	4.8	4.8 5.4	1/1
115	4.8	4.8	nd	1/1
115	nd	nd	5.4	1/1
140	4.8	nd	nd	1/1
140	4.8 5.4	nd	nd	1/1

Deletions detected by PCR in total cellular DNA. PCR was carried out for 30 cycles with denaturation at 94°C for 60 secs, anealing at 50°C for 90 secs and extension at 72°C for 150 secs. The light strand primer was situated at base position 7567 and a heavy strand primer at base position 13610. nd = not detected.

HUMAN CULTURED CELLS ARE NOT DEPENDENT UPON OXIDATIVE PHOSPHORYLATION FOR VIABILITY AND GROWTH

Morais and colleagues described in a series of papers the isolation of ρ^0 cells (completely lacking mtDNA) derived by prolonged culture of chicken embryo cells in the presence of ethidium bromide (EtBr)(Desjardins et al., 1985). Thus, under conditions of aerobiosis, such cells have no functional respiratory system. The ρ^0 cells require, for viability and growth, the obligatory supplementation of the basic growth medium with pyruvate and uridine. The pyruvate/lactate couple (redox sink) allows the re-oxidation of NADH to NAD^+ by cellular lactate dehydrogenase. Uridine is necessary to by-pass the requirement for a functional electron transport system in the pyrimidine synthesis pathway (dihydro-orotate dehydrogenase). Attardi and colleagues have extended these studies to mammalian cells, showing that the human HeLa cells can also be completely depleted of their mtDNA content by prolonged EtBr treatment when grown similarly in the presence of pyruvate and uridine (King et al., 1989). We have recently shown that it is possible to grow human cells anaerobically. The human lymphoblastoid Namalwa cell line, cultured in a RPMI-1640 and 10% serum medium under a CO_2/N_2 atmosphere, can grow anaerobically if pyruvate and uridine are also included in the medium (Vaillant et al., 1991). Such anaerobic growth occurred at the same rate as that of aerobic cultures, with a doubling time of 20 hours (Table V). A two-fold increase in glucose consumption by the cells grown under anaerobic conditions as opposed to aerobic growth was observed. Glycolytic energy production and re-oxidation of NADH by pyruvate allow the cells, under anaerobic conditions to synthesise sufficient ATP to maintain viability and a rapid growth rate. It can be calculated that, for aerobic growth of Namalwa cells, only about 12% of metabolised glucose need be completely oxidised in

Table V. Growth and glucose consumption of Namalwa cells cultured under various conditions

Gas phase	Pyruvate added	Doubling time (hours)	Glucose consumption (nmol/h/10^5 cells)
CO_2/Air	-	20	23 ± 4
CO_2/Air	+	20	18 ± 6
CO_2/N_2	-	Death	-
CO_2/N_2	+	20	46 ± 3

Adapted from Vaillant et al. (1991).

order to supply cellular ATP requirements. Thus, these studies on vertebrate cells have established that, at least in tissue culture, there is no absolute need for mitochondrial ATP production for growth and viability.

These observations may be extended to include the concept that provided NADH/NADPH can be re-oxidised at a sufficiently rapid rate by non-mitochondrial systems, then sufficient ATP

146

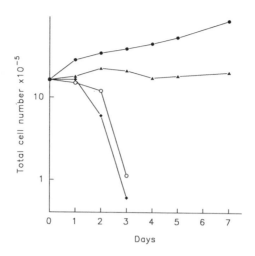

Fig. 2: Effect of added ferricyanide on EtBr-treated Namalwa cells. Namalwa cells have been maintained for 10 months in presence of EtBr (50 ng/ml in RPMI-1640 plus 10% serum, supplemented with pyruvate (1 mM) and uridine (50 μg/ml); Vaillant et al., 1991). They were transfered in a medium (RPMI 1640 plus 10% serum and uridine) containing either pyruvate (●), ferricyanide (10 μM, ▲; or 15 μM, ◆) or no addition (O). Viability was estimated every day by trypan blue exclusion on duplicate samples. Medium was changed at two days intervals.

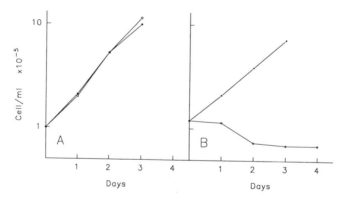

Fig. 3: Effect of pyruvate treatment on AZT treated Namalwa cells. Namalwa cells were cultured, in duplicate, in absence (Panel A) or, in presence of 400 μM AZT (Panel B). Pyruvate (1 mM) and uridine (50 μg/ml) were either added (filled symbols) or omitted (open symbols) in the growth medium (RPMI-1640 plus 10% serum). The number of viable cells was determined daily by trypan blue exclusion. Each point represents the average of two determinations.

investigated by culturing human Namalwa cells with added AZT in the presence or absence of pyruvate and uridine. Fig. 3 demonstrates that, when treated with a concentration of 400 μM AZT, the viability of human cultured Namalwa cells was greatly reduced. This effect was circumvented by the addition of pyruvate and uridine; normal growth occurs under these conditions. Pyruvate and uridine had no effect on the growth of non-treated cells. The rescue, by pyruvate and uridine, of AZT treated cells, is similar to the rescue of EtBr treated or anaerobically grown cells and demonstrates the similarity of effects of the drug on mitochondrial bioenergetic functions. Electron micrographs of AZT-treated cells (Fig. 4) showed significant modifications of the mitochondria which morphologically resembled the organelles seen in ρ^0 cells. In particular, mitochondria appeared swollen and contained few

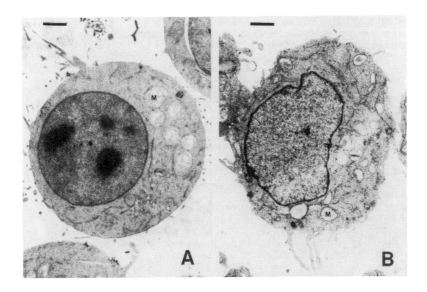

Fig. 4: Electron micrographs of Namalwa cells untreated or treated with AZT. Cells were grown for three days in RPMI-1640 plus 10% serum, supplemented with pyruvate (1 mM) and uridine (50 μg/ml). A: Untreated Namalwa cells (Bar: 2 μM). B: Namalwa cells treated with 400 μM AZT (Bar: 2 μM).

148

cristae; electron dense inclusions (micro-crystalline ?) were also seen (Fig. 4B). In comparison, non-treated cells possessed normal mitochondria with regular and lamellar cristae (Fig. 4A).

These cell culture results demonstrate that respiratory deficient cells can be rescued by redox compounds. This concept of the use of redox sinks provides a framework for the design of a general strategy to ameliorate bioenergetic defects in ageing. Amelioration therapy using redox sinks/bypass has been successfully employed with a number of mitochondrial disease patients. These therapeutic regimes have included the use of such redox compounds as vitamin C, vitamin K_3, succinate or CoQ_{10} (Ellef et al., 1984; Shoffner et al., 1989; Jinnay et al., 1990). The suggestion that the use of such a therapy in aged individuals would improve bioenergetic production and thus the physiological state and being-well of the aged, warrants serious study.

ACKNOWLEDGEMENT: This work was supported by grants from the National Heart Foundation of Australia and the National Health and Medical Research Council of Australia.

REFERENCES

Anderson, S., Bankier, A.T., Barrel, B.G., de Bruijn, M.H.L., Coulson, A.R., Drouin, J., Eperon, I.C., Nierlich, D.P., Roe, B.A., Sanger, F., Schreir, P.H., Smith, A.J.H, Staden, R., and Young, I.G. (1981) Nature 290, 457-465
Cortopassi, G.A., and Arnheim, N. (1990) Nucleic Acids Res. 18, 6927-6933
Dalakas, M.C., Illa, I., Pezeshkpour, G.H., Laukaitis, J.P., Cohen, B., and Griffin, J.L. (1990) New Eng. J. Med. 322, 1098-1105
Desjardins, P., Frost, E., and Morais, R. (1985) Mol. Cell. Biol. 5, 1163-1169
Ellef, S., Kennaway, G., Buist, N., Darley Usmar, V.M., Capaldi, R.A., Bank, W.J., and Chance, B. (1984) Proc. Natl. Acad. Sci. USA 81, 3529-3533
Ellem, K.A.O., and Kay, G.F. (1983) Biochem. Biophys. Res. Commun. 112, 183-190
Gadaleta, G., Pepe, G., De Candia, G., Quagliariello, C., Sbisa, E., and Saccone, C. (1989) J. Mol. Evol. 28, 497-516
Harman, D. (1983) Age 6, 86-94
Hayakawa, M., Torii, K., Sugiyama, S., Tanaka, M., and Ozawa, T. (1991) Biochem. Biophys. Res. Commun. 179, 1023-1029

Holt, I.J., Harding, A.E., and Morgan-Hughes, J.A. (1988) Nature 331, 717-719

Ikebe, S-I., Tanaka, M., Ohno, K., Sato, W., Hattori, K., Kondo, T., Mizuno, Y., and Ozawa, T. (1990) Biochem. Biophys. Res. Commun. 170, 1044-1048

Jinnay, K., Yamada, H., Kanda, F., Masui, Y., Tanaka, M., Ozawa, T., and Fujita, T. (1990) Eur. Neurol. 30, 56-60

King, M., and Attardi, G. (1989) Science 246, 500-504

Linnane, A.W., Marzuki, S., Ozawa, T., and Tanaka, M. (1989) Lancet i, 642-645

Linnane, A.W., and Nagley, P. (1991) In: Molecular Biology of the Myocardium (Tada, M. Ed),Japanese Scientific Societies Press, Tokyo (in press)

Linnane, A.W., Baumer, A., Maxwell, R.J., Preston, H., Zhang, C., and Marzuki, S. (1990) Biochem. Int. 22, 1067-1076

Nagley, P., Mackay, I.R., Baumer, A., Maxwell, R.J., Vaillant, F., Wang, Z-X., Zhang, C., and Linnane A.W. (1990) In: Ann. N.Y. Acad. Sci. (in press)

Obermair-Kusser, B., Müller-Höcker, J., Nelson, I., Lestienne, P., Enter, Ch., Riedele, Th., and Gerbits, K.-D. (1990) Biochem. Biophys. Res. Commun. 169, 1007-1015

Richman, D.D. et al. (1987) New Eng. J. Med. 317, 192-197

Shoffner, J.M., Lott, M.T., Voljavec, A.S., Soueidan, S.A., Costigan, D.A., and Wallace, D.C. (1989) Proc. Nat. Acad. Sci. USA 86, 7952-7956

Schon, E.A., Rizzuto, R., Moraes, C.T., Nakase, H., Zeviani, M., and DiMauro, S. (1989) Science 244, 346-349

Simpson, M.V., Chin, C.D. Keilbraugh, S.A., Lin, T.S., and Prussof, W.H. (1989) Biochem. Pharmaco. 38, 1033-1036

Vaillant, F., Loveland, B.E., Nagley, P., and Linnane, A.W. (1991) Biochem. Int. 23, 571-580

Adenine Nucleotides in Cellular Energy Transfer and Signal Transduction
S. Papa, A. Azzi & J.M. Tager (eds)
© 1992 Birkhäuser Verlag, Basel/Switzerland

STRUCTURAL AND FUNCTIONAL ALTERATIONS OF MITOCHONDRIAL $F_O F_1$-ATPSYNTHASE IN VARIOUS PATHOPHYSIOLOGICAL STATES

Ferruccio Guerrieri, Giuseppe Capozza, Franco Zanotti, Ferdinando Capuano and Sergio Papa

Institute of Medical Biochemistry and Chemistry and Centre for the Study of Mitochondria and Energy Metabolism, University of Bari, Italy

SUMMARY: The structure and function of mitochondrial $F_O F_1$ ATPsynthase in rat liver regeneration, heart of senescent rats and in rapidly growing Morris hepatoma 3924A have been studied. In the Morris hepatoma 3924A, the ATPase activity exhibited a Km for ATP considerably higher than in normal rat liver. In rat liver 24 h after partial hepatectomy and in heart of aged rats (24 month old rats) the Vmax for ATP hydrolase activity decreased and the oligomycin sensitive proton conductivity, in vesicles of the inner mitochondrial membrane, increased.
Studies with the specific inhibitor oligomycin and immunoblot analysis indicated that, in Morris hepatoma, the catalytic process in F_1 and the functional interactions between F_1 and F_O sectors are altered. During the first phase of rat liver regeneration and in heart of senescent rats the functional alteration of ATP hydrolase activity is associated with a decrease of the F_1 content in mitochondria.

INTRODUCTION

Under normal conditions more than 80% of the energy demand

of most mammalian cells is covered by ATP produced in

mitochondrial oxidative phosphorylation, the remaining being contributed by glycolysis (Papa 1989).

In certain tissues and in particular pathophysiological conditions the contribution of oxidative phosphorylation to ATP supply is smaller whilst that of glycolysis becomes more important (Uriel, 1979; Lakatta and Yim, 1982; Hansford, 1983; Sako et al. 1988; Izquierdo et al. 1990). This occurs, for exsample, in fetal (Baggetto et al. 1984; Izquierdo et al. 1990) and regenerating tissues (Uriel, 1979; Buckle et al. 1985, 1986) and in rapidly growing tumors (Uriel, 1979).

Recent observations indicate that the same shift from aerobic to fermentative energy metabolism is exhibited by tissues from senescent animals (Lakatta and Yim, 1982; Hansford, 1983; Nohl and Kramer, 1980).

Comprehension of the factors which regulate the relative contribution of fermentation and respiration to ATP supply to the cell has from time to time attracted the interest of investigators (Nohl and Kramer, 1980; Rouslin, 1983; Bukle et al. 1985, 1986; Sako et al, 1988; Capuano et al. 1989). As a contribution to this problem we begun a sistematyc study of the structural and functional characteristics of the mitochondrial F_0F_1-ATP synthase in a variety of tissues and pathophysiological states (Buckle et al. 1985, 1986; Capuano et al. 1989; Guerrieri et al. 1989a; Capozza et al. 1992).

The basic features of F_0F_1-ATP synthase are described in other papers in this volume (Nagley; Papa et al; Hatefi and Matzumo-Yagi), here experimental observations on structural and functional alterations of the mitochondrial F_0F_1-ATP synthase in rat liver regeneration in heart of aged rats and in Morris hepatoma 3924 A are reported.

RESULTS

Morris hepatoma: The ATPase activity exhibited by vesicles of the inner mitochondrial membrane (ESMP) from Morris hepatoma is considerably lower with respect to control rat liver (Capuano et al. 1989; Guerrieri et al. 1989a). Kinetic analysis by double reciprocal plots showed no difference in Vmax, whereas the Km value for ATP was considerably higher in hepatoma than in control submitochondrial particles (ESMP) (Table I).

AMP-PNP, a non hydrolyzable competitive inhibitor (Capuano et al. 1989) of the ATPase activity, showed a considerably higher K_i in ESMP from hepatoma than from control (Table I).

The oligomycin sensitive proton translocation through F_o

Table I. ATP hydrolase activity and passive proton conduction of F_oF_1 ATP synthase from rat liver and Morris hepatoma 3924A.
Vesicles of inner mitochondrial membrane (ESMP) were prepared from rat liver and Morris hepatoma 3924A as described by Capuano et al. (1989). ATP hydrolase activity and oligomycin sensitive passive proton permeability were analysed as described by Guerrieri et al. (1989b).
Where indicated ESMP were preincubated 5 min in presence of inhibitor. AMP-PNP = [ß, γ-imido] ATP. I_{50} = oligomycin concentration giving 50% of maximal inhibition. Vmax are expressed in µmol/min/mg prot. Km is expressed in mM; Ki is expressed in µM and I_{50} in µg oligomycin/mg protein.

	HYDROLASE ACTIVITY				PROTON CONDUCTION	
	Vmax	Km	Ki for AMP·PNP	I_{50}	$1/t_{\frac{1}{2}}$ (s^{-1})	I_{50}
Rat liver	1.13	0.17	1.2	0.21	2.59	0.10
Morris hepatoma	1.08	0.49	3.7	2.70	2.28	0.08

sector did not show any significant change in hepatoma submitochondrial particles respect to control particles. Oligomycin, a specific inhibitor of proton conduction by F_O, showed practically the same I_{50} for passive proton translocation in hepatoma ESMP and in control liver ESMP; however the I_{50} for oligomycin inhibition of ATP hydrolase activity was much higher in hepatoma ESMP than in control liver ESMP.

Liver regeneration: During liver regeneration the hepatocytes enter the replicative cycle characterized by a marked reorganization of gene expression (Enrich and Gahmberg, 1985). This process shows two different phases (Uriel, 1979): a) retrodifferentiation; lasting approximately 24 h after hepatectomy b) redifferentiation; which is characterized by restauration of normal order and assembly of protein complexes. Table II shows that the first phase of liver regeneration was

Table II. Changes of ATP hydrolase activity and passive proton conduction of $F_O F_1$ ATP synthase in liver submitochondrial particles 24 h after partial hepatectomy.
For preparation of ESMP from control rat liver and from regenerating rat liver, determination of ATP hydrolase activity and of oligomycin sensitive proton conduction see Buckle et al. 1986.

| | ATP HYDROLASE ACTIVITY | | PROTON CONDUCTION |
| | Vmax | Km | $1/t_{\frac{1}{2}}$ |
	(μmol/min/mg prot)	(mM)	(s^{-1})
Control Liver	1.13	0.17	2.38
Regenerating Liver (24 h)	0.28	0.19	4.00

characterized by decrease of the ATP hydrolase activity of ESMP with no significant change in Km for ATP. This was accompanied by an increase of the rate of anaerobic release of the transmembrane proton gradient.

The mitochondrial F_0F_1 ATPsynthase is structurally and functionally organized into two sectors: the F_1 catalytic sector and the proton conducting membrane sector F_0 (see Papa et al., this volume). In the enzyme normally arranged in the membrane the proton conduction by F_0 is strictly bound to catalytic activity in F_1. After displacement or removal of F_1, as it occurs in some submitochondrial particles, a passive proton conduction by F_0 can, however, be observed (Pansini et al. 1978; Guerrieri et al. 1989b).

Kinetic analysis of anaerobic H^+ release, a process which is inhibited by F_0 inhibitors oligomycin and DCCD (see Pansini et al. 1978 and Buckle et al. 1986) shows a byphasic pattern in F_1 containing particles from beef heart which changes to monophasic in F_1 depleted particles (Fig. 1A). ESMP from control liver showed also a byphasic pattern; however the ESMP from regenerating liver (24 h) showed, as F_1 depleted particles a monophasic kinetics (Fig. 1B). Immunoblot analysis, using antibody against bovine F_1 which crossreacted with rat liver α and ß subunits, showed a decrease of F_1 subunits in ESMP prepared from liver 24 h after hepatectomy respect to control liver ESMP (Fig. 2A). No change was observed in the content of the F_0 subunit OSCP.

These results suggest that during early regeneration disturbance occurs in the integration of F_1 subunits with the

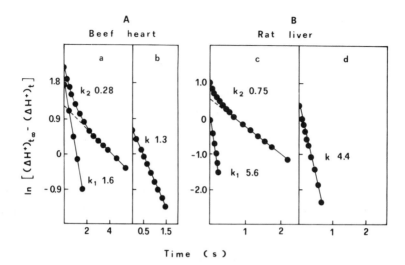

Fig. 1 Kinetic analysis of anaerobic relaxation of $\triangle\mu H^+$ set up, in ESMP, by succinate respiration. ESMP and F_1 depleted ESMP were prepared as described by Guerrieri et al. 1989 b. For preparation of ESMP from rat liver and from regenerating rat liver see Buckle et al 1986.
Mathematical analysis was carried out as in Pansini et al. 1978. Beef-heart submitochondrial particles: a) ESMP; b) F_1 depleted ESMP. Liver submitochondrial particles: c) control; d) regenerating (24 h).

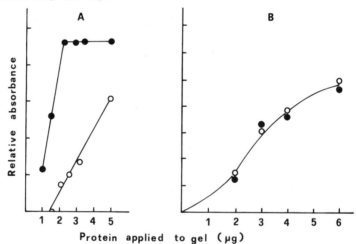

Fig. 2 Relative absorbance of immunoblots of submitochondrial particles from control liver (●) and regenerating liver (24 h) (o) with antibody against bovine F_1 (A) and bovine OSCP (B). For elctrophoresis and immunoblot procedure see Buckle et al. 1986. Redrawn from Buckle et al. 1986.

F_O membrane sector. This is transitory and, 48 h after partial hepatectomy, the ATPase activity and control of passive proton translocation by F_O are restored (Buckle et al. 1986).

Heart of Senescent rats: In ESMP prepared from heart of senescent rats (24 month old rats) the ATPase activity was considerably lower with respect to ESMP from heart of adult rats (12 months) (Table III). No change in the Km value for ATP was observed (Table III). The oligomycin sensitive proton conduction by F_O was enhanced as shown by the increase of the reciprocal value of the $t_{\frac{1}{2}}$ of the anaerobic release of the transmembrane proton gradient set up by respiration (Table III).

The increase of passive proton permeability with age was accompanied by change of the kinetic pattern from biphasic to monophasic (Fig. 3).

Semiquantitative immunoblot analysis showed that the material cross-reacting with antisera against bovine F_1 at the

Table III. Age dependent changes in ATP hydrolase activity and in passive proton conduction of F_OF_1 ATPsynthase in rat heart submitochondrial particles.
ESMP for rat heart mitochondria were isolated following the procedure used by Low and Vallin (1963) for the beef heart mitochondria. For determination of ATP hydrolase activity and passive proton conduction see Legend to Table I. The data reported are the mean of 5 experiments ±S.E.M.

AGE	ATP HYDROLASE ACTIVITY		PROTON CONDUCTION
(months)	Vmax	Km	$1/t_{\frac{1}{2}}$
	(µmol/min/mg·prot)	(mM)	(s^{-1})
12	3.18 ± 0.18	0.14 ± 0.02	0.52 ± 0.07
24	1.47 ± 0.22	0.14 ± 0.03	1.19 ± 0.09

158

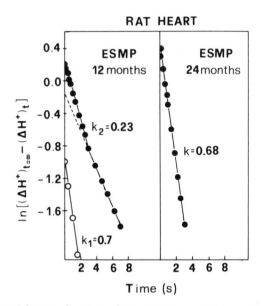

Fig. 3. Kinetic analysis of anaerobic relaxation of $\Delta\mu H^+$ set up in ESMP, from rat heart by succinate respiration.
For experimental conditions and mathematical analysis see Legend to Fig. 1.

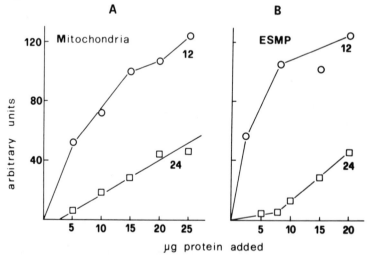

Fig. 4 Semiquantitative analysis by immunoblot of content of ß-subunit in mitochondria (A) and in ESMP (B) of heart of adult (O) and senescent (□) rats.
For Mitochondria and ESMP preparation see Legend to Table III. SDS polyacrylamide gel electrophoresis and immunoblot analysis were performed using an antiserum against bovine F_1 as described by Guerrieri et al. 1989 b.

level of β subunit decreased in ESMP, as well as in intact
mitochondria, from heart of 24 month rats respect to 12 month
rats (Fig. 4).

DISCUSSION

The results reported in this paper indicate that the
alteration of F_0F_1 ATPsynthase in ESMP from hepatoma is due to
changes in the kinetic properties of F_1 and in alteration of
F_0-F_1 interaction, as shown by the increase of I_{50} for oligomycin
inhibition of ATPase activity observed in these ESMP.

The alteration of F_0F_1 ATPsynthase complex observed during
the first phase of liver regeneration, as well as in heart
mitochondria of aged rats, can be ascribed to lower content of
F_1 subunit respect to F_0 resulting not only in lower activity
of the complex but also in faster dissipation of the
transmembrane proton gradient set up by respiration. This pattern
can explain the fact that, in these two states, the main energy
supply for the cell is provided by glycolysis.

There are various possible explanations for the presence, in
the mitochondrial membrane, of F_0 lacking bound F_1: a) repressed
expression of genes for F_1 subunits; b) defective assembly in
the inner mitochondrial membrane of F_1 subunit precursors and c)
enhanced, in vivo, proteolytic degradation of normally
synthetized F_1 subunits.

In heart of senescent rats it has been observed a deficit
at transcriptional level (Gadaleta et al., 1990) which can
reflect a generalized decrease of expression of proteins. This,
however, doesn't seem to explain the selective decrease of F_1

160

with respect to F_O (see Capozza et al., 1992).

REFERENCES

Baggetto, L., Gautheron, D.C. and Godinot, C. (1984) Arch. Biochem. Biophys 232, 670-678.

Buckle, M., Guerrieri, F. and Papa, S. (1985) FEBS Letters 188, 345-351.

Buckle, M., Guerrieri, F., Pazienza, A. and Papa, S. (1986) Eur. J. Biochem. 155, 439-455.

Capozza, G., Guerrieri, F., Kalous. M., Zanotti, F., Jirillo, E. and Papa, S. (1992) In "Biomerkers of Aging: Expression and Regulation" (Licastro, F., Caldarera, M. and Fabris, N. Eds.) in press.

Capuano, F., Stefanelli, R., Carrieri, E. and Papa, S. (1989) Cancer Res. 49, 6547-6550.

Enrich, C. and Gahmberg, C.G. (1985) FEBS Lett. 181, 12-16.

Gadaleta, M.N., Petruzzella, V., Renis, M., Fracasso, F. and Cantatore, P. (1990) Eur. J. Biochem. 187, 501-506.

Guerrieri, F., Capuano, F., Buckle, M. and Papa, S. (1989a) In "Molecular Aspects of Human Disease" (Gorrod, J.W., Albano, O. and Papa, S. Eds.) Ellis Horwood Publ. vol.1 p.108-114.

Guerrieri, F., Kopecky, J. and Zanotti, F. (1989b) In "Organelles of Eukaryotic Cells Molecular Structure and Interactions" (Tager, J.M., Azzi, A., Papa, S. and Guerrieri, F. Eds.) Plenum Pub. Co, 197-208.

Hansford, R.G., (1983) Biochim. Biophys Acta 1056, 71-73.

Izquierdo, J.M., Luis, A.M. and Cuezva J.M. (1990) Jour. Biol. Chem. 265. 9090-9097.

Lakatta, E.G. and Yim, F.C.P. (1982) Ann. J. Physiol. 242, 4927.4941.

Low, H. and Vallin, I. (1963) Biochim. Biophys. Acta 69, 361-374.

Nohl, A. and Kramer, R. (1980) Merch. Ageing. Development, 14, 137-144.

Pansini, A., Guerrieri, F. and Papa, S. (1978) Eur. J. Biochem. 92, 545-551.

Papa, S. (1989) In "Molecular Aspects of Human Disease"(Gorrod, J.W., Albano, O. and Papa, S. Eds.) Ellis Horwood Publ. vol.1, p.17-32.

Pedersen, P.L. (1977) Ann. Biochem. 80, 401-408.

Rouslin, W. (1983) Jour. Biol. Chem. 258, 9657-9661.

Sako, E.Y., Kingsly, P.B., From, A.H.L., Foker, J.E. and Ugurbi, K. (1988) Jour. Biol. Chem. 263, 10600-10607.

Walker, J.E., Lutter, R., Dupuis, A. and Runswick, M.J. (1991) Biochemistry 30, 5369-5379.

Uriel, J. (1979) Adv. Cancer Res. 29, 127-174.

Adenine Nucleotides in Cellular Energy Transfer and Signal Transduction
S. Papa, A. Azzi & J.M. Tager (eds)
© 1992 Birkhäuser Verlag, Basel/Switzerland

LONG RANGE INTRAMOLECULAR LINKED FUNCTIONS IN THE Ca^{2+} TRANSPORT ATPase OF INTRACELLULAR MEMBRANES

Giuseppe Inesi

Department of Biological Chemistry, University of Maryland School of Medicine, Baltimore, Maryland, 21201 U.S.A.

SUMMARY: The complete cycle of ATP utilization coupled to Ca^{2+} transport is operated by a single 106 kD unit of membrane bound ATPase. The cycle begins with Ca^{2+} binding which is an absolute requirement for enzyme activation. Then, phosphoryl transfer from ATP to the enzyme produces changes leading to vectorial translocation and loss of affinity of the bound Ca^{2+}. The Ca^{2+} binding and phosphorylation domains are separated by a distance of approximately 5 nanometers within the ATPase unit. Therefore the energy transduction mechanism requires long range coupling. Thapsigargin, which is a specific inhibitor of the Ca^{2+} ATPases of intracellular membranes, binds stoichiometrically and with extremely high affinity to the enzyme. The resulting inhibition affects distant functional domains, indicating that long range intramolecular linkages are also involved in the mechanism of this specific inhibition.

The catalytic and transport cycle of intracellular Ca^{2+}-ATPases:

The Ca^{2+} ATPases of intracellular membranes (Hasselbach and Makinose, 1961; MacLennan et al., 1985; Gunteski-Hemblin et al., 1988) constitute a family of highly homologous enzymes which are responsible for sequestration of cytoplasmic Ca^{2+} in intracellular stores delimited by endoplasmic reticulum. These enzymes operate against a three orders of magnitude Ca^{2+} gradient, in parallel with utilization of ATP or other phosphoryl donors (Inesi, 1985). Transient state kinetic measurements demonstrate that Ca^{2+} bound to the enzyme is displaced vectorially upon formation of a phosphorylated enzyme intermediate, and before hydrolytic cleavage

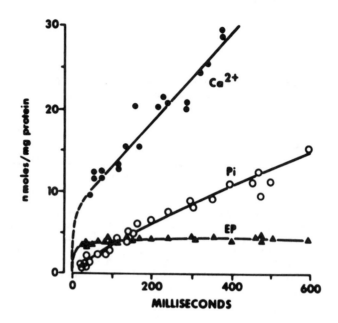

Fig 1. Demonstration of sequential enzyme phosphorylation, internalization of bound calcium and hydrolytic cleavage of Pi from the phosphoenzyme upon addition of ATP to sarcoplasmic reticulum ATPase preincubated with Ca^{2+}. Note that two moles of Ca^{2+} are internalized per mole of enzyme in the transient state. Following the first cycle, steady state Ca^{2+} uptake and Pi production proceed with a ratio of 2:1 (from Inesi et al., 1982).

of phosphate (Fig 1). A minimal number of partial reactions (Inesi et al., 1980) must then be considered for the catalytic and transport cycle, as follows:

1. $E + 2Ca^{2+}_{out} \longleftrightarrow E\ Ca_2$ $(3 \times 10^{12}\ M^{-2})$

2. $E\ Ca^2 + ATP \longleftrightarrow ATP\ E\ Ca_2$ $(1 \times 10^5\ M^{-1})$

3. $ATP\ E\ Ca_2 \longleftrightarrow ADP\ E\text{-}P\ Ca_2$ (0.3)

4. $ADP\ E\text{-}P\ Ca_2 \longleftrightarrow E\text{-}P\ Ca_2 + ADP$ $(7 \times 10^{-4}\ M)$

5. $E\text{-}P\ Ca_2 \longleftrightarrow E\text{-}P + 2Ca^{2+}_{in}$ $(3 \times 10^{-6}\ M^2)$

6. $E\text{-}P \longleftrightarrow E\ P_i$ (1)

7. $E\ P_i \longleftrightarrow E + P_i$ $(1 \times 10^{-2}\ M)$

The sequence above includes reactions that were accessed experimentally, and does not list explicitly protein conformational changes that are likely to account for the observed phenomena. We assume that the equilibrium constants derived for the partial reactions through chemical experimentation include any associated physical transition.   It is then apparent that coupling of catalytic and transport events is determined by mutual destabilization of Ca^{2+} binding and phosphorylation.   It can be shown formally (Inesi, 1985) that the free energy acquired by the enzyme through phosphoryl transfer from ATP corresponds to the free energy required for a three orders of magnitude reduction of the affinity constant of the enzyme for Ca^{2+}. Conversely, Ca^{2+} binding to the phosphoenzyme obtained with Pi in the absence of Ca^{2+}, raises its phosphorylation potential and renders formation of ATP possible when ADP is added.   The fundamental feature of the coupling mechanism is a functional linkage between phosphorylation and Ca^{2+} binding domains within the ATPase protein.

The idea of linked functions was first introduced by Jeffries Wyman (1964) to explain the reciprocal effects of oxygen and proton binding to hemoglobin. It was then defined by Gregorio Weber (1972) in terms of binding constants exhibited by different domains within a protein, and the reciprocal effects of domain occupancy by ligands. A mutual exclusion mechanism of this type is operative in the Ca^{2+} transport ATPase, where phosphorylation of the catalytic site by ATP destabilizes calcium bound to the enzyme, and occupancy of the specific calcium sites destabilizes the phosphorylated enzyme intermediate. It is then extremely useful to establish topologic relationships of calcium and phosphorylation sites, in order to evaluate whether short range events (such as proton displacements) related to the catalytic chemistry are sufficient to explain the observed destabilization of bound calcium, or whether additional conformational effects are required to explain a long range destabilization.

Relationships of structure and function: Analysis of the primary sequence of the Ca^{2+}-ATPase has yielded suggestions on the topology of the enzyme (Fig. 2), which in turn allows localization of domains. A combined approach by amino acid derivatization and site directed mutagenesis has revealed that the phosphorylation domain resides in the extramembranous region of the ATPase, while the Ca^{2+} binding domain resides within the transmembrane region where four (of the ten) spanning helices contribute residues for Ca^{2+} complexation (Inesi et al., 1990). Spectroscopic studies of fluorescent probes and their spacial relationships within the ATPase molecule indicate that phosphorylation and Ca^{2+} binding domains are separated by a distance of approximately 5 nanometers. Therefore, the functional linkage spans over this relatively long distance.

Structural considerations indicate that the four ATPase helices involved in Ca^{2+} binding are clustered in the transmembrane domain to form a rather tight channel which is stabilized by Ca^{2+} binding.

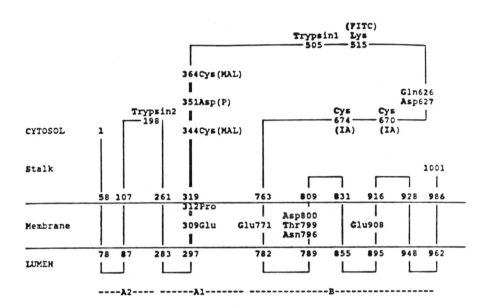

Fig 2. Topology of the Ca^{2+} ATPase of sarcoplasmic reticulum. The distribution of the ATPase sequence outside (cytosol) and inside (lumen of the sarcoplasmic reticulum) the membrane is based on the original assignment of MacLennan et al. (1985). T1 and T2 are sites of trypsin digestion that produce the peptide segments A2, A1, and B shown at the bottom of the figure. The six residues (Glu309, Glu771, Asn796, Thr799, Asp800, and Glu908) in the transmembrane region are thought to be involved in calcium linked functions (Clarke et al., 1989, Sumbilla et al., 1991). Asp351 is the residue undergoing phosphorylation at the catalytic site (Bastide et al., 1973; Degani and Boyer, 1973). The part of the sequence that shows a high degree of homology with other cation transport ATPase is shown by a double line. It is clear that this sequence intervenes between the phosphorylation and Ca^{2+} binding domains.

This transmembrane stabilization is transmitted over to the extramembranous catalytic site in the form of an activating signal permitting utilization of ATP. In turn, phosphoryl transfer from ATP to a catalytic site residue (Asp351), produces destabilization of the transmembrane helical cluster and vectorial dissociation of bound Ca^{2+}. A specific peptide segment (residues 297-359 in the Ca^{2+} transport ATPase) intervening between one of the transmembrane helices and the phosphorylation domain, is highly homologous in all cation transport ATPases and is likely to play a role in signal transmission and linkage of the cation binding and phosphorylation functions (Inesi and Kirtley, 1990).

Thapsigargin, a specific inhibitor of intracellular Ca^{2+}- ATPases: Thapsigargin (TG), a tumor promoting sesquiterpene lactone (Fig.3),

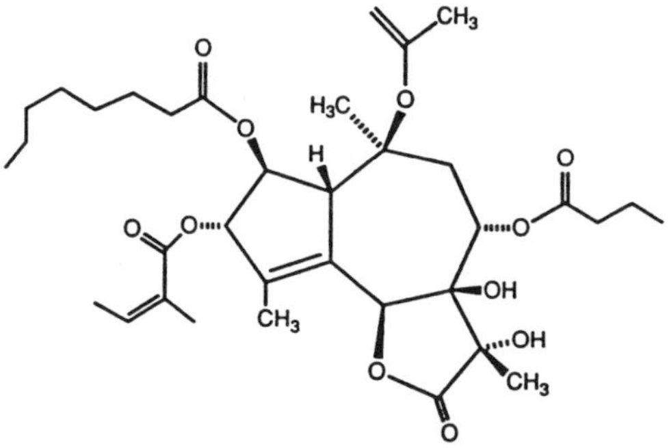

Fig 3. Structure of Thapsigargin.

has been shown to raise the intracellular concentration of Ca^{2+} in several cell lines (Jackson et al., 1988; Scharff et al., 1988). This effect was attributed to TG interference with the endoplasmic reticulum regulatory functions (Thastrup et al., 1990; Thastrup, 1990). Subsequently, we found that the Ca^{2+} transport and ATPase activities of SR vesicles purified from skeletal muscle are inhibited by very low concentrations of thapsigargin, in quantities stoichiometrically equivalent to the enzyme present in the assay (Sagara and Inesi, 1991). We also found that the inhibition is produced equally well on various isoforms of intracellular Ca^{2+} transport ATPases (Campbell et al., 1991; Fig 4) obtained by transfecting mammalian cells with specific cDNAs. TG is much less effective on the plasma Ca^{2+} ATPase and other cation transport ATPases. For these reasons, TG may be considered to be a specific inhibitor of the intracellular Ca^{2+} transport ATPases. Owing to its specificity, TG is of high value as a tool for mechanistic studies of the catalytic and transport cycle, as well as for intracellular perturbations of Ca^{2+} dependent regulatory functions.

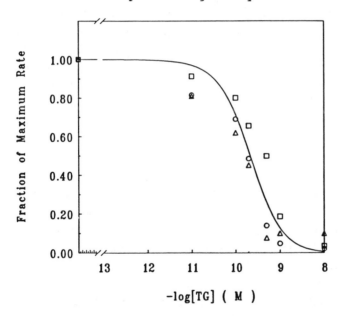

Fig 4. Inhibition of ATP dependent Ca^{2+} transport by Thapsigargin. The different symbols refer to experiments carried out with different ATPase isoforms (SERCA1, SERCA2a and SERCA2b) obtained by transfecting Cos1 cells with appropriate cDNA. (from Campbell et al., 1991).

As for localization of the inhibitory effect in the partial reactions comprising the catalytic and transport cycle of the SR ATPase, we found (Sagara and Inesi, 1991) that Ca^{2+} binding to the activating sites of the enzyme (reaction 1 in the sequence above) and, consequently, the Ca^{2+} dependent enzyme phosphorylation with ATP (reaction 3, formation of the phosphorylated intermediate) are inhibited.

Ca^{2+} independent enzyme phosphorylation with Pi (reaction 6) is also inhibited by TG. This reaction, which is the reversal of the phosphoenzyme hydrolytic cleavage, is Ca^{2+} independent and, in fact, requires dissociation of bound calcium from the enzyme (Masuda and de Meis, 1973). Therefore, our experiments show that two ATPase partial reactions (such as Ca^{2+} binding and enzyme phosphorylation with Pi) which can be studied under conditions rendering them independent of each other, are both inhibited by TG. It is of interest that binding of a single molecule of thapsigargin affects both Ca^{2+} binding and the phosphorylation reactions which occur in rather distant domains. It is then apparent that the mechanism of this specific inhibition involves long range intramolecular linkages.

CONCLUSIONS

Studies of structure and function indicate that coupling of catalysis and cation transport, as well as specific inhibition of these activities, are mediated by mechanisms spanning relatively long distances within the ATPase protein. Homologous polypeptide strands connecting catalytic and cation binding domains in various cation transport ATPases may have a role in signal transmission and energy transduction. It is apparent that long range linked functions are common features of several energy transducing enzymes such as myosin (actin binding and catalytic domains) and ATP synthase (transmembrane H+ channel and catalytic domain).

REFERENCES

Bastide, F., Meissner, G., Fleischer, S. & Post, R.L. (1973) J.Biol. Chem. **248**, 8385-8391

Campbell, A.M., Kessler, P.D., Sagara, Y., Inesi, G. & Fambrough, D.M. (1991) J. Biol. Chem. **266**, 16050-16055

Clarke, D.M., Loo, T.W., Inesi, G. & MacLennan, D.H. (1989) Nature **339**, 476-478

Gunteski-Hamblin, A.-M., Greeb, J. & Shull, G.E. (1988) J. Biol. Chem. **263**, 15032-15040

Hasselbach, W. & Makinose, M. (1961) Biochem. Z. **333**, 518-528

Inesi, G. (1985) Annu. Rev. Physiol. **47**, 573-601

Inesi, G. & Kirtley, M.E. (1990) J. Membr. Biol. **116**, 1-8

Inesi, G., Kurzmack, M., Coan, C. & Lewis, D. (1980) J. Biol. Chem. **255**, 3025-3031

Inesi, G., Sumbilla, C. & Kirtley, M.E. (1990) Physiol. Rev. **70**, 749-760

Inesi, G., Watanabe, T., Coan, C. & Murphy, A.J. (1982) Ann. NY Acad. Sci. **402**, 515-534

Jackson, T.R., Patterson, S.I., Thastrup, O. & Hanley, M.R. (1988) Biochem. J. **253**, 81-86

MacLennan, D.H., Brandl, C.J., Korczak, B. & Green, N.M. (1985) Nature **316**, 696-700

Masuda, H. & DeMeis, L. (1973) Biochemistry **12**, 4581-4585

Sagara, Y. & Inesi, G. (1991) J. Biol. Chem. **266**, 13503-13506

Scharff, O., Foder, B., Thastrup, O., Hofman, B., Miler, J., Ryder, L.P., Jacobsen, K.D., Langhoff, E., Dickweiss, E. & Christensen, S.B. (1988) Biochim. Biophys. Acta **972**, 257-264

Sumbilla, C., Cantilina, T., Collins, J.H., Malak, H., Lakowicz, J.R. & Inesi, G. (1991) J. Biol. Chem. **266**, 12682-12689

Thastrup, O. (1990) Agents and Actions **29**, 8-15

170

Thastrup, O., Cullen, P.J., Drobak, B.K., Hanley, M.R. & Dawson, A.P. (1990) Proc. Natl. Acad. Sci. **87**, 2466-2470

Weber, G. (1972) Biochemistry **11**, 864-878

Wyman, J. (1964) Adv. Prot. Chem. **19**, 223-286

Adenine Nucleotides in Cellular Energy Transfer and Signal Transduction
S. Papa, A. Azzi & J.M. Tager (eds)
© 1992 Birkhäuser Verlag, Basel/Switzerland

SYNTHESIS AND CHARACTERISATION OF RADIOLABELED AZIDO-
DERIVATIVES OF P^1,P^5-DI(ADENOSINE-5')PENTAPHOSPHATE FOR
MAPPING STUDIES OF ADENYLATE KINASE

Jérôme GARIN and Pierre VIGNAIS
DBMS/Biochimie (CNRS/UA 1130). Centre d'Etudes Nucléaires
85X 38041 Grenoble CEDEX (France)

SUMMARY : The precise localisation of the adenylate kinase
nucleotide binding sites is still a matter of debate. P^1,P^5-
di(adenosine-5')pentaphosphate (Ap5A) is a potent inhibitor of
adenylate kinase. In this paper we describe the original
synthesis of two radiolabeled azido-derivatives of Ap5A,
namely 2-azido-[^{32}P]Ap5A and 8-azido-[^{32}P]Ap5A. These
photoprobes should be useful tools to map the adenylate kinase
catalytic sites.

The localisation of the nucleotide binding sites in
nucleotide binding proteins has largely benefited from the use
of the technique of photoaffinity labeling (for general
reviews, see Chowdhry, 1979 ; Czarnecki et al., 1979 ; Potter
& Haley, 1983; Vignais & Lunardi, 1985). The photoaffinity
analogs have two major advantages. First, they remain
chemically unreactive until light irradiation has occurred,
which allows the study of their reversible interaction with
the protein in the dark. Second, the high reactivity of the
photogenerated group (nitrene group for 2-azido-ADP/ATP and 8-
azido-ADP/ATP) results in the covalent binding of the probe to
amino acid residues located at the ligand binding site
(nucleotide binding site for 2-azido-ADP/ATP and 8-azido-

ADP/ATP) regardless of the chemical nature of these residues. In fact, photolabeling by 2-azido-ADP/ATP and 8-azido-ADP/ATP of leucyl, isoleucyl, glutamyl, prolyl, lysyl, cysteinyl and tryptophanyl residues have been reported (Hollemans et al., 1983 ; Knight and McEntee, 1985 ; Garin et al., 1986 ; Kuwayama and Yount, 1986 ; Hegyi et al. 1986).

Until recently, results obtained by photolabeling with azido-ADP/ATP had not been confronted with models of protein structure deduced from NMR, cristallographic, or directed mutagenesis analysis. A recent careful study of mutants of *Escherichia coli* H^+-ATP synthase (Wise, 1990) showed that the aromatic ring of a tyrosyl residue (β-Tyr$_{331}$) does play a critical role in catalysis. Interestingly, the equivalent tyrosyl residue on the bovine mitochondrial H^+-ATP synthase (β-Tyr$_{345}$) was previously reported to be the amino acid residue photolabeled by 2-azido-ADP (Garin et al., 1986). Another interesting correlation between the results of photolabeling and those of protein crystals X-ray analysis was recently reported in the case of actin. A model of the atomic structure of rabbit skeletal muscle actin at 2.8 Å resolution has been determined by X-ray analysis (Kabsch et al., 1990). Localization of the adenine base of the nucleotide bound to actin as revealed by X-ray analysis was consistent with the results of photolabeling studies conducted with 2-azido-ADP (Kuwayama & Yount, 1986) and 8-azido-ATP (Hegyi et al., 1986).

Adenylate Kinase (ATP-AMP phosphotransferase, EC 2.7.4.3.) is a monomeric enzyme of small size (M_w 21700 for the rabbit muscle enzyme) that catalyses the reaction $Mg^{2+}ATP + AMP <=> Mg^{2+}ADP + ADP$. Through the use of a number of physicochemical approaches it was shown that adenylate kinase has two nucleotide binding sites, one for a magnesium-bound nucleotide and the other for a free nucleotide. The amino acid sequences of adenylate kinase from *Escherichia coli*, yeast, pig, human, rabbit, calf muscle and beef heart mitochondria have been reported. Six highly conserved regions have been identified, indicating strong similarities between adenylate kinases from different species (Schulz et al., 1986).

3D structure models of adenylate kinase have been early reported. Those studies based on the use of X-ray cristallography (Pai et al., 1977) and NMR (Fry et al., 1985, 1986 and 1987) led to different localizations of the substrates binding sites. The combination of recent results obtained from X-ray diffraction (Egner et al., 1987 ; Müller & Schulz, 1988 ; Dreusicke et al., 1988 ; Diederichs & Schulz, 1990), from NMR (Rösch et al., 1989 ; Vetter et al., 1990 ; Reinstein et al., 1990), and from site-directed mutagenesis studies (Reinstein et al., 1988 and 1989 ; Kim et al., 1989 and 1990 ; Yan et al., 1990 ; Liang et al.,1991 ; Yan & Tsai, 1991) led to a new model of adenylate kinase structure (for a critical review, see Tsai & Yan, 1991). This model still leaves some questions opened, including the localisation of the ATP binding site. It was therefore thought that the photolabeling approach could be useful to get an insight in the spatial arrangement of the peptide chain of adenylate kinase. To our knowledge, the only work reported in this field is that of Chuan et al. (1989) who synthesized a 8-azido fluorescent analog of ATP (8-azido-2'-O-dansyl-ATP). This probe was found to bind to Leu_{115}, Cys_{25}, and probably His_{36}. This result was in good agreement with the "NMR model" of the adenylate kinase structure proposed by Fry et al. (1986). However, this adenylate kinase model has been criticized (Vetter et al., 1990 and 1991 ; Tsai & Yan, 1991).

To map the adenylate kinase nucleotide binding sites, we decided to synthesize specific photoactivable ligands of the enzyme derived from P^1,P^5 di(adenosine-5')pentaphosphate (Ap5A), a potent and specific inhibitor of adenylate kinase (Lienhard & Secemski, 1973).

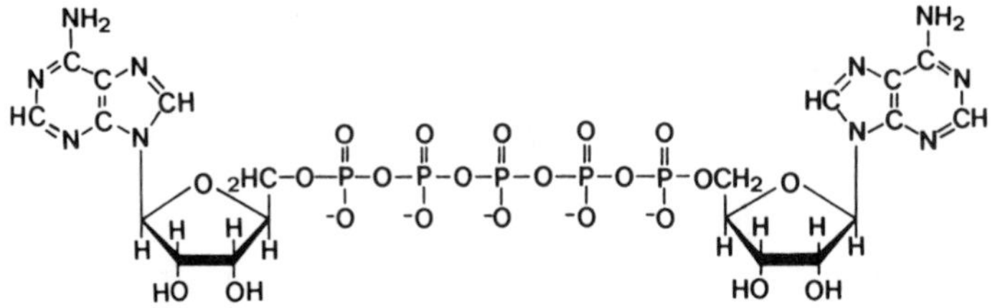

174

Figure 1 : Schematic representation of Ap5A

Ap5A consists of ATP and AMP linked by a phosphoryl group (figure 1), and it binds simultaneously to both nucleotide binding sites of adenylate kinase (Vetter et al. 1990). Using a fluorescent analog of Ap5A (mAp5Am) the dissociation constant for Ap5A to *E. coli* adenylate kinase was recently estimated to be 15 nM (Reinstein et al., 1990). For comparison, the dissociation constant for ATP (Mg^{2+}) is 85 µM. Two photoactivable analogs of Ap5A were synthesized in their radiolabeled form, namely 2-azido-[^{32}P]Ap5A and 8-azido-[^{32}P]Ap5A.

MATERIALS AND METHODS

Chemicals : AMP, ADP, and ATP were from Boehringer (Mannheim), [$\alpha^{32}P$]ATP was from New England Nuclear. 2-chloro-adenosine, and 8-azido-ADP were purchased from Sigma. Silica gel 60F254 aluminium plates were from Merck. All reagents used in the synthesis of 2-azido-adenosine were of the purest grade commercially available. The solvents were redistilled.

Preparation of Adenylate Kinase : before use, rabbit muscle adenylate kinase from Boehringer (Mannheim) was repurified. Fifteen mg of enzyme (ammonium sulfate suspension) were resuspended in 3 mL buffer A (Tris-HCl 20 mM, NaCl 100 mM, DTT 5 mM, EDTA 1 mM (final pH 7.5). The solution was loaded on a Sephacryl S-100 HR (Pharmacia) column (100 cm x 2.5 cm) equilibrated with buffer A. The fractions corresponding to the main peak of absorbancy at 280 nm were pooled and ammonium sulfate was added (85% final). Analysis of the purified enzyme by HPLC revealed the removal of five minor contaminants representing about 25% of the total protein content (not shown).

Kinetic studies : solutions for the coupled colorimetric assay of the forward reaction were as follows : 2 mM $MgCl_2$, 100 mM Tris-HCl (pH 7.5), 200 μM NADH, 400 μM PEP and 80 mM KCl (Berghäuser and Schirmer, 1978) . AMP and ATP concentrations were 150 μM and 200 μM respectively. The final volume in the test cuvette was 1 mL (temperated to 25°C) containing 10 units from each of the helper enzymes lactate deshydrogenase and pyruvate kinase. The amount of adenylate kinase added to start the reaction was between 50 and 100 ng per reaction.

RESULTS AND DISCUSSION

Synthesis and Purification of 2-azido-[^{32}P]Ap5A : 2-azido-ADP was synthesized from 2-chloro-adenosine as described by Boulay et al. (1985). 2-azido-[^{32}P]Ap5A was obtained by a modification of the procedure described by Ng and Orgel (1987) for the synthesis of Ap2A, Ap4A, and Ap6A. 1-ethyl-3-(3-dimethylaminopropyl)carbodiimide (EDC) was used as the coupling agent between 2-azido-ADP and [^{32}P]ATP. The reaction developed in a Hepes buffer in the presence of Mg^{2+} ions. The presence of divalent metal ions was reported to facilitate the nucleophilic attack of one phosphate group on the activated derivative of the other (Burton & Krebs, 1953).
2-azido-[^{32}P]Ap5A was separated from other mononucleotides (unreacted 2-azido-ADP and [^{32}P]ATP) and dinucleotides (2-azido-[^{32}P]Ap3A, di(2-azido)-Ap4A, and [^{32}P]Ap6A) by a two

step HPLC procedure using a Spherisorb C_{18} ODS_2 column and a poly F column (Du Pont de Nemours). The purity of the probe was checked by thin layer chromatography on silica gel plates (60F254) in dioxane / 2-propanol / 20% NH_4OH / H_2O (40/20/50/30). A single radioactive and UV-absorbing spot, with a R_f of 0.60 was revealed (figure 2). On the basis of radioactivity measurement, the yield of 2-azido-[a^{32}P]Ap5A recovered was estimated to be about 30 % with respect to the added [^{32}P]ATP. The UV spectrum of the photoprobe in 0.1 N HCl showed maxima at 266 nm and 235 nm (figure 3A). From the [^{32}P]ATP radioactive specific activity and the 2-azido-[^{32}P]Ap5A radioactivity and UV absorbancy, a molar extinction coefficient of 24850 $M^{-1}cm^{-1}$ was calculated.

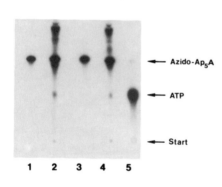

1 2 3 4 5

Figure 2 : Analysis of the synthesis products by chromatography on silica gel plates followed by autoradiography.
Lane 1 : pure 2-azido-[^{32}P]Ap5A obtained after HPLC.
Lane 2 : products of the EDC coupling between 2-azido-ADP and [^{32}P]ATP.
Lane 3: pure 8-azido-[^{32}P]Ap5A obtained after HPLC.
Lane 4 : products of the EDC coupling between 8-azido-ADP and [^{32}P]ATP.
Lane 5 : [^{32}P]ATP.

Synthesis and Purification of 8-azido-[^{32}P]Ap5A : the procedure was similar to that described for 2-azido-[^{32}P]Ap5A except that 2-azido-ADP was replaced by 8-azido-ADP. 8-azido-[^{32}P]Ap5A UV spectrum in 0.1 N HCl (figure 3B) revealed a shoulder around 290 nm and a maximum at 262 nm with a molar extinction coefficient of 24450 $M^{-1}cm^{-1}$.

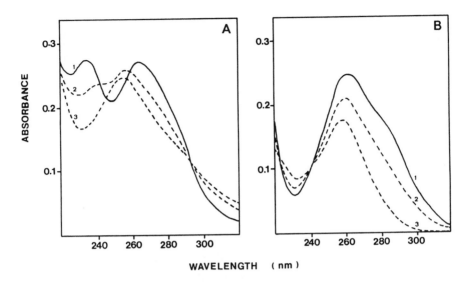

Figure 3 : Absorbance spectrum and photoreactivity of 2-azido-Ap5A (panel A) and 8-azido-Ap5A (panel B). Panel A : a solution of 2-azido-Ap5A (10 μM in 0.1 N HCl) in a quartz cuvette of 1-cm pathway was submitted to successive flashes of light delivered by a VL 6L lamp placed 5 cm from the cuvette. Trace 1 : control. Trace 2 : 15-s flash. Trace 3 : 30-s flash. Panel B : a solution of 8-azido-Ap5A (10 μM in 0.1 N HCl) was submitted to light flashes as previously described.

2-azido-Ap5A and 8-azido-Ap5A are potent inhibitors of adenylate kinase. Adenylate kinase activity is markedly inhibited in the presence of low concentrations of Ap5A (Lienhard & Secemski, 1973). The results presented in figure 4 show that azido-Ap5A probes are also potent inhibitors of adenylate kinase : in our experimental conditions, the ligand concentrations resulting in a 50% inhibition of adenylate kinase activity (I50) were 40 nM Ap5A, 120 nM 2-azido-Ap5A, and 400 nM 8-azido-Ap5A.

178

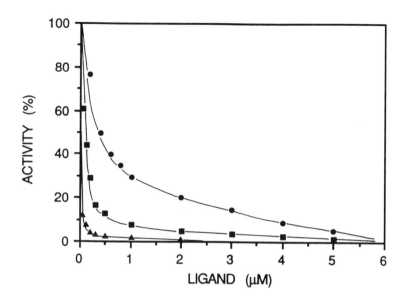

<u>Figure 4</u> : Dose-effect curves relative to the adenylate kinase inhibition caused by Ap5A (▲), 2-azido-Ap5A (■) or 8-azido-Ap5A (●). The kinetic studies were conducted as described in Materials and Methods. The final concentration of added dinucleotide varied from 0.02 μM to 5.0 μM. Percentage of remaining activity was calculated using as reference the same enzyme without addition of dinucleotide.

From these results, it is clear that the presence of the azido group on azido-Ap5A probes slightly pertubed the recognition of the ligands by adenylate kinase but did not prevent the binding of these ligands and the subsequent inhibition. The difference between I_{50} values of 2-azido-Ap5A and 8-azido-Ap5A might be explained by the different conformations of the probes. In solution, 2-azido-ADP is believed to have a preferential "anti" conformation (Czarnecki, 1984) typical of natural adenine nucleotides (Davies & Danyluk, 1974), whereas the presence of the azido group at the C8 position of the purine favors the "syn" conformation (Sarma et al., 1974). As the nucleotides were reported to bind to adenylate kinase in the "anti" conformation (Fry et al., 1987), the lower affinity of 8-

azido-Ap5A for adenylate kinase could be due to the minor fraction of the reactive "anti" conformation of the "8-azido-ADP side" of 8-azido-Ap5A. Such a direct effect of the conformations of 2-azido-ADP and 8-azido-ADP on their affinity for mitochondrial H^+-ATPase has been demonstrated (Garin et al., 1988).

In summary, 2-azido and 8-azido derivatives of Ap5A prove to be interesting probes to map the catalytic sites of adenylate kinase. Work is in progress to identify the amino acid residues photolabeled by 2-azido-Ap5A and 8-azido-Ap5A.

REFERENCES

Bergäuser J. & Schirmer R.H. (1978) *Biochim. Biophys. Acta* *537*, 428-435.
Boulay F., Dalbon P., and Vignais P.V. (1985). *Biochemistry* *24*, 7372-7379.
Chowdhry V. (1979).*Ann. Rev. Biochem. 48*, 293-325.
Chuan H., Lin J. and Wang J.H. (1989). *J. Biol. Chem. 264*, 7981-7988.
Czarnecki J., Geahlen R., and Haley B. (1979). *Methods Enzymol. 61*, 642-653.
Czarnecki J.J. (1984). *Biochim. Biophys. Acta 800*, 41-51.
Davies D.B., and Danyluk S.S. (1974). *Biochemistry 13*, 4417--4434.
Diederichs K. & Schulz G.E. (1990). *Biochemistry 29*, 8138--8144.
Dreusicke D., Karplus A.K., and Schulz G.E. (1988). *J. Mol. Biol. 199*, 359-371.
Egner U., Tomasselli A.G., and Schulz G.E. (1987). *J. Mol. Biol. 195*, 649-658.
Fry D.C., Kuby S.A., and Mildvan A.S. (1985). *Biochemistry 24*, 4680-4694.
Fry D.C., Kuby S.A., and Mildvan A.S. (1986). *Proc. Natl. Acad. Sci. USA 83*, 907-911.
Fry D.C., Kuby S.A., and Mildvan A.S. (1987). *Biochemistry 26*, 1645-1655.
Garin J., Boulay F., Issartel J.P., Lunardi J., and Vignais P.V. (1986). *Biochemistry 25*, 4431-4437.
Garin J., Vignais P.V., Gronenborn A.M., Clore G.M., Gao Z., and Bauerlein E. (1988). *FEBS Lett. 242*, 178-182.
Hegyi G., Szilagyi L., and Elzinga M. (1986). *Biochemistry,* *25*, 5793-5798.
Hollemans M., Runswick M.J., Fearnley I.M., and Walker J.E.

(1983). *J. Biol. Chem. 258*, 9307-9313.

Kabsch W., Mannherz H.G., Suck D., Pai E.F., and Holmes K.C. (1990). *Nature, 347*, 37-44.

Kim H.J., Nishikawa S., Tanaka T., Uesugi S., Takenaka H.,` Hamada M., and Kuby S.A. (1989). *Protein Engineering 2*, 379-386.

Kim H.J., Nishikawa S., Tokutomi Y., Takenaka H., Hamada M., Kuby S.A., and Uesugi S. (1990). *Biochemistry 29*, 1107--1111.

Knight K.L. & McEntee K. (1985). *J. Biol. Chem. 260*, 10185--10191.

Kuwayama H. & Yount R.G. (1986). *Biophys. J., 49*, W-34.

Liang P., Phillips G.N., and Glaser M. (1991). *Struct. Funct. Genet. 9*, 28-36.

Lienhard G.E. & Secemski I.I. (1973). *J. Biol. Chem. 248*, 1121-1123.

Müller C.W. & Schulz G.E. (1988). *J. Mol. Biol. 202*, 913--915.

Ng K.E. & Orgel L.E. (1987). *Nucl. Acids Res. 15*, 3573-3580.

Pai E.F., Sachsenheimer W., Schirmer R.H., and Schulz G.E. (1977). *J. Mol. Biol. 114*, 37-45.

Potter R.L. & Haley B.E. (1983). *Methods Enzymol. 91*, 613--633.

Reinstein J., Brune M., and Wittinghofer A. (1988). *Biochemistry 27*, 4712-4720.

Reinstein J., Gilles A., Rose T., Wittinghofer A., Saint Girons I., Bârzu O., Surewicz W.K., and Mantsch H.H. (1989). *J. Biol. Chem. 264*, 8107-8112.

Reinstein J., Schlichting I., and Wittinghofer A. (1990). *Biochemistry 29*, 7451-7459.

Rösch P., Klaus W., Auer M., and Goody R.S. (1989). *Biochemistry 28*, 4318-4325.

Sarma R.H., Lee C.H., Evans F.E., Yathindra H., and Sundaralingam M. (1974). *J. Am. Chem. Soc. 96*, 7337-7348.

Schulz G.E., Schiltz E., Tomasselli A.G., Frank R., Brune M., Wittinghofer A., and Schirmer R.H. (1986). *Eur. J. Biochem. 161*, 127-132.

Tsai M.D.& Yan H. (1991). *Biochemistry 30*, 6806-6818.

Vetter I.R., Reinstein J., and Rösch P. (1990). *Biochemistry 29*, 7459-7467.

Vetter I.R., Konrad M., and Rösch P. (1991). *Biochemistry 30*, 4137-4142.

Vignais P.V. & Lunardi J. (1985). *Ann. Rev. Biochem. 54*, 977-1014.

Wise J.G. (1990). *J. Biol. Chem. 265*, 10403-10409.

Yan H., Shi Z., and Tsai M.D. (1990). *Biochemistry 29*, 6385--6392.

Yan H., and Tsai M.D. (1991). *Biochemistry 30*, 5539-5546.

Adenine Nucleotides in Cellular Energy Transfer and Signal Transduction
S. Papa, A. Azzi & J.M. Tager (eds)
© 1992 Birkhäuser Verlag, Basel/Switzerland

PROTEIN PHOSPHORYLATION AND THE REGULATION OF SUGAR TRANSPORT IN GRAM-NEGATIVE AND GRAM-POSITIVE BACTERIA

Milton H. Saier, Jr., Jonathan Reizer and Josef Deutscher

Department of Biology, University of California at San Diego, La Jolla, CA 92093-0116

SUMMARY:
Sugar uptake and cytoplasmic inducer generation appear to be regulated by the phosphenolpyruvate:sugar phosphotransferase system (PTS) by distinct mechanisms in Gram-negative versus Gram-positive bacteria. In Gram-negative bacteria, the free form of the glucose-specific IIA protein of the PTS, which can be phosphorylated on a histidyl residue by PEP and the PTS energy coupling proteins, inhibits non-PTS permease activities. In Gram-positive bacteria, the phosphorylated form of the energy coupling HPr protein of the PTS, which can be phosphorylated on a seryl residue by ATP and a protein kinase, appears to inhibit non-PTS permease activities. In this summary article, the current status of these two PTS-mediated regulatory mechanisms will be evaluated.

Carbohydrates are transported into both Gram-negative and Gram-positive bacteria by a multiplicity of mechanisms, one of which depends on the catalytic activities of the protein constituents of the phosphoenolpyruvate (PEP):sugar phosphotransferase system (PTS). In these organisms the PTS is known to control the activities of certain non-PTS carbohydrate permeases (i.e., those for lactose and maltose in Escherichia coli; those for gluconate and glucitol in Bacillus subtilis). Recent evidence suggests that the regulatory mechanisms by which the PTS controls non-PTS sugar permeases in these two major groups of prokaryotes are different and involve two distinct heat stable phosphocarrier proteins of the PTS as well as two different phosphorylation mechanisms. In E. coli a sugar-specific PTS protein, IIAglc, is phosphorylated by a PEP-dependent mechanism on a histidyl residue, and only the free (dephospho) form of IIAglc binds to the allosteric site on the cytoplasmic side of the permease protein to inhibit its activity. In contrast, in B. subtilis, a non sugar-specific PTS protein, HPr, is phosphorylated by an ATP-dependent mechanism on a seryl residue, and only the phosphorylated product, HPr(ser-P), appears to bind to the target permeases to allosterically inhibit their activities. Recently, the 3-dimensional structures of these two allosteric regulatory proteins of the PTS have been determined both by X-ray crystallography and by 2-dimensional NMR. In this brief review,

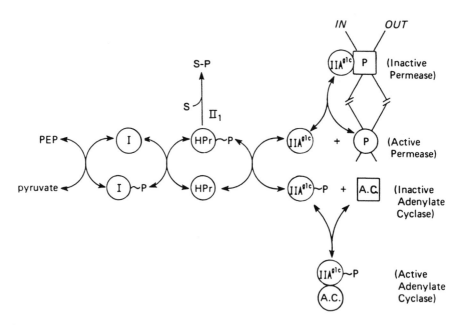

Fig. 1: Regulatory scheme involving the reversible phosphorylation of the IIAglc protein of the PTS in enteric bacteria such as E. coli and S. typhimurium. The abbreviations are as follows: I, Enzyme I; H, HPr; IIAglc, the Enzyme IIAglc protein; P, permease; AC, adenylate cyclase. (Modified from Saier, 1977, with permission.)

the current status of research on these two regulatory mechanisms will be presented.

The involvement of the PTS as a protein phosphorylating system in the regulation of gene transcription and metabolism in bacteria has been discussed in earlier reviews (Saier, 1989a,b; Saier et al., 1990). The regulatory mechanism by which the IIA^{glc} protein of enteric bacteria controls the activities of non-PTS permeases such as the lactose and maltose permeases, as well as the intracellular enzymes glycerol kinase and adenylate cyclase, was reviewed in detail in 1989 (Saier, 1989a). At that time the essential features of the regulatory mechanism had already been established. The IIA^{glc} protein (Saier and Reizer, 1990, 1991) which is reversibly phosphorylated by PEP, Enzyme I and HPr of the PTS, is believed to allosterically control all of these proteins. Adenylate cyclase is apparently allosterically activated by phosphohistidyl IIA^{glc} while the permeases and glycerol kinase are allosterically inhibited by free IIA^{glc} (Fig. 1). Two recent advances in the definition of the detailed regulatory mechanism have been (1) to identify residues in the target permeases which when mutated abolish binding of IIA^{glc} to the allosteric site, and (2) to determine the 3-dimensional structure of IIA^{glc}. The results of the former studies (Wilson et al., 1990; Dean et al., 1990; Kuhnau et al., 1991) identified regions in the LacY and MalK proteins of the lactose and maltose permeases, respectively, which are believed to be involved in IIA^{glc} binding. These regions in the two permease proteins show a surprising degree of sequence similarity considering that LacY and MalK are non-homologous proteins, one, LacY, being the sole integral membrane constituent of the lactose permease, the other, MalK, being a peripheral membrane protein which is only one of several constituents of the maltose permease. The probability that the degree of sequence similarity in the regions of these two permeases shown in Figure 2 (Dean et al., 1990) arose by chance is less than one in 10,000. This fact leads to the suggestion that the sequence similarity observed arose by

184

convergent evolution as a result of a need for a common function
(i.e., binding of IIA^{glc}).

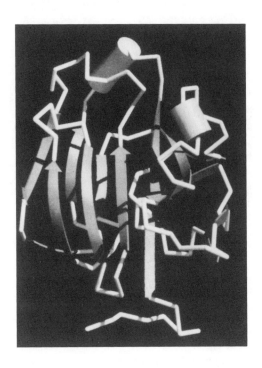

Fig. 2: The three-dimensional structure of IIA^{glc} of B. subtilis.
This protein is homologous with IIA^{glc} of enteric bacteria and can
replace it in the regulatory system depicted in Fig. 1.   The
structure shown is an anti-parallel β-barrel with Greek key and
jelly roll topological features. (Reproduced from Liao et al.,
1991 with permission.)

The three dimensional structure of IIA^{glc} is shown in Figure 3
(Liao et al., 1991; Fairbrother et al., 1991).   This heat stable
protein which can be autoclaved (121°) without complete loss of
activity (Sutrina et al., 1990; Reizer et al., 1991) possesses
the essence of an anti-parallel β-barrel with Greek key and jelly

roll topological features (Liao et al., 1991; Branden and Tooze, 1991). Knowledge of this 3-dimensional structure should allow precise definition of the regions of the protein which interact with target systems. Sequence analyses of mutant IIA^{glc} proteins which have lost the capacity to bind and regulate the target systems should be highly informative.

MalK   275 Val Gln · · · Val Gly Ala Asn Met Ser Leu · · · Gly Ile Arg Pro
              .   .   :   :   :   :   :   .   .   .
LacY   198 Ala Asn Ala Val Gly Ala Asn His Ser Ala Phe Ser Leu Lys Leu

Fig. 3: Sequence comparison between regions of the MalK and LacY proteins which include residues (indicated in bold print) which when mutated abolish the binding of IIA^{glc} to the maltose and lactose permeases of E. coli, respectively. The two sequences exhibit striking similarity as shown. The probability that this degree of sequence identity arose by chance is less than 10^{-4}. Since these two proteins are not homologous, convergent evolution for a common function (binding of IIA^{glc}) probably was responsible for the sequence similarity observed. (Modified from Dean et al., 1990 with permission.)

A schematic view of the Gram-positive regulatory system is presented in Figure 4. HPr, another heat stable protein, can be phosphorylated either on a histidyl residue (H$_{15}$) by PEP and Enzyme I or on a seryl residue (S$_{46}$) by ATP and an HPr(ser) kinase. The latter enzyme is allosterically activated by cytoplasmic intermediates of carbohydrate metabolism such as fructose-1,6-diphosphate and gluconate-6-phosphate (Deutscher and Engelmann, 1984; Reizer et al., 1984). Phosphorylation of the seryl residue inhibits phosphorylation of the histidyl residue about 100-fold. Thus, phosphorylation of serine 46 potentially reduces phosphoryl transfer via the PTS to less than 1% of the normal rate.

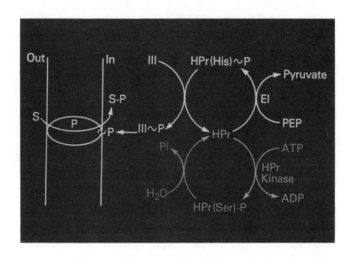

<u>Fig. 4:</u> Scheme illustrating the two modes of HPr phosphorylation in Gram-positive bacteria such as <u>B</u>. <u>subtilis</u>. Top (in white): phosphorylation of the active site histidyl residue, H_{15}, by PEP and Enzyme I. Bottom (in grey): phosphorylation of the regulatory site seryl residue, S_{46}, by ATP and the HPr(ser) kinase; The two processes are antagonistic, but not mutually exclusive. (Modified   from Reizer, 1989 with permission.)

   In <u>vivo</u> studies (Reizer et al., 1989; Sutrina et al., 1990; Reizer et al., 1991) did not provide evidence for the suggestion (Deutscher et al., 1984) that this mechanism serves as a device for regulating the activity of the PTS. This fact, as well as the recent demonstration of HPr, HPr(ser) kinase and HPr(ser-P) phosphatase in heterofermentative lactobacilli (Reizer et al., 1988; Reizer, 1989) and in <u>Acholeplasma</u> <u>laidlawii</u> (Reizer et al., in preparation) that lack a functional PTS prompted us to consider an alternative physiological function for the HPr(ser) kinase - HPr(ser-P) phosphatase system (Deutscher et al., in

Table I. Catabolite repression of gluconate kinase synthesis by PTS sugars in B. subtilis strains.

Sugar(s) present during growth	Gluconate Kinase Activity		
	GM122	SA003	GM808
None	<1	<1	<1
Gluconate	24±4	24±6	25±4
Gluconate + glucose	2±1	21±6	2±1
Gluconate + mannitol	4±2	20±3	4±2
Gluconate + glucitol	23±3	21±4	24±4

The three strains used were: column 1: GM122 (wild-type); column 2: SA003 (an isogenic mutant in which the regulatory seryl residue in the chromosomally-encoded HPr is replaced by an alanyl residue); column 3: GM808 (a primary site revertant). Enzyme activities are expressed in nmoles of product formed per min. per mg. protein at 37°C. Experimental conditions will be described in a forthcoming publication (Deutscher et al., in preparation).

preparation). As shown in Table I, synthesis of gluconate kinase in a wild-type Bacillus subtilis strain is inducible by growth in the presence of gluconate and strongly repressed by inclusion of glucose or mannitol, both PTS sugars in B. subtilis, in the growth medium. The non-PTS sugar, glucitol, was not repressive, showing that the repressive phenomenon exhibits specificity for sugar substrates of the PTS. In the second column in Table I, data are presented for an isogenic B. subtilis strain, which has a single mutation in the chromosomal structural gene for HPr, a mutation (S46A) which changes the regulatory seryl residue to alanine so that ATP-dependent phosphorylation of this protein cannot occur. The consequence of this mutation is that the repressive effect of the PTS sugars on gluconate kinase synthesis is essentially abolished. When the S46A mutation was reversed (primary site reversion), the wild-type repressive effect was restored, establishing that the mutation was responsible for the observed in vivo response. Very similar behavior was observed for another catabolic enzyme, glucitol dehydrogenase, involved in the catabolism of the non-PTS sugar, glucitol (data not shown).

The mechanism of the repressive effect reported in Table I was further investigated (Reizer et al., manuscript in preparation). The same strains utilized for the experiment in Table I were examined with respect to inhibition of the uptake of [^{14}C]gluconate in cells induced by growth in the presence of both gluconate and the PTS sugar, glucose or mannitol. The PTS sugar inhibited gluconate uptake into intact cells if synthesis of the uptake system for that PTS sugar had been induced prior to initiation of the uptake experiment. This result provided evidence that the apparent catabolite repression was in fact due to exclusion of the inducer, gluconate, from the cytoplasm of the cell. An inducer exclusion mechanism is thus implied.

In contrast to the situation discussed above for the PTS-mediated regulatory system in Gram-negative enteric bacteria, HPr(ser-P) (rather than IIA^{glc}) may be the species which binds to the allosteric regulatory site on the cytoplasmic side of the gluconate permease in B. subtilis to inhibit its activity. However, since phosphorylation of the active histidyl residue in HPr inhibits HPr(ser) phosphorylation as noted above, the binding of this seryl phosphorylated PTS protein to the non-PTS permease may be functionally equivalent to binding of the free form of the PTS protein, IIA^{glc}, in E. coli and other enteric bacteria. These preliminary results therefore suggest that Gram-positive and Gram-negative bacteria have evolved two functionally equivalent but mechanistically distinct processes in order to allow the PTS to regulate cytoplasmic levels of non-PTS sugar inducers. These mechanisms provide the bacteria with the capacity to sense the availability of various carbohydrates in their environment and respond by the selection of the preferred carbon and energy source.

The three dimensional structure of the B. subtilis HPr protein has recently been determined (Wittekind et al., 1989; Herzberg et al., in preparation). This small, heat stable protein consists of an open face β-sandwich formed by four antiparallel β-strands serving as the underlying bread packed against three α-helices serving as the overlying spread (Fig. 5). As revealed by the X-ray structure, the β-sheet curls back on itself so that the regulatory seryl residue (S$_{46}$) is close to the active site histidyl residue (H$_{15}$). Thus, the presence of the negatively

charged phosphoryl group at position 46 may inhibit introduction
of the phosphoryl group at the active site by electrostatic
repulsion or by inhibition of Enzyme I binding.   The same
explanation applies to inhibition of seryl residue$_{46}$
phosphorylation by prior phosphorylation of histidyl residue$_{15}$.
These 3-dimensional structural studies therefore provide the
first information leading to an understanding of the regulatory
interactions in molecular detail.   Further studies will be
required to establish the details of the interactions by which
allosteric regulation of the target permeases is effected.

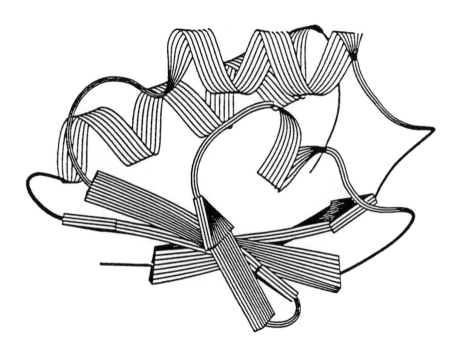

Fig. 5:   Three dimensional structure of HPr from B. subtilis.
The structure is that of an open face β-sandwich with four β
strands below three α-helices as discussed in the text.
(Reproduced from Herzberg et al., 1991 with permission.)

190

Acknowledgments: We thank Mary Beth Hiller for expert assistance in the preparation of this manuscript, the Alexander von Humboldt Foundation of Germany for financial support to MHS during the preparation of this manuscript, and Dr. Matthias Müller for his generous hospitality. The work reported in this review was supported by Public Health Service grants 5RO1 AI21702 and 2RO1 AI14176 from the National Institute of Allergy and Infectious Diseases.

REFERENCES

Branden, C. and Tooze, J. (1991) Introduction to Protein Structure. Garland Publishing, Inc., New York & London

Dean, D.A., Reizer, J., Nikaido, H., and Saier, M.H., Jr. (1990) J. Biol. Chem. 265, 21005-21010Deutscher, J. and Engelmann, R. (1984) FEMS Microbiol. Lett. 23, 157-162

Deutscher, J., Kessler, U., Alpert, C.A., and Hengstenberg, W. (1984) Biochemistry 23, 4455-4460

Fairbrother, W.J., Cavanagh, J., Dyson, H.J., Palmer, A.G., III, Sutrina, S.L., Reizer, J., Saier, M.H., Jr., and Wright, P.E. (1991) Biochemistry. 30, 6896-6907

Kuhnau, S., Reyes, M., Sievertsen, A., Shuman, H.A. (1991) J. Bacteriol. 173, 2180-2186

Liao, D.-I., Kapadia, G., Reddy, P., Saier, M.H., Jr., Reizer, J., and Herzberg, O. (in press) Biochemistry

Reizer, J. (1989) FEMS Microbiol. Rev. 63, 149-156

Reizer, J., Novotny, M.H., Hengstenberg, W., and Saier, M.H., Jr. (1984) J. Bacteriol. 160, 333-340

Reizer, J., Peterkofsky, A., and Romano, A.H. (1988) Proc. Natl. Acad. Sci. USA 85, 2041-2045

Reizer, J., Sutrina, S.L., Saier, M.H., Jr., Stewart, G.C., Peterkofsky, A., and Reddy, P. (1989) EMBO J. 8, 2111-2120

Reizer, J., Sutrina, S., Wu, L.-F., Reddy, P., Deutscher, J., and Saier, M.H., Jr. (in press) J. Biol. Chem.

Saier, M.H., Jr. (1977) Bacteriol. Revs. 41, 856-871.

Saier, M.H., Jr. (1989a) Microbiol. Revs. 53, 109-120

Saier, M.H., Jr. (1989b) Res. Microbiol. 140, 349-352

Saier, M.H., Jr. and Reizer, J. (1990) Res. Microbiol. 141, 1033-1038

Saier, J.H., Jr., Reizer, J., J. Bacteriol., in press.

Saier, M.H., Jr., Wu, L.-F., and Reizer, J. (1990) Trends Biochem. Sci. 178, 391-395

Sutrina, S.L., Reddy, P., Saier, M.H., Jr., and Reizer, J. (1990) J. Biol. Chem. 265, 18581-18589

Wilson, T.H., Yunker, P.L., and Hansen, C.L. (1990) Biochim. Biophys. Acta 1029, 113-116

Wittekind, M., Reizer, J., Deutscher, J., Saier, M.H., Jr. and Klevit, R.E., (1989) Biochemistry 28, 9908-9912.

Adenine Nucleotides in Cellular Energy Transfer and Signal Transduction
S. Papa, A. Azzi & J.M. Tager (eds)
© 1992 Birkhäuser Verlag, Basel/Switzerland

SIGNAL TRANSDUCTION OF THE PHOSPHATE REGULON IN ESCHERICHIA COLI MEDIATED BY PHOSPHORYLATION

Kozo Makino, Mitsuko Amemura, Soo-Ki Kim, Hideo Shinagawa, and Atsuo Nakata

Department of Experimental Chemotherapy, Research Institute for Microbial Diseases, Osaka University, 3-1 Yamadaoka, Suita, Osaka 565, Japan

SUMMARY: In Escherichia coli, expression of genes involved in the phosphate(*pho*) regulon is regulated by phosphate concentration of the medium. Proteins PhoR and PhoB, which are regulatory factors for the transcriptional regulation of the *pho* genes, are known to belong to a family of two-component regulatory factors that respond to a variety of environmental stimuli in bacteria. PhoB is the actual activator protein, which binds to the *pho* gene promoters. PhoR is most likely a transmembrane protein and signal transducer, which exhibits both PhoB-specific phosphorylation and dephosphorylation activities. The phosphorylation of PhoB protein occurs concurrently with the acquisition of the ability to activate transcription from the *pho* promoters. Cross talk by protein phosphorylation appears to occur from PhoM, a PhoR like protein, to PhoB. Mutational analyses of PhoB protein indicated that the two domains, which are consisted of phosphate-acception and transcriptional activation domains, are separable and the non-phosphorylated domain blocks the transcriptional activation domain.

INTRODUCTION

In Escherichia coli, a number of genes are involved in the transport and assimilation of phosphate and are inducible by phosphate starvation of the medium. These include *phoA*, *phoE*, *pstS*, and *ugpB*, which code for alkaline phosphatase, outer membrane porin e, phosphate-binding protein, and sn -glycerol-3-phosphate-binding protein, respectively. They constitute a single phosphate(*pho*) regulon, and are under the same physiological and genetic control(Torriani and Ludtke, 1985; Shinagawa et al., 1987). Genetic evidence suggests that the product of *phoB* is the direct transcriptional activator for the *pho* regulon (Makino et al.,1982, 1986;

Shinagawa et al., 1983; Guan et al., 1983). Regulation of the *pho* regulon by PhoB requires the function of the *phoR* gene. PhoR protein regulates the *pho* regulon negatively with excess phosphate, and positively with limited phosphate. The phosphate levels in the medium are monitored by the phosphate binding protein(the *pstS* gene product) and the signal is transduced by the function of the *pst-phoU* operon to PhoR, which in turn modulates the PhoB function(Nakata et al., 1987; Shinagawa et al., 1987). The PhoR/PhoB pair is structurally as well as functionally similar to many two-component regulators of bacteria that respond to environmental changes(Makino et al., 1986a, b; Ronson et al., 1987). The PhoR-like proteins are sensors or transducers which modulate the activator function of PhoB-like proteins in response to environmental signals. Based on results of extensive biochemical studies on the two-component regulatory systems, particularly, Ntr-regulon and bacterial chemotaxis, it has been originally suggested that a common cascade mechanism, i. e., the transfer of a phosphoryl group(phosphotransfer) between the two components, is involved in the regulatory systems(Stock et al.,1989 and references therein). We have also indicated that the PhoR/PhoB system is a paradigm of such adaptive response systems in prokaryotes(Makino et al., 1989).

In PhoR-defective strains, the *pho* regulon is expressed constitutively, and the expression is dependent on the functions of *phoM* and *phoB* (Wanner and Latterell, 1980). Therefore, PhoM replaces the positive regulatory function of PhoR in the mutants by cross talking to PhoB.

In this article, we describe our recent studies on the signal transduction and regulatory mechanism of the *pho* regulon.

RESULTS AND DISCUSSION

Structure of the phoB-phoR Operon

The nucleotide sequence of the 3-kilobase chromosomal DNA containing the *phoB* and *phoR* genes has been determined(Makino et al., 1986a, b). In the regulatory region of the operon, we found a consensus nucleotide sequence shared by the regulatory regions of the *phoA*, *pstS*, and *phoE* genes, which we proposed to name the phosphate(*pho*) box(Fig. 1). Since all of these genes are positively regulated by the *phoB* gene product, this suggested that transcription of the *phoB-*

```
phoA:  CAGTAAAAAGTTAATCTTTTCAACAGCTGTCATAAAGTTGTCACGGCCGAGACTTATAGTCGCTTTGTTTTTA-
                                                           (-10)        -mRNA
       TTTTTTAATGTATTTGTACATGGAGAAAATAAAGTG
                         (SD)           Met


phoS:  CTCTCTGTCATAAAACTGTCATATTCCTTACATATAACTGTCACCTGTTTGTCCTATTTTGCTTCTCGTAGCC-
                                                           (-10)        -mRNA
       AACAAACAATGCTTTATGAATCCTCCCAGGAGACATTATG
                          (SD)         Met


phoE:  TACCACATTTTAAGAATATTATTAATCTGTAATATATCTTTAACAATCTCAGGTTAAAAACTTTCCTGTTTTC-
                                                           (-10)        -mRNA
       AACGGGACTCTCCCGCTGAATATTCGCGCGTTAATTAAAATCAGGAATGAAAATG
                                                (SD)        Met


phoB:  ATAACCTGAAGATATGTGCGACGAGCTTTTCATAAATCTGTCATAAATCTGACGCATAATGACGTCGCATTAA-
                                                           (-10)        -mRNA
       TGATCGCAACCTATTTATTACAACAGGGCAAATCATG
                          (SD)        Met
```

Figure 1. Comparison of the nucleotide sequence of the *phoA*, *phoS*, *phoE*, and *phoB* promoter regions. The possible -10 sequences, Shine-Dalgarno sequences, transcriptional initiation sites, and translational initiation codons are shown by boldface type. Eighteen nucleotides with a highly conserved sequence(*pho* box) of each gene are underlined.

phoR operon is also regulated positively by its own products, in agreement with the finding of Guan et al.(1983). The open reading frame corresponding to the *phoB* gene can code for a protein with 229 amino acid residues with an Mr of 26,161. Immediately distal to the coding region of *phoB*, a sequence whose transcript can form a stem-and-loop structure was found. This sequence may serve as a transcriptional attenuator since expression of *phoB* is more efficiently than that of *phoR*. No promoterlike structure has been found in the intercistronic region between *phoB* and *phoR*, which is in agreement with the previous finding that the transcription of the *phoR* gene is initiated at the promoter of the *phoB* gene(Tommassen et al., 1982; Makino et al., 1985). The open reading frame corresponding to the *phoR* gene was identified about 60 base pairs downstream of the termination codon of the *phoB* gene, and it can code for a protein consisting of 431 amino acid residues with an M of 49,666. This value agrees well with the M of the PhoR protein determined by the minicell(Tommassen et al., 1982) and maxicell methods(unpublished data). Since the PhoR protein deduced from the nucleotide sequence of the gene revealed a long stretch of hydrophobic residues(ca. 60 residues) in the amino-terminal region, it may be a membrane-bound protein.

A possible stem-and-loop structure of the transcript, followed by several Us characteristic of Rho-independent terminators, was found distal to the *phoR* coding region. Therefore, the *phoB* and *phoR* genes are likely to constitute an operon.

Purification of PhoB and PhoR Proteins

To study the functions of the PhoB and PhoR proteins, we purified the proteins and examined their biochemical properties in vitro. A PhoB'-'LacZ fusion protein was purified from the cells carrying the fusion gene by monitoring ß-galactosidase activity of the fusion protein, and an antiserum was prepared(Sugita et al., 1985). The PhoB protein was purified from cells carrying a PhoB-overproducing plasmid by immunologically monitoring the protein fractions during the purification steps(Makino et al., 1988).

We had difficulty in purifying PhoR protein in a functional form for biochemical studies, since rather small amounts of the protein were synthesized even in induced cells and it was associated with the membrane fraction, which was not solubilized by non-ionic detergents. PhoR protein may be bound to the membrane by the amino-terminal hydrophobic region. To solve these problems, we constructed a plasmid with a mutant *phoR* gene(*phoR1084*) placed under a strong inducible promoter P*tac*, which encoded PhoR protein lacking the amino-terminal 83 residues(Makino et al., 1989; Yamada et al., 1990). Cells carrying the plasmid produced alkaline phosphatase constitutively, indicating that PhoR1084 is locked in a form to activate PhoB. PhoR1084, which was found in the soluble fraction, was purified with over 95% purity.

Autophosphorylation of PhoR1084

The purified PhoR1084 autophosphorylated in the presence of ATP using its γ-position of phosphate(Makino et al., 1989). The phosphorylation reached a maximum with incorporation of 0.6 mol phosphate per mol of PhoR1084 monomer. To infer the phosphorylated residue in PhoR1084, the chemical stability of phospho-PhoR1084 was examined under various conditions. The phsphoryl-protein linkage in PhoR1084 was stable at neutral to alkaline pH and unstable in acid and pyridine or hydroxylamine. These results are similar to those for phospho-CheA(Wylie et al., 1988) and phospho-NR$_1$(Ninfa et al., 1988), and suggest that the phosphate in PhoR1084 is likely to be N-linked, presumably phosphohistidine. Radiolabeled

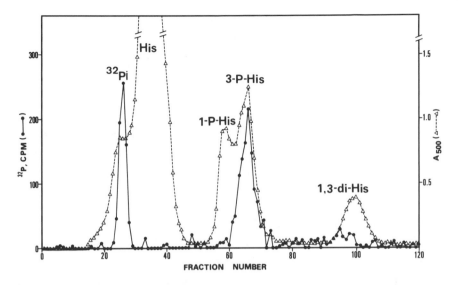

Figure 2. Cochromatography of phosphohistidine in phospho-PhoR1084 with 3-phosphohistidine on Dowex 1 column chromatography. The [32P]phospho-PhoR1084 was hydrolyzed in 3 M KOH at 120°C for 6 h. A mixture 1-, 3- , and 1,3-phosphohistidine was prepared as described by Amemura et al.(1990). The mixture of phosphohistidines and hydrolysate of phospho-PhoR1084 was adsorbed on a column of Dowex 1 equilibrated with HCO_3-. The column was eluted with a linear gradient of 0 to 1.0 M $KHCO_3$(900 ml), and the radioactivity of each fraction was counted in a liquid scintillation counter. Histidine and phosphohistidines were asaayed as described by Amemura at al.(1990).

phospho-PhoR1084 was hydrolyzed at pH 13, and the hydrolysates were analyzed by chromatography on Dowex 1 with a mixture of the histidines phosphate linked at the N-1 and/or N-3 position. The peak of the radioactivity coincided with that of 3-phosphohistidine(Fig. 2). Thus, the phosphoryl group of phospho-PhoR1084 is linked at the N-3 position of the imidazole ring in a histidine moiety in PhoR1084.

Phosphorylation of PhoB by Phospho-PhoR1084

The genetic evidence suggests that the wild-type PhoR protein modulates the PhoB function in response to phosphate levels in the medium(Wanner, 1987; Shinagawa et al., 1987). We examined the ability of PhoR1084 to catalyze phosphorylation of PhoB(Makino et al., 1989). PhoB was phosphorylated only in the presence of PhoR1084 and ATP. To examine whether phosphate was transferred from phospho-PhoR1084, we purified phospho-PhoR1084 and incubated it with PhoB. PhoB protein was phosphorylated by the phospho-PhoR1084. The reaction proceeded very rapidly with over 95% phosphotransfer in less than 30 seconds, but phospho-PhoR1084 in the absence of PhoB was rather stable and most of the isotope remained attached to PhoR1084 even after incubation for 60 minutes. We also examined the effects of exposing phospho-PhoB to 0.1 M-HCl, 0.25 M-NaOH, or 0.8 M-NH$_2$OH for ten minutes at 37°C. Approximately 60% of the phospho-PhoB survived the treatment with acid, but only 25% survived the treatment with NH$_2$OH, and none survived the treatment with NaOH. These result suggest that the phosphate is linked to an acyl group of glutamate or aspartate. Furthermore, biochemical experiment using a borohydride reduction method indicated that the phosphate of PhoB is linked to an aspartate(unpublished data).

Properties of Phosphorylated PhoB

To elucidate the biochemical consequence of PhoB phosphorylation, We compared the in vitro DNA-binding ability of the phosphorylated form of PhoB with that of the non-phosphorylated form, by means of DNase I footprinting analysis (Makino et al., 1989). The phosphorylated form of PhoB was prepared in vitro by incubating it with PhoR1084 in the presence of ATP. Equivalent amounts of DNA fragments with the *pstS* promoter region containing two *pho* boxes were mixed and incubated with either the phosphorylated or non-phosphorylated forms of PhoB. Upon the incubation of PhoB with PhoR1084 in the presence of ATP, the DNA-binding ability of the former to the *pho* boxes was enhanced about 5 times than that of non-phosphorylated PhoB. Since PhoR1084 and/or ATP did not protect the *pho* boxes from DNase I digestion, the enhancement of protection by PhoB required both PhoR1084 and ATP. These results indicate that phosphorylation of PhoB mediated by PhoR enhances the affinity of PhoB for the *pho* boxes of the *pho* genes. In vitro transcription experiments, using RNA polymerase containing a major sigma factor, the sigma 70, indicated that PhoB incubated with PhoR1084 and ATP greatly stimulated transcription from the *pstS* promoter by RNA polymerase, while neither PhoB nor PhoR1084 alone stimulated transcription. This result shows that phospho-PhoB formed by PhoR1084 and ATP is the active form that promotes transcription from the *pho* promoters.

PhoR has Dephosphorylation Activity

Genetic studies suggest that PhoR protein has another form with excess phosphate of the medium(Shinagawa et al., 1987). To elucidate this possibility, we isolated a mutant *phoR* gene(*phoR4222*), which is mutated at position 213(His to Tyr). The wild type cells containing the *phoR4222*-plasmid did not express the *pho* genes with limited phosphate, indicating that the mutant locks in a form that makes PhoB inactivate. To determine the biochemical properties of the PhoR4222 protein, we purified the protein and performed in vitro experiments. The PhoR4222 did not autophosphorylate in the presence of ATP even after long incubation, but dephosphrylated phospho-PhoB. These data indicate that another form of PhoR in vivo is the form to dephosphorytate phospho-PhoB.

The *Pho* Box as a Consensus Sequence for the Promoter of the *Pho* Regulon

The genes in the *pho* regulon, including *phoA*, *pstS*, *phoE*, and *phoB*, have been shown to be regulated physiologically by phosphate in the medium and genetically by *phoB* and several regulatory genes such as *phoR*, *phoM*, and the genes in the *pst-phoU* operon(Nakata et al., 1989; Shinagawa et al., 1987). Therefore, they are likely to share a common regulatory element in the promoter region. In agreement with this idea, a consensus sequence, which we named the *pho* box, was found in the promoter regions of all of them as shown in Fig. 1(Makino et al., 1986a). A well-conserved 18-nucleotide sequence, CTG/TTCATAA/TAA/TC TGTCAC/T, was found 10 nucleotides upstream of the putative -10 regions in all cases. The *pho* box is duplicated in tandem in the regulatory region of *pstS* as shown in Fig. 1. Mutational analysis of the *pstS* promoter region was performed with in vivo and in vitro experiments(Kimura et al., 1989). Deletions extending into the upstream *pho* box but retaining the downstream *pho* box greatly reduced promoter activity, but the remaining activity was still regulated by phosphate levels in the medium and by the PhoB protein, indicating that each *pho* box is functional. No activity was observed in deletion mutants which lacked the remaining *pho* box or the -10 region. The phosphorylated PhoB protein binding region in the *pstS* regulatory region was studied with the deletion plasmids by a gel-mobility retardation assay(Kimura et al., 1989) and DNase I footprinting(Makino et al., 1988, 1989). These results indicate that the protein binds to each *pho* box on the *pstS* promoter in vivo and in vitro.

Cross Talk to the Phosphate Regulon by PhoM Protein

In the absence of the *phoR* function, the genes in the phosphate regulon are expressed constitutively and the expression is dependent on the functions of *phoM* and *phoB*. A DNA fragment containing the *phoM* operon was cloned(Makino et al., 1984) and the nucleotide sequence was determined(Amemura et al ., 1986). The nucleotide sequence of the *phoM* region showed an open reading frame(*phoM-orf2*), immediately upstream of *phoM*, that is highly homologous to PhoB. Therefore, the PhoM and PhoM-open reading frame 2(ORF2) pair appears to constitute a two-component regulatory system that may respond to unknown signals and regulate a set of genes. We constructed a plasmid with *lacZ '-'phoM* fusion gene, which encoded a hybrid protein(PhoM1206) in which the hydrophobic amino-terminal half of the native PhoM was replaced by ß-galactosidase(Amemura et al., 1990). The *phoM1206* gene could complement the *phoM* mutation in vivo. We purified PhoM1206 from the overproducing strain carrying the plasmid. In the presence of ATP, the protein was autophosphorylated at the N-3 position of imidazole ring of a histidine residue, and the phospho-PhoM1206 phosphorylated PhoB. PhoM1206 could also transphosphorylate the product of *phoM-orf2*, but PhoR1084 could not phosphorylate the ORF2. Therefore, cross talk by protein phosphorylation appears to occur from PhoM to PhoB not from PhoR to PhoM-open reading frame 2.

Mutational Analyses of the PhoB protein

The transcriptional activator PhoB is composed of 229 amino acids, and postulated to contain at least three functional domains(Makino et al., 1989); (I) phosphate-accepting domain from phospho-PhoR, (II) DNA(the *pho* box)-binding domain, and (III) a domain interacting with RNA polymerase holoenzyme. To elucidate each domain, we constructed a series of *phoB* mutants, including deletions in N-terminal or C-terminal region and point mutations, and analyzed transcription-enhancing, phosphate-accepting, and DNA-binding activities. The truncated proteins containing 127 N-terminal amino acids were fully competent as the substrate for phosphorylation by phospho-PhoR1084(domain I). Analysis of phosphorylation by phospho-PhoR1084 in vitro revealed that the Asp53 of PhoB is the phosphate-accepting residue and the Thr83 plays an important role for the phosphorylation. Domains (II) and (III) are involved in the region of 90 C-terminal amino acids. DNA-binding activity of PhoB was abolished, if any one of three Arg, of two Thr, or of two Gly in the C-terminal region was replaced by other amino acid. A mutant, Glu177 to Lys, did not enhance transcription but was normal in DNA-binding and phosphate-accepting activities. The region containing Glu177 of PhoB may be related to the interacting domain with RNA polymerase.

Relationship between PhoR and Pst-PhoU Proteins

The *pho* regulon is regulated by multiple regulatory genes in a very complex manner. These genes seem to constitute a cascade regulatory network(Nakata et al 1987). The work presented here clearly indicates that the product of phoR has two forms, kinase and phosphatase forms. Since mutations in the *pst-phoU* operon cause constitutive expression of the *pho* genes, the PhoR protein may not be a direct sensor for the phosphate concentrations of the medium and function as an informational mediator that receives a phosphate signal transmitted by the function of the products of the *pst-phoU* operon. The operon is consisted of five genes, whish include *pstS*, *pstC*, *pstA*, *pstB*, and *phoU* (Amemura et al., 1985; Surin et al., 1985). Since *pst* genes are all related to the transport of phosphate and the *phoU* gene is not(Zuckier and Torriani, 1981), PhoU may interact directly with the Pst complex or catalyze formation of an effector molecule from a phosphate derivative that is produced from phosphate by the function of the Pst complex. PhoU that has received the phosphate signal or the effector molecule may modify PhoR to the phosphatase form that inactivates the transcriptional function of PhoB. The nature of the signal is not known at present.

Acknowledgements: This work was supported by a Grant-in-Aid for Scientific Research from the Ministry of Education, Science, and Culture of Japan.

REFERENCES

Amemura, M., Makino, K., Shinagawa, H., Kobayashi, A., and Nakata, A. (1985) J. Mol. Biol. 184, 241-250

Amemura, M., Makino, K., Shinagawa, H., and Nakata, A. (1986) J. Bacteriol. 168, 293-302

Amemura, M., Makino, K., Shinagawa, H., and Nakata, A. (1990) J. Bacteriol. 172, 6300-6307

Guan, C.-D., Wanner, B., and Inoyue, H. (1983) J. Bacteriol. 156, 710-717

Kimura, S., Makino, K., Amemura, M., Shinagawa, H., and Nakata, A. (1989) Mol. Gen. Genet. 215, 374-380

Makino, K., Shinagawa, H., and Nakata, A. (1982) Mol. Gen. Genet. 187, 181-186

Makino, K., Shinagawa, H., and Nakata, A. (1984) Mol. Gen. Genet. 195, 381-390

Makino, K., Shinagawa, H., and Nakata, A. (1985) J. Mol. Biol. 184, 231-240

Makino, K., Shinagawa, H., Amemura, M., and Nakata, A. (1986a) J. Mol. Biol. 190, 37-44

Makino, K., Shinagawa, H., Amemura, M., and Nakata, A. (1986b) J. Mol. Biol. 192, 549-556

Makino, K., Shinagawa, H., Amemura, M., Kimura, S., Nakata, A., and Ishihama, A. (1988) J. Mol. Biol. 203, 85-95

Makino, K., Shinagawa, H., Amemura, M., Kawamoto, T., Yamada, M., and Nakata, A. (1989) J. Mol. Biol. 210, 551-559

Nakata, A., Amemura, M., Makino, K., and Shinagawa, H. (1987) In: Phosphate Metabolism and Cellular Regulation in Microorganisms(A. Torriani-Gorini, F. G. Rothman, S. Silver, A. Wright, and E. Yagil, Eds), A. S. M., Washington, DC., pp. 150-155

Ninfa, A. J., Ninfa, E. G., Lupas, A. N., Stock, A., Magasanik, B., and Stock, J. (1988) Proc. Nat. Acad. Sci., USA. 85, 5492-5496

Ronson, C. W., Nixson, B. T., and Ausubel, F. M. (1987) Cell 49, 579-581

Shinagawa, H., Makino, K., and Nakata, A. (1983) J. Mol. Biol. 168, 477-488

Shnagawa, H., Makino, K., Amemura, M., and Nakata, A.(1987) In: Phosphate Metabolism and Cellular Regulation in Microorganisms(A. Torriani-Gorini, F. G. Rothman, S. Silver, A. Wright, and E. Yagil, Eds), A. S. M., Washington, DC., pp. 20-25

Stock, J. B., Ninfa, A. J., and Stock A. M. (1989) Microbiol. Rev. 53, 450-490

Surin, B. P., Rosenberg, H., and Cox. G. B. (1985) J. Bacteriol. 161, 189-198

Tommassen, J., De Geus, P., Lugtenberg, B., Hackett, J., and Reeves, P. (1982) J. Mol. Biol. 157, 265-274

Torriani, A., and Ludtke, D. N. (1985) In: The molecular biology of bacterial growth (M. Schaechter, F. C. Neidhart, J. Ingraham, and N. O. Kjeldgaard, Eds), J & B publishers, Boston, pp. 224-242

Wanner, B. L., and Latterell, P. (1980) Genetics 96, 353-366

Wylie, D., Stock, A., Wong, C.-Y., and Stock, J. (1988) Biochem. Biophys. Res. Commun. 151, 891-896

Yamada, M., Makino, K., Shinagawa, H., and Nakata, A.(1990) Mol. Gen. Genet. 220, 366-374

Zuckier, G., and Torriani, A. (1981) J. Bacteriol. 145, 1249-1256

Adenine Nucleotides in Cellular Energy Transfer and Signal Transduction
S. Papa, A. Azzi & J.M. Tager (eds)
© 1992 Birkhäuser Verlag, Basel/Switzerland

THE PROTEIN KINASE C FAMILY IN SIGNAL TRANSDUCTION AND CELLULAR REGULATION

Yasuo Fukami and Yasutomi Nishizuka

Laboratory of Molecular Biology, Biosignal Research Center, Kobe University, Nada, Kobe 657, Japan

The receptor-mediated hydrolysis of inositol phospholipids is now generally accepted to be a common mechanism for transducing various extracellular signals into the cell, such as those from certain hormones, neurotransmitters, antigens, some growth factors, and many other biologically active substances (1, 2). When a ligand binds to certain receptors on the cell surface, inositol phospholipids are hydrolyzed by phospholipase C, producing diacylglycerol and inositol phosphates. The primary effect of the diacylglycerol is to activate an ubiquitous enzyme, protein kinase C (PKC), which in turn phosphorylates a wide range of cellular proteins. Thus, PKC has been postulated to play key roles in signal transduction and cellular regulation mediated by the receptor function. PKC also serves as the major receptor for tumor-promoting phorbol esters, such as phorbol 12-myristate 13-acetate, which can directly activate PKC *in vitro* and *in vivo*. Since phorbol esters are metabolically more stable than diacylglycerols, they have been used as a long-lived PKC activator to show the importance of PKC activation in a variety of cellular responses in various cell systems (3).

Although PKC was originally thought to be a single entity, it is now clear that PKC is a large family of proteins with multiple subspecies (4). The current question is, therefore, as to the biological roles of individual PKC subspecies. What kind of PKC subspecies is(are) responsible for a certain kind of cellular function? To address this question, several lines of investigation are in progress. Some of the recent studies on the PKC family are briefly summarized below.

PROPERTIES OF THE NEWLY ISOLATED PKC SUBSPECIES

Molecular cloning and enzymological analysis of PKC has revealed the existence of multiple subspecies of the enzyme in a number of mammalian tissues (4, 5). At present, 8 subspecies of PKC have been identified (Table 1). Initially, four cDNA clones which encode α-, βI-,

Table 1. PKC Subspecies Structurally Identified in Mammalian Tissues

	Amino acid residues	Molecular weight	Tissue expression
α	672	76,799	Universal
βI	671	76,790	Some tissues
βII	673	76,933	Many tissues
γ	697	78,366	Brain only
δ	673	77,517	Many tissues ?
ε	737	83,474	Brain only ?
ζ	592	67,740	Brain, liver etc. ?
η(L)	683	77,972	Lung, skin, heart

βII-, and γ-subspecies were isolated from the rat brain cDNA library. These subspecies have been enzymologically well characterized and are now termed 'classical PKC subspecies'. Subsequently, another group of cDNA clones encoding δ-, ε-, and ζ-subspecies were identified using the same library. An additional cDNA clone encoding η(L)-subspecies was isolated from the mouse epidermis (6) and human keratinocytes (7). These newly isolated subspecies are structurally distinct from the classical subspecies in that they lack a putative Ca^{2+} binding domain. In contrast to the classical subspecies, enzymological properties of the newly isolated subspecies have not been well studied. Recently, the δ- and ε-subspecies of PKC were purified from the rat brain by the aid of subspecies-specific antibodies, and their enzymological properties have been revealed (8, 9). As expected, these subspecies were indeed independent of Ca^{2+} concentrations, when activated with diacylglycerol and phosphatidylserine, suggesting that they are regulated in a distinct fashion from that of the classical PKC subspecies. Immunocytochemical analysis and *in situ* hybridization analysis being undertaken may reveal a unique tissue distribution and subcellular localization of the newly identified subspecies individually.

Undoubtedly, there may be several additional PKC subspecies to be identified in mammalian tissues. Isolation and characterization of these forthcomers will be an important task for the next few years.

PKC SUBSPECIES IN LOWER EUKARYOTES

In mammals described above, PKC subspecies seem to have specialized roles in cellular functions of particular tissues and cell types. However, in lower eukaryotes or in germ cells, they might be responsible for general and fundamental functions such as cell growth and development. To find out such general and fundamental roles of PKC, and also to overlook the PKC family from the phylogenetic standpoint, it is important to carry out the studies not only with mammals but also with lower vertebrates or lower eukaryotes, such as shown in Table 2. Lower eukaryotic organisms are especially usefull when one wants to carry out a genetical analysis. For example, Levin *et al.* (15) isolated a cDNA clone, *PKC1*, which belongs to the classical PKC group, from *S. cerevisiae*. By using gene disruption technique, this *PKC1* gene has been shown to be required for the regulation of the yeast cell cycle. Possible involvement of the PKC family in cell cycle control is now becoming a very exciting field. However, there is little biochemical evidence concerning the mechanism of the involvement of PKC subspecies.

Table 2. PKC Subspecies in Various Lower Eukaryotes

Species	PKC-related genes*	
Mammals (rat, mouse, human)	α, βI, βII, γ	δ, ϵ, ζ, η
Xenopus laevis	XPKC I, XPKC II	-
Drosophila melanogaster	dPKC53E(br) dPKC53E(ey)	dPKC98F
Caenorhabditis elegance	-	*tpa-1*
Saccharomyces cerevisiae	*PKC1*	-

*References; α to ϵ (reviewed in ref. 5), ζ (10), η (6, 7), XPKC I and XPKC II (11), dPKC53E(br) (12,13), dPKC53E(ey) and dPKC98F (13), *tpa-1* (14), and *PKC1* (15).

Recently, one yeast PKC isozyme was isolated and characterized (16). The yeast PKC showed quite different enzymological properties from those of mammalian PKCs. First of all, it did not respond to the phorbol ester, phorbol 12-myristate 13-acetate (PMA), although it was activated by diacylglycerol in the presence of phosphatidylserine. It was also very different from the mammalian PKCs in its substrate specificity. It phosphorylated myelin basic protein very well, but H1 histone or protamine was not a good substrate, while these proteins were substantially phosphorylated by the α-subspecies of rat PKC. Phosphoamino acid analysis revealed that the yeast PKC preferentially phosphorylates threonine residues in the substrate proteins, whereas the mammalian α–PKC phosphorylates predominantly serine residues. Thus, the yeast PKC appears to have a distinct substrate recognition from that of the mammalian PKC. Since activity of the yeast PKC did not depend on Ca^{2+}, this enzyme seems not to belong to the classical group of the PKC family, and therefore, it is different from the one that has been cloned and shown to be involved in the cell cycle. The function of this yeast PKC has not been clarified yet.

In addition to the yeast system, we recently started to investigate the system of *Xenopus* oocytes. In *Xenopus* oocytes, the cell cycle is physiologically arrested at the G_2/M boundary of the first meiotic division in the process of oocyte maturation. Several maturation-promoting stimuli, such as progesterone, insulin and the tumor-promoting PMA, can induce the oocyte maturation. The PMA-induced oocyte maturation and some other observations suggest that PKC is involved in the maturation process. Although two PKC cDNA clones which resemble those encoding the classical mammalian PKCs have been isolated (11), enzymological characterization of the *Xenopus* PKC has not been carried out. Very recently, we isolated two distinct *Xenopus* PKCs from oocytes and characterized their properties (Sahara *et al.*, manuscript submitted). The two *Xenopus* PKCs were very similar to the mammalian α–PKC in their enzymological and immunological properties, but they could be discriminated from the α–, β–, or γ–PKC. Whether the two PKCs correspond to the two cDNA clones is to be determined. Although the two PKCs were equally activated by PMA *in vitro*, they responded differently in PMA-stimulated oocytes. Namely, one PKC disappeared rapidly upon PMA treatment of oocytes, while the other PKC remained unchanged during the oocyte maturation.

ACKNOWLEDGMENTS

Some of the studies reported here were supported in part by research grants from the Special Research Fund of the Ministry of

Education, Science and Culture, Japan; and by research grants from the Muscle Dystrophy Association U.S., the Juvenile Diabetes Foundation International U.S., the Yamanouchi Foundation for Research on Metabolic Disorders; the Sankyo Foundation of Life Science; the Merck Sharp & Dohme Research Laboratories; the Biotechnology Laboratories of Takeda Chemical Industries; the New Lead Research Laboratories of Sankyo Company; and Osaka Cancer Research Fund.

REFERENCES

1. Nishizuka, Y. (1984) Nature **308**, 693-698.
2. Berridge, M. J., and Irvine, R. F. (1984) Nature **312**, 315-321.
3. Nishizuka, Y. (1986) Science **233**, 305-312.
4. Nishizuka, Y. (1988) Nature **334**, 661-665.
5. Kikkawa, U., Kishimoto, A., and Nishizuka, Y. (1989) Ann. Rev. Biochem. **58**, 31-44.
6. Osada, S-I., Mizuno, K., Saido, T. C., Akita, Y., Suzuki, K., Kuroki, T., and Ohno, S. (1990) J. Biol. Chem. **265**, 22434-22440.
7. Bacher, N., Zisman, Y., Berent, E., and Livneh, E. (1991) Mol. Cell. Biol. **11**, 126-133.
8. Ogita, K., Miyamoto, S-I., Yamaguchi, K., Koide, H., Fujisawa, N., Kikkawa, U., Sahara, S., Fukami, Y., and Nishizuka, Y. Proc. Natl. Acad. Sci. USA, in press.
9. Koide, H., Ogita, K., Kikkawa, U., and Nishizuka, Y. Proc. Natl. Acad. Sci. USA, in press.
10. Ono, Y., Fujii, T., Ogita, K., Kikkawa, U., Igarashi, K., and Nishizuka, Y. (1989) Proc. Natl. Acad. Sci. USA **86**, 3099-3103.
11. Chen, K.-h., Peng, Z-g., Lavu, S., and Kung, H-f. (1988-89) Second Messengers and Phosphoproteins **12**, 251-260.
12. Rosenthal, A., Rhee, L., Yadegari, R., Paro, R., Ullrich, A., and Goeddel, D. V. (1987) EMBO J. **6**, 433-441.
13. Schaeffer, E., Smith, D., Mardon, G., Quinn, W., and Zuker, C. (1989) Cell **57**, 403-412.
14. Tabuse, Y., Nishiwaki, K., and Miwa, J. (1989) Science **243**, 1713-1716.
15. Levin, D. E., Fields, O., Kunisawa, R., Bishop, J. M., and Thorner, J. (1990) Cell **62**, 213-224.
16. Ogita, K., Miyamoto, S-I., Koide, H., Iwai, T., Oka, M., Ando, K., Kishimoto, A., Ikeda, K., Fukami, Y., and Nishizuka, Y. (1990) Proc. Natl. Acad. Sci. USA **87**, 5011-5015.

Adenine Nucleotides in Cellular Energy Transfer and Signal Transduction
S. Papa, A. Azzi & J.M. Tager (eds)
© 1992 Birkhäuser Verlag, Basel/Switzerland

STRUCTURE-FUNCTION STUDIES ON THE PROTEIN KINASE C FAMILY MEMBERS

*D.J. BURNS, *P. V. BASTA, *W. D. HOLMES, *L.M. BALLAS,
*N. B. RANKL, *J. L. BARBEE, ^R.M. BELL, AND *C. R. LOOMIS

*Sphinx Pharmaceuticals Corporation, Durham, N.C. 27717 and ^Department of
Biochemistry, Duke University Medical Center, Durham N.C. 27710

Introduction

The phospholipid-dependent, diacylglycerol or phorbol ester activated serine-threonine protein kinase first described by Nishizuka and colleagues is a major cellular regulatory enzyme (Nishizuka 1988). This protein kinase, protein kinase C (PKC) has been implicated in a variety of cellular processes including secretion, modulation of gene expression, proliferation and tumor promotion. PKC has also been identified as the major intracellular receptor for the tumor-promoting phorbol esters (Nishizuka 1989). The enzyme has a requirement for anionic phospholipids (i.e. phosphatidyl-L-serine) as cofactors (Takai et al. 1979), and also requires phorbol esters or the neutral lipid, sn-1,2 diacylglycerol (Kaibuchi et al. 1981) for maximal activation. sn-1,2 diacylglycerols are the naturally occurring activators of this enzyme.

Molecular cloning analysis of tissue from several mammalian species has identified eight structurally related members of the protein kinase C family (Figure 1) (Parker et al. 1989 and Bell and Burns 1991). All of the PKC family members have regions of shared homology (C or conserved regions), and regions of diversity (V or variable regions) (Figure 1). The PKC family members can be divided into two distinct classes based on the presence or absence of one particular structural domain (Huang 1989). This domain, C2 or conserved domain 2, is present in the α, βI, βII, and γ family members, but absent in the δ, ε, ζ, and η family members (Ono et al. 1988). The C2 region appears to confer Ca^{2+} sensitivity to the PKC family members (Ono et al. 1989a); perhaps by serving as a Ca^{2+}-dependent phospholipid binding domain.

208

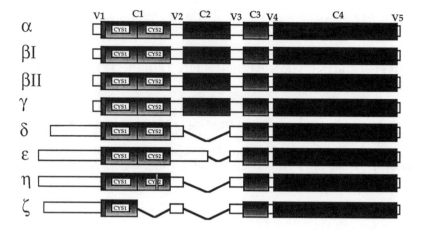

Figure 1: **Molecular Structures of the Protein Kinase C Family Members.** The eight members of the protein kinase C family are presented: 1) the α, βI, βII, and γ family members are dependent on Ca^{2+} for maximal activation and contain an extra conserved domain (C2); 2) the δ, ε, ζ, and η family members do not contain a C2 domain and do not require Ca^{2+} for activation. V1-V5, variable or non-conserved regions of the PKC molecule. C1-C4, conserved regions of the PKC molecule.

Structurally, the protein kinase C molecule can be divided into two functionally distinct domains (Lee and Bell 1986, and Ono et al. 1989b) (Figure 2). The amino-terminal portion of the molecule is known as the regulatory domain. This region contains the single or dual cysteine-rich, zinc-finger-like motif(s) which have been implicated as phorbol ester binding sites (Kaibuchi et al. 1989, Ono et al. 1989b, and Burns and Bell, 1991). The regulatory domain also contains the pseudosubstrate site; a conserved amino acid sequence that closely resembles protein kinase C substrate phosphorylation sites (House and Kemp 1990). The carboxy-terminal portion of the protein resembles other serine/threonine kinases, and contains consensus sequences for ATP and substrate binding sites (Hanks et al. 1988) This region of the PKC molecule is referred to as the catalytic domain.

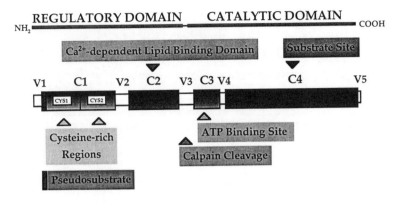

Figure 2: **Structural Features of Protein Kinase C.** The general features of the protein kinase C molecule are highlighted.

Materials and Methods

Human protein kinase C clones were isolated from cDNA libraries using established methods (Ono et al. 1988). All of the various protein kinase C constructs were expressed in the insect-cell baculovirus expression system and characterized for protein kinase C activity and [3H]PDBu binding as previously described (Burns and Bell 1991).

Cloning and Expression of Human PKC-δ and -ε Clones

Two independent clones coding for the human PKC-δ have been isolated from a single cDNA library; clone 1 and 2 (Figure 3). The predicted amino acid sequences of both of the human PKC-δ clones are 676 amino acids and have about a 90% sequence identity to the rat and mouse PKC-δ 's. There are three nucleotide differences between the two δ clones that translate into two conservative amino acid changes in the catalytic portion of the enzyme (amino acid number 375 (Ser versus Phe) and amino acid number 593 (Met versus Val). Amino acid residues 375 and 593

Human δ Clones

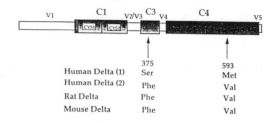

	375	593
Human Delta (1)	Ser	Met
Human Delta (2)	Phe	Val
Rat Delta	Phe	Val
Mouse Delta	Phe	Val

Human ε Clones

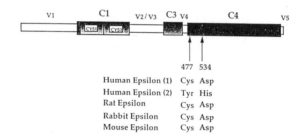

	477	534
Human Epsilon (1)	Cys	Asp
Human Epsilon (2)	Tyr	His
Rat Epsilon	Cys	Asp
Rabbit Epsilon	Cys	Asp
Mouse Epsilon	Cys	Asp

Figure 3: **Amino Acid Differences Between the Two PKC Delta and Epsilon Clones.** The important amino acid residue differences between the human delta and epsilon clones are compared to each other, as well as, to clones isolated from other mammalian species.

from PKC delta 2 are identical to the corresponding residues in the rat and mouse clones. When the cDNA clones are expressed in the insect-cell baculovirus expression system, the resulting proteins display protein kinase C activity independent of Ca^{2+}, bind radioactive phorbol esters, and are recognized by a δ-specific antiserum (Figure 4 and data not shown). Histone III-S, myelin basic protein, and peptides derived from the δ pseudosubstrate site (Figure 4) all serve as good substrates for PKC-δ. When assayed in the presence of PS/Triton X-100 mixed

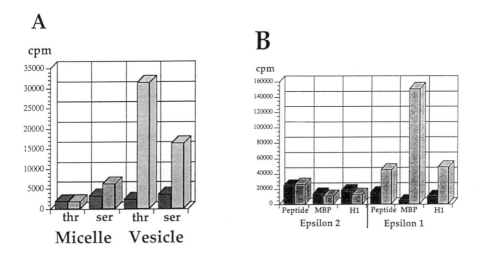

Figure 4. **(A) Peptides as Substrates for PKC Delta and (B) Protein Kinase C Activities of Epsilon 1 and 2.** (A) Modified threonine and serine peptides based on the pseudosubstrate site of PKC delta were tested as substrates for human PKC delta using both phosphatidylserine/Triton X-100 mixed micelles and phosphatidylserine vesicles. **(thr)** TMNRRGTIKQAKI; **(ser)** TMNRRGSIKQAKI. (B) A modified pseudosubstrate peptide (peptide), myelin basic protein (MBP), and histone III-S (H1) were all tested as substrates for epsilon 1 and epsilon 2 using a standard vesicle assay to measure PKC activity. For both (A) and (B), darker shaded boxes represent activity in the absence of lipid cofactors, while the lighter boxes represent activity in the presence of phosphatidylserine and diacylglycerol.

micelles, however, the activity of PKC-δ is much reduced relative to PS vesicles with all tested substrates (see Figure 4). The reason for the behavior of PKC-δ in the two assay systems is not well understood, but may reflect an inhibition of the enzyme by the detergent.

Two independent cDNA clones have also been isolated for human PKC-ε which also differ by two amino acids in the catalytic domain of the enzyme (C4) (Figure 3). PKC epsilon clone 2 was isolated from a human temporal cortex cDNA library, and then used to screen a human frontal cortex library for PKC epsilon 1. The human brain cDNA libraries were derived from the same individual. A comparison of the two human PKC-ε sequences with the deduced amino acid sequences of the rat, rabbit and mouse ε sequences reveals a high degree of homology. The two human clones have between 98% and 99% identity with those sequences. A comparison of the human ε clones reveals two nucleotide differences which translate into two amino acid differences at position 477 (Cys versus Tyr) and 534 (Asp versus His). Amino acid residues 477 and 534 from PKC epsilon 1 are identical to the corresponding residues in the rat, rabbit, and mouse clones. The two amino acid substitutions present in the epsilon 2 clone are quite distinct and represent non-conservative amino acid changes. The first substitution of a tyrosine for a cysteine at position 477 in epsilon 2 is in a region of the protein kinase C family that is not well conserved, however, the second substitution of a histidine for an aspartic acid residue is in an area of highly conserved amino acid residues. This aspartic acid residue is stringently conserved in all of the identified protein kinase C family members, and more strikingly, in all known serine/threonine and tyrosine kinases (Hanks and Hunter 1988). The histidine residue of epsilon 2 corresponds to the aspartic acid residue 166 of the cyclic AMP-dependent protein kinase. Asp[166] in conjunction with other invariant residues has been implicated in ATP binding; it has been postulated that the Asp[166] and Asp[184] of the cyclic AMP-dependent protein kinase may interact with the phosphate groups of ATP through Mg^{2+} salt bridges (Hanks and Hunter). Thus, the substitution of a positively charged histidine residue in place of a negatively charged aspartic acid residue at position 534 in epsilon 2 could potentially alter many of the catalytic properties of the enzyme. In fact, the mutation of a single amino acid residue in the ATP-binding site of protein kinase C renders the enzyme catalytically inactive (Ohno et al. (1990) and Freisewinkel et al. 1991).

When the epsilon 1 and 2 cDNA's were expressed in the insect-cell baculovirus expression system, the resulting proteins were able to bind phorbol esters and were recognized by two antipeptide antibodies directed against two distinct PKC-ε epitopes (data not shown). Only epsilon 1, however, had a functional kinase activity that was dependent upon phosphatidylserine and diacylglycerol (Figure 4). As would be expected for Ca^{2+}-independent PKC family members, the kinase activity of epsilon 1 is not enhanced by the addition of Ca^{2+} (data not shown). Myelin basic protein also serves as a good substrate for human epsilon 1. In contrast, epsilon 2 did not have any significant protein kinase C activity. The *in vivo* significance of a non-functional protein kinase C family member is not readily apparent. One possible explanation of these findings is that the correct activation conditions or appropriate substrates for epsilon 2 have not been found. Alternatively, a second possibility is that epsilon 2 plays a role in PKC signal attenuation. Evidence for this type of regulation has recently been demonstrated for a growth factor receptor that does not display tyrosine kinase activity (Chou and Hayman 1991).

PKC Contains Two Phorbol Binding Sites

All of the protein kinase C family members contain a single or tandem cysteine-rich repeat (C1 region) that is required for phorbol ester binding (Ono et al. 1989 and Kaibuchi et al. 1989). To investigate the structure/function relationships of this region of the PKC regulatory domain, a series of deletion and truncation mutants were created and expressed in the insect-cell baculovirus expression system (Figure 5) (Burns and Bell 1991). The constructs were designed to determine the location and number of phorbol ester binding domains. As shown in Figure 5, all of the PKC-derived constructs that contained single or tandem cysteine-rich region(s) bound phorbol esters ([3H]PDBu).

In general, mutants containing the second cysteine-rich region bound phorbols to a slightly greater extent than those that contained only the first cysteine-rich region; construct 11 is an exception (Figure 5). Part of this may be reflective of the

214

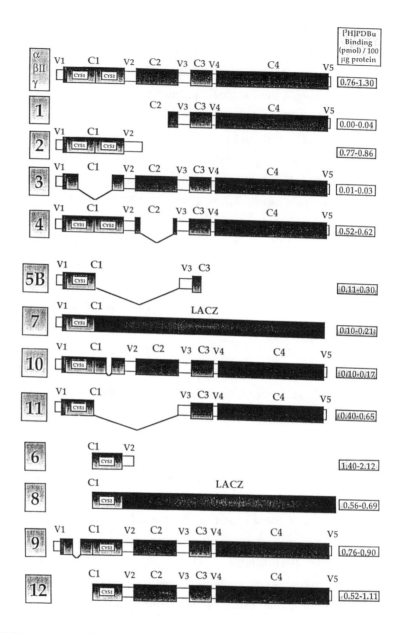

Figure 5. **Structures and [3H]PDBu Binding Activities of Several Protein Kinase C Constructs.** The structural features of the full-length protein kinase C clones and the PKC deletion and truncation constructs are depicted. The range of [3H]PDBu activities observed from several independent experiments is presented to the right of each construct.

levels of expressed protein. Constructs containing only the first cysteine-rich region appeared to be present in smaller quantities than those containing the second cysteine-rich region as determined by Western blotting with PKC-specific antisera (data not shown). However, to fully assess phorbol binding potentials, individual binding affinities were determined for several of these constructs. Scatchard analyses indicated that all of the tested constructs bound phorbol esters with nanomolar affinities: construct 5B, K_d of 31 ± 4 nM; construct 10, K_d of 26 nM; and construct 11, K_d of 29 ± 8 (first cysteine-rich region) versus construct 6, K_d of 59 ± 13 nM; construct 9, K_d of 29 ± 4; and construct 12, K_d of 28 ± 9 nM (second cysteine-rich region). Thus, one protein kinase C molecule contains two separate domains capable of binding phorbol esters with equivalent affinities. In addition, the minimal domain structure of the PKC molecule required for phorbol ester binding is beginning to be delineated: a polypeptide of 146 amino acid residues (construct 5B) from the first cysteine-rich region, as well as, a polypeptide of 86 amino acid residues (construct 6) from the second cysteine-rich region were both able to bind phorbol esters ([3H]PDBu). Thus, all of the structural information necessary for interactions with phospholipids and phorbol esters must be present in these polypeptides.

References

Bell, R.M. and Burns, D.J. (1991) "Lipid activation of protein kinase C" J. Biol. Chem. 266, 4461-4464.

Burns D.J., and Bell, R.M. (1991) "Protein kinase C contains two phorbol ester binding domains" J. Biol. Chem. 266, 18330-18338

Chou, Y.-H., and Hayman, M.J. (1991) "Characterization of a member of the immunoglobin gene superfamily that possibly represents an additional class of growth factor receptor" Proc. Natl. Acad. Sci. U.S.A. 88, 4897-4901

Freisewinkel, I., Riethmacher, D., and Stabel, S. (1991) "Downregulation of protein kinase C-γ is independent of a functional kinase domain" FEBS Lett. 280, 262-266

Hanks, S.K., Quinn, A.M., and Hunter, T. (1988) "The protein kinase family: conserved features and deduced phylogeny of the catalytic domains" Science 241, 42-52

House, C., and Kemp, B. (1990) "Protein kinase C pseudosubstrate prototype: structure-function relationships" Cell. Signalling 2, 187-190.

Huang, K.-P. (1989) "The mechanism of protein kinase C activation" TINS 12, 425-432.

Kaibuchi, K., Takai, Y., and Nishizuka, Y. (1981) "Cooperative roles of various membrane phospholipids in the activation of calcium-activated, phospholipid-dependent protein kinase" *J. Biol. Chem.* **256**, 7146-7149.

Kaibuchi, K., Fukumoto, Y., Oku, N., Takai, Y., Arai, K., and Muramatsu, M. (1989) "Molecular genetic analysis of the regulatory and catalytic domains of protein kinase C" J. Biol. Chem. 264, 13489-13496.

Lee, M-H., and Bell, R.M. (1986) "The lipid binding, regulatory domain of protein kinase C" J. Biol. Chem. 261, 14867-14870.

Nishizuka, Y. (1988) "The molecular heterogeneity of protein kinase C and its implication for cellular regulation" Nature 334, 661-665

Nishizuka, Y. (1989) "The family of protein kinase C for signal transduction" JAMA 262, 1826-1833.

Ohno, S., Konno, Y., Akita, Y., Yano, A., and Suzuki, K. "A point mutation at the putative ATP-binding site of protein kinase C abolishes the kinase activity and renders it down-regulation-insensitive" J. Biol. Chem. 265, 6296-6300

Ono, Y., Fujii, T., Ogita, K., Kikkawa, U., Igarashi, K., and Nishizuka, Y. (1988) "The structure, expression, and properties of additional members of the protein kinase C family" J. Biol. Chem. 263, 6927-6932.

Ono, Y., Fujii, T., Ogita, K., Kikkawa, U., Igarashi, K., and Nishizuka, Y. (1989a) "Protein kinase C ζ subspecies from rat brain: its structure, expression, and properties" Proc. Natl. Acad. Sci. USA 86, 3099-3103.

Ono, Y., Fujii, T., Igarashi, K., Kono, T., Tanaka, C., Kikkawa, U., and Nishizuka, Y. (1989b) "Phorbol ester binding to protein kinase C requires a cysteine-rich zinc-finger-like sequence" Proc. Natl. Acad. Sci. USA 86, 4868-4871.

Parker, P.J., Kour,G., Marias,R., Mitchell, F., Pears, C., Schaap, D., Stabel, S., and Webster, C. (1989) "Protein kinase C - a family affair" Mol. Cell. Endrocrinol. 65, 1-11.

Takai, Y., Kishimoto, A., Iwasa, Y., Kawahara, Y., Mori, T., Nishizuka, Y., Tamura, A., and Fujii, T. (1979) "A role for membranes in the activation of a new multifunctional protein kinase system" *J. Biochem.* (Tokyo) **86**, 575-578.

Adenine Nucleotides in Cellular Energy Transfer and Signal Transduction
S. Papa, A. Azzi & J.M. Tager (eds)
© 1992 Birkhäuser Verlag, Basel/Switzerland

MECHANISM OF PROTEIN KINASE C-MEDIATED SIGNAL TRANSDUCTION

Kuo-Ping Huang, Freesia L. Huang, Hiroki Nakabayashi,
Charles W. Mahoney and Kuang-Hua Chen

Section on Metabolic Regulation, Endocrinology and Reproduction
Research Branch, National Institute of Child Health and Human
Development, National Institutes of Health, Bethesda, MD 20892

SUMMARY: Protein kinase C (PKC) plays a pivotal role in the
regulation of numerous cellular functions in response to a variety
of external stimuli that cause the increase in $[Ca^{2+}]_i$, sn-1,2-
diacylglycerol (DAG), and cis-unsaturated fatty acids. In vivo, a
majority of PKC is believed to be loosely associated with membrane
components. Among the various phospholipids tested, the
polyphosphoinositides appear to be the best candidate as anchoring
sites for PKC, which readily interacts with these phospholipids
under the physiological concentrations of Mg^{2+}. PKC anchored to
phosphatidylinositol-4, 5-bisphosphate can further interact with
Ca^{2+} and phosphatidylserine to generate a partially active enzyme.
In the presence of DAG, the kinase becomes fully active. The
activated PKC is susceptible to proteolysis to generate
constitutively active PKM, which exhibits a broader substrate
specificity than PKC. Phosphorylation of a group of calmodulin
(CaM)-binding proteins by PKC provides a link between PKC-mediated
responses and other signaling pathways that make use of CaM as
activator. Phosphorylation of one of those CaM-binding proteins,
neurogranin (RC3), reduces its affinity for CaM and thus renders
CaM available for other enzymes. Activation of PKC in the nucleus
is important in controlling gene expression. Phosphorylation of
CCAAT/enhancer-binding protein at a serine residue within the basic
DNA-binding region attenuates the binding of this protein to
specific DNA probes. Control of PKC gene expression has been linked
to the regulation of cellular growth and differentiation.
Structural features of PKCγ gene promoter region have been
characterized and potential regulatory elements identified. This
article illustrates several possible mechanisms that confer the
functional diversity of PKC.

INTRODUCTION

Members of the protein kinase C (PKC) gene family have been implicated in diverse cellular regulation (Nishizuka,1988; Huang, 1990). At least eight PKC family members have been identified by molecular cloning (for review see Bell & Burns, 1991). These serine/threonine protein kinases consist of two groups of isozymes distinguishable by their requirements for Ca^{2+} when assayed in the presence phosphatidylserine (PS) and diacylglycerol (DAG). Structurally, the group B PKCs (δ, ϵ, ζ and η) are distinct from the group A PKCs (α, βI, βII, and γ) in the regulatory domain where the conserved C2 region is absent in the former. The Ca^{2+} requirements among the group A PKCs are distinguishable when cis-unsaturated fatty acids are used as activators (Naor et al., 1988). The emerging data on the tissue, cellular, and subcellular localizations and the distinct biochemical characteristics of these enzymes support the contention that each member of this enzyme family has distinct function. In this article we briefly discuss several possible mechanisms that can confer the functional diversity of these enzymes.

RESULTS

Activation of PKC associated with polyphosphoinositides: A majority of the group A PKCs are believed to be associated with the membrane fractions in vivo. Association of PKCs with membrane phospholipids is a prerequisite for activation of this enzyme. Among the various phospholipids tested, PS is most effective in supporting the kinase activity in the presence of Ca^{2+} and DAG. Analysis of PKC and phospholipid interaction by monitoring the changes in the intrinsic fluorescence of the enzyme revealed that PKC interacts most favorably with polyphosphoinositides than any other phospholipids either in the presence or absence of Ca^{2+}. Interaction of PKC with polyphosphoinositides is greatly facilitated in the presence of Mg^{2+}. In contrast, interaction of PKC with PS is facilitated in the presence of Ca^{2+} but not with Mg^{2+}. Thus, it is likely that PKC is preferentially associated with polyphosphoinositides under the basal physiological conditions when [Ca^{2+}]$_i$ is low and Mg^{2+} is in the millimolar range. Among the three major PKC isozymes (PKC α, β, and γ) tested, PKCγ appears to have the highest affinity toward these phospholipids. It is intuitive to speculate that anchoring of PKCγ to cellular membrane rich in polyphosphoinositides may contribute to the enrichment of this isozyme in the dendrites of several neurons (Huang et al., 1988).

PKC is a relatively flexible molecule which undergoes extensive conformational changes upon interaction with its activators as demonstrated by its ability to undergo intramolecular autophosphorylation at multiple sites located in both the regulatory and catalytic domains (Huang et al., 1986). In addition, PKC appears to interact with both phosphatidylinositol-4,5-bisphosphate (PIP$_2$) and PS simultaneously. This is illustrated by the changes in the intrinsic fluorescence of PKC following sequential addition of PS and PIP$_2$ (Fig. 1). The characteristic changes in the fluorescence resulting from interactions of PKC with PS and PIP$_2$ persisted regardless of the order of addition of these two phospholipids. In addition, PKC associated with PIP$_2$ micelles

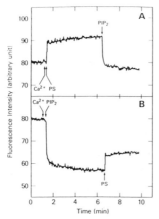

Fig. 1. Changes in the intrinsic fluorescence of PKC following interaction with PS and PIP$_2$ in the presence of Ca^{2+}. Panel A, Ca^{2+} (1 mM), PS (12 μM), and PIP$_2$ (6 μM) were added to PKC (3 μg/ml in 20 mM Tris-Cl buffer, pH 7.5, containing 0.5 mM DTT, 0.25 mM EDTA, 0.25 mM EGTA, 5% glycerol) in a sequential order at the indicated time points and the changes in fluorescence were monitored. Panel B, the order of addition of these components was Ca^{2+}, PIP$_2$, and PS.

can further interact with PS vesicles to form a larger aggregate as analyzed by gel filtration chromatography (Huang & Huang, 1991). These results suggest that these two phospholipids may have distinct binding regions on the PKC molecule. In vivo, attachment of PKC to membrane phosphoinositides places the kinase in close proximity to DAG generated from the hydrolysis of these phospholipids by PI-specific phospholipase C. In addition, binding of PKC to PIP_2 could also confer moderate stimulation of the kinase at increased $[Ca^{2+}]_i$ even without DAG (Chauhan & Brockerhoff, 1988). It is envisioned that an increase in $[Ca^{2+}]_i$ results in an interaction of the PIP_2-associated PKC with PS to form a partially active enzyme. Upon the generation of DAG following hydrolysis of phosphoinositides by phospholipase C, PKC becomes fully active. The activated form of the kinase can be converted into the membrane-inserted (Bazzi & Nelsestuen, 1988) or membrane protein-associated form (Gopalakrishna et al.,1986).

Altered substrate specificity of proteolytically activated PKC: Prolonged activation of PKC frequently results in a proteolytic degradation of the kinase most likely by endogenous Ca^{2+}-activated proteases. In vitro, PKC can be degraded by trypsin or calpain to generate a 40-50 kDa catalytic and 30-40 kDa regulatory fragments (Huang and Huang,1986; Kishimoto et el. 1989). The various PKC isozymes have differential sensitivity to proteolysis both in vitro and in phorbol ester-treated cells (Huang et al.,1989). Incubation of human platelets with thrombin or PMA has been shown to cause the phosphorylation of myosin light chain (LC) at two sets of sites recognized separately by PKC and myosin light chain kinase (MLCK) under the in vitro assay conditions (Naka et al. 1988). MLCK, a Ca^{2+}/CaM-activated protein kinase, phosphorylates LC at Ser^{19} and Thr^{18} resulting in an increase in the actin-activated Mg^{2+}-ATPase activity (Sellers et al., 1981). In contrast, PKC phosphorylates Thr^9 and Ser^1 or Ser^2 with a resulting decrease in the Mg^{2+}-ATPase activity of myosin that has already been activated by MLCK (Nishikawa et al., 1984). The PMA-induced phosphorylation of myosin LC in human platelets could be due to activation of both PKC and

MLCK or due to PKM generated from protease-degraded PKC (Tapley & Murray, 1985). Under the _in vitro_ assay conditions, incubation of myosin LC with PKM resulted in the phosphorylation of both sets of sites recognized by PKC and MLCK as analyzed by isoelectric focusing peptide mapping (Fig. 2). The PKM-catalyzed phosphorylation of LC is not greatly affected by MLCK inhibitor ML-9 nor by the activators of MLCK, Ca^{2+} and CaM. Furthermore, a synthetic peptide corresponding to the Ser^{19} site of LC recognized by MLCK is also readily phosphorylated by PKC and PKM (Nakabayashi et al., 1991), suggesting that Ser^{19} is a potential phosphorylation site of PKC. The broadening of substrate specificity of PKM may be due to the relief of steric constraint imposed by the regulatory domain. Recently, it has been shown that PKM also possess PI kinase activity (Tusupov et al., 1991), and proteolytic degradation of PKC-ϵ increases its kinase activity toward histone (Schaap et al., 1990). These findings suggest that proteolytic degradation of PKC could amplify the kinase activity by forming a spontaneously active enzyme with a broader substrate specificity.

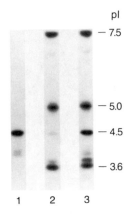

Fig. 2. Isoelectric focusing analysis of the ^{32}P-labeled tryptic peptides derived from LC phosphorylated by MLCK, PKC, and PKM. Tryptic phosphopeptides from LC phosphorylated by MLCK (lane 1), PKC (lane 2), and PKM (lane 3) were analyzed by isoelectric focusing on polyacrylamide gel. The apparent isoelectric points of the ^{32}P-labeled peptides are as indicated.

224

Amplification of the PKC-mediated responses by phosphorylation of
CaM-binding protein: Several CaM-binding proteins, such as MARCKS
(Stumpo et al., 1989), neuromodulin (GAP43, B50) (Alexander et al.,
1987), and neurogranin (RC3) (Baudier et al., 1991; Watson et al.,
1990), are specific substrates of PKC. Neuromodulin and neurogranin
are Ca^{2+}-independent and MARCKS is a Ca^{2+}-dependent CaM-binding
protein. Rat brain neurogranin purified by heat treatment, DEAE-
cellulose and hydroxylapatite column chromatography has an amino
acid sequence identical to that deduced from the rat RC3 cDNA
clone. The rat brain protein exhibits extensive sequence identity
to the bovine brain neurogranin. Phosphorylation of neurogranin by
PKC or PKM resulted in a stoichiometric incorporation of phosphate
into Ser^{36}. The PKM-catalyzed phosphorylation of neurogranin is
inhibited by CaM and the inhibition is relieved by Ca^{2+}, suggesting
that Ca^{2+}/CaM complex is ineffective for inhibition. Indeed,
neurogranin does bind to a CaM-Sepharose column in the absence of
Ca^{2+} and can be eluted by buffer containing Ca^{2+} (Fig. 3).
Phosphorylation of neurogranin by PKM prevented its interaction
with CaM-Sepharose column even in the absence of Ca^{2+}. These results

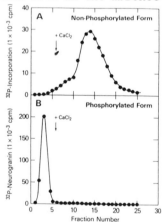

Fig. 3. Chromatography of phosphorylated and non-phosphorylated
neurogranin on CaM-Sepharose. Non-phosphorylated neurogranin (panel
A) and PKC-phosphorylated neurogranin were separated on CaM-
Sepharose equilibrated with 20 mM Tris-Cl buffer, pH 7.5,
containing 1 mM DTT, 0.5 mM EDTA, 0.5 mM EGTA, and 10% glycerol.
For elution, 3 mM $CaCl_2$ was included in the buffer.

indicate that phosphorylation of Ser[36] located near the predicted CaM-binding domain controls the binding affinity of neurogranin to CaM. Storm and coworkers (Liu & Storm, 1990) suggest that this group of PKC substrates sequester CaM under low $[Ca^{2+}]_i$ and release it at an increased level of $[Ca^{2+}]_i$ and/or by phosphorylation with PKC. Thus, neurogranin, a postsynaptic enriched protein (Represa et el., 1990), may function as a regulator to control the level of available CaM in response to activation of PKC γ, which appears to co-localize with this protein in many neurons in CNS (Huang et al., 1988).

Regulation of nuclear transcriptional factor function by PKC: Many of the long term events regulated by PKC involve the regulation of gene transcription as is the case with genes containing phorbol ester-response elements (Angel et al., 1987). The mechanism of signal transduction from the phorbol esters to the transcriptional process remains unknown. Nuclear membrane is rich in phosphoinositides (Cocco et al., 1990) and contain functional IP_3 receptors (Malviya et al., 1990) and ATP-stimulated Ca^{2+} uptake system (Nicotera et al.,1989). In addition, nuclear extract contains PKC β (Rogue et al., 1990) and DAG kinase, a key enzyme in attenuating PKC activity. It appears that nuclear PKC can function by phosphorylation of transcriptional factors upon an increase in Ca^{2+} and DAG. CCAAT/enhancer-binding protein (C/EBP), a member of the bZIP family of proteins characterized by a leucine repeat-dimerization interface and an upstream basic region containing a DNA site-specific binding domain, was used as a model to test the functional effect of PKC-catalyzed phosphorylation. Several truncated forms of the C/EBP containing the leucine zipper and basic regions were phosphorylated by PKC at multiple sites dependent on Ca^{2+}, lipid activators, and isozyme type. Both PKC α and β are more active than PKC γ in phosphorylating C/EBP. Phosphorylation of Ser[299], located at a basic region of the predicted DNA binding site (Vinson et al., 1989) resulted in a weakening of binding of C/EBP to DNA as measured by gel mobility shift assay and footprinting analysis. These findings can be explained by the "scissors-grip" model for the interaction of C/EBP with its specific recognition sequence (Vinson et al., 1989). This model predicts a dimerized set of α-helices (leucine zipper) which bifurcates in such a manner as to project positively charged regions in opposite directions into the major groove of DNA. According to this model, Ser^{299} is positioned within the major groove of DNA. One might therefore predict that phosphorylation of this amino acid residue would interfere with DNA binding by either interfering with the recognition of the specific DNA binding site or simply by ionic charge repulsion (Fig. 4).

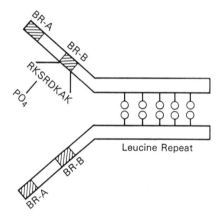

Fig. 4. Schematic representation of the phosphorylation of C/EBP by PKC. C/EBP containing a leucine zipper and upstream basic region for DNA binding was phosphorylated by PKC at multiple sites. The amino acid sequence surrounding Ser^{299}, a crucial site for affecting DNA binding, is shown.

Control of PKC γ gene expression: In the central nervous system the expression of PKCγ is under developmental control; it is most active during synaptogenesis and in the adult (Huang et al., 1990). Recently, the promoter region of rat PKC γ gene was cloned, sequenced and mapped by transient expression assays (Chen et al., 1990). This promoter contains several nuclear protein binding sites homologous to previously characterized protein binding sites for SP1, AP2, adenovirus type 2 major late promoters, and progesterone receptor as well as sites unique for the PKC γ promoter as determined by footprinting analysis (Fig. 5). Among them, a binding site (site M) appears to be specific for nuclear proteins from those tissues that do not express PKC γ, such as fetal brain, spleen, and liver. The nuclear extracts of adult brain appear to be deficient in a protein for this site and thus, may be considered as a negative regulatory element. Based on the known regulatory mechanisms of previously characterized nuclear protein binding elements, we predict that PKC γ gene expression may be responsive to cAMP and phorbol ester (due to AP2) and progesterone or other steroid hormones. In fetal brain, the level of PKC γ is very low in comparison with those of PKC α and β. Thus, activation of the latter two kinases, which usually coexist with PKC γ in the same neuron, may provide a positive signal for the expression of PKC γ gene. This control mechanism may contribute to the delayed increase of PKC γ as compared to PKC α and β during brain development.

Binding Site	Position	Sequence
A	+168/+205	TGGCCCCTATCGCCGTCCACCTGTTTCCTCAGAAAAA
B (Sp1)	+112/+149	GCGCTGCGCACCGTTAGTGCCCTGCCCCTGTCCTTCCG
C (Sp1)	+84/+96	GGTGCCGGGGGGT
D (AP2)	+45/+66	GCTCTGACCCCACCCGCTTTCT
E (AP2)	-58/-35	TGGCAAGCCTACCCCCACCCCCGA
F (Sp1)	-86/-61	AAACTGAAACCCCGCCCCTTGGTGCT
G (AP2)	-108/-92	TTGGGGGTGGGGGAACG
H	-154/-122	GGGAGTGTGCACGTGGAGAGGAGGGAGGGGCAA
I	-248/-229	GTGGGAGCCAAGATAACAGA
J	-260/-251	AGGAAGATGG
K	-297/-272	GATTCTGGCAAGACCCACACAAGCGG
L	-465/-442	GGCATGGCACCAAGCCCAGAGGCT
M	-679/-669	GAATTAATAGG
N	-741/-728	CCCCCTTGCGACAG
O (MLP)	-1360/-1338	GGTGGGGCGTGACCTGGGGGAGG
P	-1418/-1394	GAACGGCTTCCTGCCAGAATGGACC
Q (PR)	-1517/-1490	AGTTCTCCGTTGCGTTCCAAGAATAGAG

Fig. 5. Schematic representation of the nuclear protein binding sites within the PKC γ promoter. The nucleotide sequences mapped by DNase I footprinting analysis are given below the diagram. Asterisk indicates the hypersensitive sites. Nuclear protein for site M was present only from those tissues that do not express PKC γ.

Conclusions

The PKC gene family consists of at least eight members, α, βI, βII, γ, δ, ε, ζ, and η, which are responsible for numerous normal as well as pathological cellular responses. These kinases are stimulated by Ca^{2+}, PS, and DAG; however, other mechanisms of activation have been uncovered so that the classification of PKC as Ca^{2+}/PS/DAG-stimulated enzymes becomes less definite. The kinase anchored at the membrane polyphosphoinositides may become active in the presence of Ca^{2+} provided that PS, a stable membrane phospholipid constituent, is available for the kinase. Conversion of the loosely membrane bound PKC into the membrane-inserted form (Bazzi & Nelsestuen, 1988), and perhaps the membrane protein or cytoskeletal protein associated form, changes the kinase into an activator-independent enzyme. Free cis-unsaturated fatty acids can stimulate PKC either in the presence or absence of Ca^{2+} dependent upon the isozyme under study. In addition, these fatty acids can stimulate PKC in synergy with DAG at low Ca^{2+} concentrations (Shinomura et al., 1991). Thus, the various PKC isozymes may be activated under different physiological conditions that account for the diverse cellular responses. Activation of PKC frequently results in a proteolytic degradation of this enzyme. The proteolyzed enzyme, PKM, exhibits a broader substrate specificity than the native enzyme and thus could amplify the kinase activity. This mechanism may be functioning in the PMA-treated cells such as human platelets (Tapley & Murray, 1985) and neutrophils (Pontremoli et al.,1986). The "cross talk" between PKC and other Ca^{2+}-mediated as well as cAMP-mediated responses may result from phosphorylation of CaM-binding proteins by PKC. CaM released from these phosphorylated proteins becomes available for other CaM-dependent enzymes such as Ca^{2+}/CaM-dependent protein kinases, cAMP-phosphodiesterase, and adenylate cyclase. The signaling mechanism that controls gene expression could be mediated by the phosphorylation of transcriptional factors. Several members of the bZIP family of transcriptional factors, such as GCN4, v-Jun, CPC-1, HBP-1, TGA-1, and Opaque 2, have homologous amino acid sequences to the Ser^{299} site of C/EBP and are likely to be regulated by PKC. The mechanism of activation of PKC inside the nucleus has yet to be defined. Finally, the amplification of PKC-mediated responses could be modulated by the level of the kinase. Distinct tissue-specific and developmental stage-controlled PKC isozymes expression has been demonstrated; the mechanism of transcriptional control of these enzymes is not yet understood. Characterization of the genomic structures of these genes will likely lead to the identification of important regulators for the expression of these enzymes.

REFERENCES

Angel, P., Imagawa, M., Chiu, R., Stein, B., Imbra, R.J.,Rahmsdorf, H.J., Jonat, C., Herrlich, P., and Karin, M. (1987) Cell 49, 729-739

Alexander, K.A., Cimler, B.M., Meier, K.E., and Storm, D.R. (1987) J. Biol. Chem. 262, 6108-6113.

Baudier, J., Deloulme, J.C., VanDorsselaer, A., Black, D., and Mathes, H. (1991) J. Biol. Chem. 266, 229-237.

Bazzi, M.D. and Nelsestuen, G.L. (1988) Biochem. Biophys. Res. Commun. 152, 366-343.

Bell, R.M. and Burns, D.J. (1991) J. Biol. Chem. 266,4661-4664.

Chauhan, V.P.S., and Brockerhoff, H. (1988) Biochem. Biophys. Res. Commun. 155, 18-23

Chen, K.-H., Widen, S.G., Wilson, S.H., and Huang, K.-P. (1990) J. Biol. Chem. 265, 19961-19965.

Cocco, L., Capitani, S., Martelli, A.M., Irvine, R.F., Gilmour, R.S.,Maraldi, N.M., Barnabei, O., and Manzoli, F.A. (1990) Adv. Enzyme Regul. 30, 155-172.

Gopalakrishna, R., Barky, S.H., Thomas, T.P., and Anderson, W.B. (1986) J. Biol. Chem. 261, 16438-16445.

Huang, F.L., Yoshida, T., Nakabayashi, H., Young, W.S.III, and Huang, K.-P. (1988) J. Neurosci. 8, 4734-4744.

Huang, F.L., Yoshida, Y., Melo-Cunha, J.R., Beaven, M.A., Huang, K.-P. (1989) J. Biol. Chem. 264, 4238-4243.

Huang, F.L., Young, W.S.III., Yoshida, Y., and Huang, K.-P. (1990) Dev. Brain Research 52, 121-130.

Huang, F.L., and Huang, K.-P. (1991) J. Biol. Chem. 266, 8727-8733.

Huang, K.-P., Chan, K.-F.J., Singh, T.J., Nakabayashi, H., and Huang, F.L. (1986) J. Biol. Chem. 261, 12134-12140.

Huang, K.-P. and Huang, F.L. (1986) Biochem. Biophys. Res. Commun. 139, 320-326.

Huang, K.-P. (1990) BioFactors 2, 171-178.

Kishimoto, A., Mikawa, K., Hashimoto, K., Tasuda, I., Tanaka, S., Tominaga, M., Kuroda, T., and Nishizuka, T. (1989) J. Biol. Chem. 264, 4088-4092.

Liu, Y. and Storm, D.R. (1990) Trends Pharmacol. Sci. 11, 107-111.

Malviya, A.N., Rogue, P., and Vincendon, G. (1990) Proc. Natl. Acad. Sci. USA 86, 453-457.

Naka, M., Saitoh, M., and Hidaka, H. (1988) Arch. Biochem. Biophys. 261,235-240.

Nakabayashi, H., Sellers, J.R., and Huang, K.-P. (1991) FEBS Lett. in press.

Naor, A., Shearman, M.S., Kishimoto, A. and Nishizuka, Y. (1988) Mol. Endocrinol. 2, 1043-1048.

230

Nicotera, P., McConkey, D.J., Jones, D.P., and Orrenius, S. (1989)
    Proc. Natl. Acad. Sci. USA 86, 453-457.
Nishikawa, M., Sellers, J.R., Adelstein, R.S., and Hidaka, H.
    (1984) J. Biol. Chem. 259, 8808-8814.
Nishizuka, Y. (1988) Nature 334,661-665.
Pontremoli, S., Melloni, E., Michetti, M., Sacco,O., Salamino, F.,
    Sparatore, B., and Horecker, B.L. (1986) J. Biol. Chem. 261,
    8309-8313.
Represa, A., Deloulme, J.C., Sensenbrenner, M., Ben-Ari, Y., and
    Baudier, J. (1990) J. Neurosci. 10, 3782-3791.
Rogue, P., Labourdette, G., Masmoudi, A., Yoshida, T., Huang, F.L.,
    Huang, K.-P., Zwiller, J., Vincendon, G., and Malviya, A.N.
    (1990) J. Biol. Chem. 265, 4061-4065.
Schaap, D., Hsuan, J., Totty, N., and Parker, P.J. (1990) Eur. J.
    Biochem. 191, 431-435.
Sellers, J.R., Pato, M.D., and Adelstein, R.S. (1981) J. Biol.
    Chem. 256, 13137-13142.
Shinomura, T., Asaoka, Y., Yoshida, K., and Nishizuka, Y. (1991)
    Proc. Natl. Acad. Sci. USA 88, 5149-5153.
Stumpo, D.J., Graff, J.M., Albert, K.A., Greengard, P., and
    Blackshear, P.J. (1989) Proc. Natl. Acad. Sci. USA 86, 4012-4016.
Tapley, P.M. and Murray, A.M. (1985) Eur. J. Biochem. 151, 419-423.
Tusupov, O.K., Severin, S.E., and Shuets, V.I. (1991) Biochem.
    Biophys. Res. Commun. 176, 1007-1013.
Vinson, C.R., Sigler,P.B., and McKnight, S.L. (1989) Science 246,
    911-916.
Watson, J.B., Battenberg, E.F., Wong, K.K., Bloom, F.E., and
    Sutcliffe, J.G. (1990) J. Neurosci. Res. 26, 397-408.

Adenine Nucleotides in Cellular Energy Transfer and Signal Transduction
S. Papa, A. Azzi & J.M. Tager (eds)
© 1992 Birkhäuser Verlag, Basel/Switzerland

STRUCTURAL STUDIES ON THE PROTEIN KINASE C GENE FAMILY

Cathy Pears, Nigel Goode and Peter J. Parker

Imperial Cancer Research Fund, Lincolns Inn Fields, London WC2A 3PX

SUMMARY.

Protein kinase C (PKC) consists of a family of at least nine members
which can be subdivided into two main groups on the basis of sequence
homology. Modification is required for the enzyme to be functional as
the primary translation product (~77kD) appears to express no kinase
activity. PKC isolated from tissue sources is a phosphoprotein of a
higher molecular weight (~80kD). Use of a kinase deficient mutant of
PKC indicates that the primary translation product undergoes an
initial phosphorylation by a separate kinase with subsequent
autophosphorylation to generate the "mature" protein.   The primary
translation products expressed in E. coli bind phorbol esters with
apparently normal affinity, consistent with this family forming a
major class of phorbol ester receptors. Isolation of the individual
isotypes of PKC from tissue sources or after expression of cDNA
sequences in mammalian cells has revealed differences in substrate
specificity and effector dependence of kinase activity. The fusion of
regulatory and catalytic domains from different isotypes has revealed
that the regions of the protein responsible for the limited substrate
range of PKC-ε in vitro lie within the regulatory domain of the
enzyme, suggesting a role for this region in limiting access to the
active site. The substrate selectivity may be extended towards a
physiologically relevant substrate. Activated versions of the various
isotypes of PKC are being constructed by mutagenesis in order to study
the functions of the individual isotypes following expression in
mammalian cells.

Protein kinase C (PKC) consists of a family of serine/threonine

protein kinases which play a major role in signal transduction in

response to the production of diacylglycerol (reviewed in Nishizuka,

1988; Parker et al., 1989). The primary translation product, as

generated in E. coli or in a reticulocyte lysate system, is a 76

kilodalton (kD) protein which displays apparently normal phorbol ester

binding activity (Cazaubon et al., 1990; Stabel and Parker,

unpublished observations) but which is devoid of kinase activity. The active kinase, as purified from tissue sources, is usually a mixture of PKC-α, -β and -γ and has an apparent molecular mass of 80 kD. This is consistent with the primary translation product undergoing post-translational modification in the cell to activate the kinase activity and to cause an apparent increase in molecular weight. This is confirmed by pulse-chase experiments in which the initial 76 kD product is chased into a higher molecular weight form over a period of about 30 minutes (Pears et al., 1991a). Purified PKC is a phosphoprotein and treatment with potato acid phosphatase causes an increase in migration rate on gel electrophoresis (Woodgett et al., 1987) consistent with the post-translational alteration in molecular mass being due to the incorporation of phosphate into PKC. However, phosphatase treatment does not significantly alter the kinase activity of purified PKC-α towards exogenous substrate (figure 1). Inhibition of the phosphatase and incubation of the 76 kD form of PKC with effectors leads to a reversal of this molecular weight shift (Pears et al., 1991a), indicating that the apparent size increase in molecular mass is due to autophosphorylation. This is confirmed by pulse-

Effect of potato acid phosphatase on PKC activity

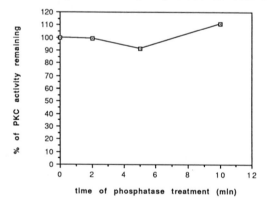

Figure 1. PKC-α was dephosphorylated with potato acid phosphatase, diluted and assayed in the presence of effectors and histone III S as substrate. The percentage of PKC activity remaining is plotted against the time of phosphatase treatment.

labelling experiments of sf9 insect cells overexpressing PKC-α using the baculovirus expression system. When cells expressing wild-type PKC-α are labelled for 10 minutes with ^{35}S-methionine, the initial translation product has an apparent molecular mass of 76 kD. Chasing with an excess of unlabelled methionine leads to an apparent increase to an 80 kD form (figure 2). When a point mutation is introduced to replace a conserved lysine residue found at the ATP binding site of PKC-α the resulting protein displays no detectable kinase activity (Pears and Parker, 1991). Similar pulse-chase experiments with cells expressing this mutant version of PKC-α reveal no apparent increase in molecular mass confirming that this is due to autophosphorylation (figure 2).

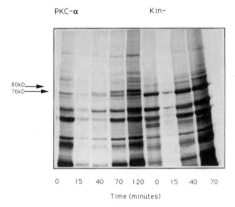

PKC-α   Kin-

80kD
76kD

0   15   40   70   120   0   15   40   70

Time (minutes)

Figure 2. Sf9 insect cells, 2 days after infection with recombinant baculovirus expressing either wild-type or kinase deficient (kin-) PKC-α were labelled with ^{35}S-methionine for 10 minutes and then chased with an excess of cold methionine for the times shown. Total protein was then resolved by electrophoresis and visualised by autoradiography. The position of the primary and mature PKC products is marked with arrows.

Evidence that the activation of PKC is due to a transphosphorylation reaction comes from expression of the kinase deficient PKC-α in COS cells. The cells were transfected with plasmid encoding PKC-α or the kinase deficient version and incorporation of ^{32}P-orthophosphate was visualized by immunoprecipitation and autoradiography. This reveals that both versions of PKC are

234

phosphoproteins (Pears et al., 1991a). As the mutant form has no intrinsic kinase activity, it must be the substrate for another kinase present in the cell, although it is theoretically possible that the endogenous PKC-α in the COS cells is responsible. However, if ^{35}S-methionine pulse-chase experiments are carried out in the presence of phorbol esters the rate of shift in molecular weight is decreased (Pears et al., 1991a). As phorbol esters stimulate the kinase activity of PKC the initial phosphorylation reaction responsible for activating the intrinsic PKC activity must be carried out by another kinase whose activity is inhibited in the presence of TPA.

All this data is, therefore, consistent with the scheme illustrated in figure 3. The primary translation product of PKC-α is a 76 kD protein which is phosphorylated by a "PKC kinase". This leads to activation of the PKC kinase activity and subsequent autophosphorylation generates the 80 kD mature protein. This initial transphosphorylation reaction identifies a potentially new control point in the generation of active PKC as the activity is repressed in

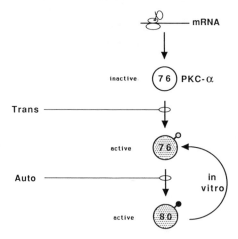

Figure 3. Schematic representation of the post-translational modifications of PKC.

the presence of PKC activators. The accumulation of an inactive precursor form of PKC-β on phorbol ester treatment of cells has been reported (Borner et al., 1988). It is not known whether the inhibition of this kinase reaction is due to a direct feed-back mechanism wherein activation of PKC leads to a decrease in the activity of this second kinase or whether this decreased rate is due to an indirect effect, such as sequestration of the PKC to the membrane. A kinase activity which will activate bacterially expressed PKC-γ has been identified (Cazaubon and Parker, unpublished observations) allowing these questions to be addressed. This system will also be used to identify the sites of auto- and transphophorylation of PKC and their subsequent mutation will reveal the role of these modifications in the biological function of the enzyme.

PKC consists of a family of closely related proteins (reviewed in Nishizuka, 1988; Parker et al., 1989) with at least six genes identified to date and alternative splicing increasing the total number of gene products (Parker et al., 1986; Coussens et al., 1986; Ono et al., 1987; Coussens et al., 1987; Ono et al., 1989a; Osada et al., 1990). Comparison of the amino acid sequences deduced from cDNA clones has revealed the presence of four highly conserved regions (C_{1-4}) interspersed by five more variable ones. The C_3 and C_4 regions contain sequences conserved in all protein kinases (Hanks et al., 1988) and this C-terminal region of the protein encodes the catalytic activity. The C_1 domain contains two copies of a cysteine rich repeat or so-called zinc finger motif and encodes the phorbol ester binding activity (Ono et al., 1989b; Cazaubon et al., 1990). PKC-ζ only contains a single copy of this repeat and has yet to be shown to respond to phorbol esters (Ono et al., 1989a; Ways et al., submitted). The C_2 region is only found in PKC-α, -β and -γ where it is thought to confer the calcium dependency of these isotypes. Similar regions are conserved in other proteins and a homologous sequence in phospholipase A_2 has been shown to be responsible for the calcium-dependent membrane translocation of this protein (Clark et al., 1991). PKC-ϵ, -δ, -ζ and -

η have not as yet been purified from tissue sources but biochemical analysis of the products of expression of the cDNA sequences of PKC-δ and -ε confirm that they are independent of calcium for activation (Schaap et al., 1989; Ono et al., 1989; Olivier and Parker, 1991). The arrangement of these conserved sequences in PKC has led to the postulation of a two domain structure for the protein with the N-terminal regulatory domain normally acting to inhibit the C-terminal kinase domain. The two domains are linked by the V3 hinge region which can be cleaved by partial proteolysis with trypsin or calpain *in vitro* to generate a 30 kD phorbol ester binding protein and a 50 kD constitutively active kinase (Lee and Bell., 1986; Huang et al., 1989; Young et al., 1988; Kishimoto et al., 1989).

The inhibition of the constitutive kinase activity of PKC in its unstimulated state is thought to be mediated by the interaction of a short stretch of peptide sequence found just upstream of the C_1 domain with the active site. This so-called pseudosubstrate sequence has the distribution of basic residues found surrounding sites known to be phosphorylated by PKC, but contains an alanine residue instead of a serine or threonine. A peptide spanning this region is a potent competitive inhibitor of PKC-α *in vitro* (House and Kemp, 1987) and the inclusion of a serine in place of the alanine generates an effective substrate. This theory was tested by the mutagenesis of this sequence in PKC-α to include a charged glutamic acid residue (Pears et al., 1990) to generate E25PKC-α. A peptide with this alteration has a greatly reduced ability to act as a competitive inhibitor and inclusion of this residue in the intact protein causes increased effector-independent activity towards both histone and a peptide substrate compared to wild-type PKC-α (figure 4). This confirms the importance of this sequence in maintaining PKC in its inactive state.

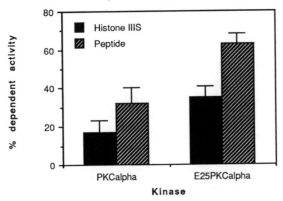

Figure 4. PKC activity was determined using two independent preparations of PKC-α and E25PKC-α partially purified after overexpression using the baculovirus system, for both histone III S and a peptide based on the pseudosubstrate site of PKC-α as substrate. The activity is expressed as a percentage of that seen towards the same substrate in the presence of optimal concentrations of effectors.

The effector independence is greater for the smaller substrate suggesting that other regions are involved in the interaction between the two domains. Mutagenesis of the pseudosubstrate site does leave the active site free to allow substrate to bind but a smaller peptide substrate can diffuse more easily into the active site than a polypeptide substrate (figure 4). This increased effector independence was confirmed in the intact cell by determining the ability of this mutant version of PKC to increase expression of a reporter constuct when co-transfected into mammalian cells. Deletion of seven amino acids spanning the pseudosubstrate site also leads to increased constitutive activity when assayed in this way (Pears et al., 1990).

The role of the activation of the various isotypes of PKC in the reported biological effects of phorbol esters treatment is being investigated using mutant versions of all the isotypes of PKC activated by mutagenesis of their pseudosubstrate sites (Ways, Kiley, Pears and Parker, unpublished observations). Mammalian cells often contain more than one isotype in various amounts and phorbol ester treatment could lead to selective activation or down-regulation. The identification of other proteins which bind phorbol esters (Ahmed et al., 1990; Maruyama and Brenner, 1991) also raises the possibility that phorbol ester treatment may alter their functions, which are as yet undefined, and some of the biological effects may be independent of PKC activation. The introduction of individual or combinations of activated PKC isotypes into cells should allow the roles of the isotypes to be distinguished and the effects of PKC activation to be investigated in the absence of phorbol ester treatment.

Isolation of the individual isotypes of PKC from tissue sources or after overexpression of cDNA sequences has revealed differences in effector dependence and substrate specificity of kinase activity (Marais and Parker, 1989; Schaap et al., 1989; Ono et al., 1989a; Olivier and Parker, 1991). In particular PKC-ε and -δ show a greatly reduced *in vitro* substrate range and are unable, for example, to phosphorylate histone. The structural basis for this substrate selectivity was investigated by creating a chimaeric protein containing the regulatory sequences of PKC-ε fused to the catalytic domain of PKC-γ. The substrate preference was investigated after transient expression in COS cells (Pears et al., 1991b). PKC-ε and PKC-ε/γ were separated from the endogenous PKC-α by ammonium sulphate fractionation and their ability to phosphorylate histone relative to a peptide substrate known to be efficiently phosphorylated by both parent isotypes and the chimaera was determined (figure 5). The inability of this chimaeric protein to use histone as a substrate indicates that the selectivity of PKC-ε is determined by the regulatory domain and suggests that, even after activation, the regulatory domain still interacts with the catalytic domain so as to monitor entry into the active site. This is confirmed by the partial proteolysis of PKC-ε *in vitro* which cleaves in the $V_{2/3}$ region and which generates a constitutively active kinase domain fragment which is capable of efficiently phosphorylating histone (Schaap et al., 1990). The most likely region for conferring this substrate limitation is the extended V_1 (or so-called V_0) region found upstream of the C1 domain in PKC-ε and -δ.

Relative activity of PKC subtypes

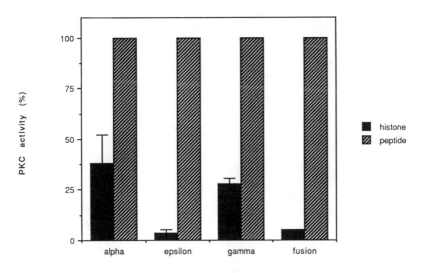

Figure 5. The ability of the fusion protein PKC-ε/γ to phosphorylate histone is compared to the ability of PKC-ε and -γ. After transient expression in COS cells the PKC activities were separated from the endogenous PKC-α by ammonium sulphate fractionation. The activity towards histone is plotted as a percentage of that shown towards a peptide substrate which is efficiently phosphorylated by both isotypes.

240

This selective substrate range may extend to a physiological substrate. Glycogen synthase kinase-3 (GSK-3) may be responsible for the phosphorylation of the proto-oncogene c-jun at sites which inhibit its ability to bind to DNA and hence to increase transcription of phorbol ester responsive genes (Woodgett, 1991). Phosphorylation of GSK-3 by PKC-α leads to the incorporation of up to 1.5 moles of phosphate per mole of GSK-3 and this reaction inhibits the ability of GSK-3 to phosphorylate c-jun (Goode, Hughes, Woodgett and Parker, manuscript in preparation). The ability of PKC to phosphorylate GSK-3 is isotype dependent, so perhaps identifying a physiologically relevant divergence in function between different PKC isotypes.

Ahmed, S., Kozma, R., Monfries, C., Hall, C., Lim, H.H., Smith, P and Lim, L. (1990) Biochem. J. 272, 767-773

Borner, C., Eppenberger, U., Wyss, R. and Fabbro, D. (1988) Proc. Natl. Acad. Sci. USA 85, 2110-2114

Cazaubon, S., Webster, C., Camoin, L., Strosberg, A.D. and Parker, P.J. (1990) Eur. J. Biochem. 194, 799-804

Clark, J.D., Lin, L.L., Kriz, R.W., Ramesha, C.S., Sultzman, L.A., Lin, A.Y., Milona, N. and Knopf, J. (1991) Cell 65, 1043-1052

Coussens, L., Parker, P.J., Rhee, L., Yang-Feng, T.L., Chen, E., Waterfield, M.D., Franke, U. and Ullrich, A. (1986) Science 233, 859-866

Coussens, L., Rhee, L., Parker, P.J. and Ullrich, A. (1987) DNA 6, 389-394

Hanks, S.K., Quinn, A.M. and Hunter, T. (1988) Science 241, 42-52

House, C. and Kemp, B.E. (1987) Science 238, 1726-1728

Huang, F.L., Yoshida, Y., Cunha-Melo, J.R., Beaven, M.A. and Huang, K.-P. (1989) J. Biol. Chem. 264, 4238-4243

Kishimoto, A., Mikawa, K., Hashimoto, K., Yasuda, I., Tanaka, S., Tominaga, M., Kuroda, T. and Nishizuka, Y. (1989) J. Biol. Chem. 264, 4088-4092

Lee, M.-H. and Bell, R.M. (1986) J. Biol. Chem. 261, 14867-14870

Marais, R.M. and Parker, P.J. (1989) Eur. J. Biochem. 182, 129-137

Maruyama, I.N. and Brenner, S. (1991) Proc. Natl. Acad. Sci. USA 88, 5729-5733

Nishizuka, Y. (1988) Nature 334, 661-665

Olivier, A. and Parker, P.J. (1991) Eur. J. Biochem. in press

Ono, Y., Fujii, T., Ogita, K., Kikkawa, U., Igarashi, K. and Nishizuka, Y. (1987) FEBS Letts 226, 125-128

Ono, Y., Fujii, T., Ogita, K., Kikkawa, U., Igarashi, K. and Nishizuka, Y. (1989a) Proc. Natl. Acad. Sci. USA 86, 3099-3103

Ono, Y., Fujii, T., Igarashi, K., Kuno, T., Tanaka, C., Kikkawa, U. and Nishizuka, Y. (1989b) Proc. Natl. Acad. Sci. USA 86, 4868-4871

Osada, S., Mizuno, K., Saido, T.C., Akita, Y., Suzuki, K., Kuroki, T. and Ohno, S. (1990) J. Biol. Chem. 265, 22434-22440

Parker, P.J., Coussens, L., Totty, N., Rhee, L., Young, S., Chen, E., Stabel, S., Waterfield, M.D. and Ullrich A. (1986) Science <u>233</u>, 853-859

Parker, P.J., Kour, G., Marais, R.M., Mitchell, F., Pears, C.J., Schaap, D., Stabel, S. and Webster, C. (1989) Mol. Cell. Endocrinol. <u>65</u>, 1-11

Pears, C.J., Kour, G., House, C., Kemp, B.E. and Parker, P.J. (1990) Eur. J. Biochem. <u>194</u>, 89-94

Pears, C.J., Stabel, S. and Parker, P.J (1991a) Manuscript submitted

Pears, C.J. and Parker, P.J. (1991) FEBS Letts. <u>284</u>, 120-122

Schaap, D., Parker, P.J., Bristol, A., Kriz, R and Knopf, J. (1989) FEBS Lett. <u>243</u>, 351-357

Schaap, D., Hsuan, J., Totty, N. and Parker, P.J. (1990) Eur. J. Biochem. <u>191</u>, 431-435

Woodgett, J. (1991) TIBS <u>16</u>, 177-181

Woodgett, J. and Hunter, T. (1987) J. Biol. Chem. <u>262</u>, 4836-4843

Young, S., Rothbard, J. and Parker, P.J. (1988) Eur., J. Biochem. <u>173</u>, 247-252

Adenine Nucleotides in Cellular Energy Transfer and Signal Transduction
S. Papa, A. Azzi & J.M. Tager (eds)
© 1992 Birkhäuser Verlag, Basel/Switzerland

α-TOCOPHEROL INHIBITS SMOOTH MUSCLE CELL PROLIFERATION AND PROTECTS AGAINST ARTERIOSCLEROSIS

Angelo Azzi, Daniel Boscoboinik, Gianna-M. Bartoli and Eric Chatelain

Institut für Biochemie und Molekularbiologie Universität Bern, CH-3012 Bern, Switzerland.

SUMMARY: The target of α-tocopherol action has been investigated in smooth muscle cells. The inhibitory effect of α-tocopherol on protein kinase C correlated with the inhibition of cell proliferation. A specific arrest of growth by α-tocopherol could be localized at the G_1/S boundary and was characterized by both an inhibition of protein kinase C activity and an increase of kinase C expression in the late G_1 phase.

Vascular muscle smooth cell (VSMC) proliferation is a central phenomenon in the onset of arteriosclerosis and hypertension (1-4). Serum, platelet-derived growth factor (PDGF), epidermal growth factor (EGF), angiotensin , interleukin-1 and endothelin (5-10) can all stimulate VSMC growth by producing molecular effects such as increase in cytosolic calcium and pH, phosphatidylinositol hydrolysis and synthesis of diacylglycerol with consequent protein kinase C activation, increased formation of *c-fos* mRNA. etc. These secondary events activate DNA synthesis and cell division (11). Heparin (12), prostaglandins (13), staurosporine (14), nisoldipin (15), and TMB-8 (16) have been shown to inhibit cell proliferation, by acting at some stage of the signal transduction pathway. α-Tocopherol as well, at physiological concentrations, inhibits smooth muscle cell proliferation possibly through a modulation of protein kinase C activity (17). Tocopherols are known to affect cellular functions (18) through their general radical scavenging activity (19), but they have also been shown to stabilize membranes (20), to inhibit growth in several types of cells (21-23) and to play a role in prostaglandin synthesis (24).

We have studied now the point of action of α-tocopherol in relation with the cell cycle (25,26) and the implication of PKC in the inhibition of cell proliferation (27-28).

MATERIALS AND METHODS

Tissue culture plastics were purchased from Falcon Labware (Becton Dickinson & Co.) and growth media and serum for cell culturing were obtained from Gibco Laboratories (Grand Island, NY). [Methyl-^{3}H]thymidine (25 Ci/mmol) and [γ-^{32}P]ATP (10 Ci/mmol) were from Amersham International. d-α-, and d-β-tocopherol were generous gifts from Hoffmann La Roche & Co (Basel, Switzerland) and Henkel Co (La Grange, IL, USA). The structural formulae of the compounds are shown in Fig. 1

A specific peptide (Bachem) Pro-Leu-Ser-Arg-Thr-Leu-Ser-Val-Ala-Ala-Lys-Lys, was used as substrate for the assay of PKC activity (29). Streptolysin-O (25,000 units) and PMA were from Calbiochem. A7r5 cells (rat aortic smooth muscle cell line, VSMC, clone DB1X from American Type Culture Collection) were grown in Dulbecco's modified Eagle medium (DMEM) containing 25 mM-bicarbonate, 60 U/ml penicillin, 60 µg/ml streptomycin, and 10% fetal calf serum (FCS). Cells were seeded into 100-mm dishes and grown to confluence at 37°C in a humidified atmosphere of 5% CO_2. Culture media were changed every 3 days. Synchronization at the G_1/S boundary was obtained by serum deprivation and hydroxyurea treatment [20]. Exponentially growing cells (in 6-well plates containing 1.5 ml of DMEM with 2% FCS), were made quiescent (G_o) by treatment with 0.2% FCS for 48 h. After re-stimulation by 2% FCS during 8 h, 1.5 mM hydroxyurea was added to each plate. After 14 h of treatment cells were washed with PBS and transferred to the fresh complete medium (DMEM, 2% FCS).Onset and duration of the S-phase was determined by pulse labelling with [^{3}H]thymidine. A7r5 VSMC cell number remained unchanged during 22 h from the stimulation and then cell divided, thus making a cycle in around 24 h. Addition of tocopherols to cell cultures was made according to the following procedure. Given amounts of ethanolic α- and β-Tocopherol solutions were transferred to a test tube, the ethanol was evaporated by a stream of dry nitrogen, serum was added and vigorously stirred in a Vortex mixer. Aliquots of this serum corresponding to given tocopherol amounts were added to the tissue cultures. Alternatively tocopherols were added as ethanol solution. The final ethanol concentration during cell growth never exceeded 0.5% and the same amount of ethanol was added to control cells. Compounds were added when restimulating the cells with serum following 48 h deprivation. Cells were trypsinized and counted in a hemocytometer in triplicate after the completion of the cell cycle (approximately 24 h). Viability was assessed by the trypan blue dye exclusion method. Measurements of PKC activity in permeabilised smooth muscle cells were performed according to the procedure of Alexander et al. [21] with minor modifications. A7r5 cells in the late G_1 phase of the cycle, preincubated for 8 h in the presence of the indicated tocopherol homologue, were washed twice in PBS, resuspended in intracellular buffer (5.16 mM-$MgCl_2$, 94 mM-KCl, 12.5 mM-

Hepes, 12.5 mM-EGTA, 8.17 mM-$CaCl_2$, pH 7.4) and aliquoted in 220 μl portions ($1.5x10^6$ cells/ml). Assays were started by adding [γ-^{32}P]ATP (40 cpm/pmol, final concentration 240 μM), peptide substrate (final concentration 250 μM) and Streptolysin-O (0.6 I.U.).

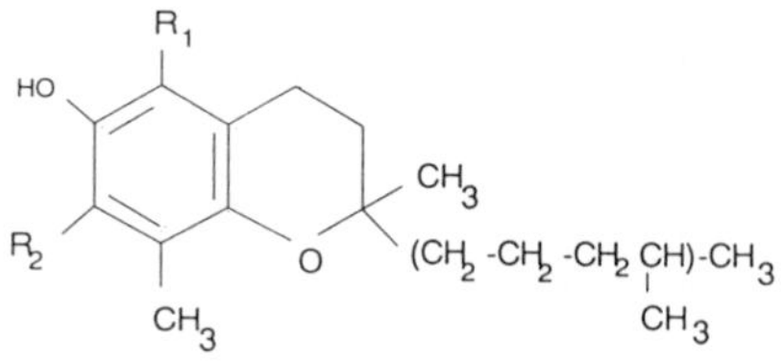

	R_1	R_2
α-tocopherol	CH_3	CH_3
β-tocopherol	CH_3	H

Fig 1: Structural formulae of α-tocopherol and β-tocopherol

The reaction mixtures were incubated at 37°C for 5 min and the reaction was stopped by adding 100 μl of 25% (w/v) trichloroacetic acid in 2 M-acetic acid. After 10 min on ice, samples were centrifuged for 5 min and spotted on P81 ion-exchange chromatography paper (Whatman International) which were then washed several times with 30% (v/v) acetic acid containing 1% (v/v) H_3PO_4 and once with ethanol. The P81 papers were dried, and the bound radioactivity was counted in a liquid scintillation analyzer. To estimate the background phosphorylation of the peptide due to a kinase activity other than PKC, assays were performed in cells treated for 24 h with 1 μM PMA.

Determination of the uptake of α- and β-, tocopherol in VSMC: Cells were incubated 24 h in complete medium with 100 μM of α- or β-tocopherols. After washing, measurements of tocopherol content were performed by reverse phase HPLC using a C-18 column (Waters, Inc. with an in-line electrochemical detector essentially as described earlier [25]. The eluent was 20 mM-lithium perchlorate in methanol-ethanol (1:9) (v.v).

Synchronization of cultured smooth muscle cells: Vascular smooth muscle cells growing exponentially in media containing FCS were made quiescent by incubating them for 48 h in low-serum media. Under these conditions, little amount of [³H]-thymidine was incorporated into DNA (17) but cells could be restimulated to growth when medium was supplemented with FCS or growth-factors were added.

To determine the onset and duration of the different phases of the cell cycle, cells were pulsed for 1 h with [³H]thymidine at several times after releasing from the quiescent state (G_0). [³H]thymidine incorporation into DNA started to increase approximately 11 h after addition of the stimulus and reached the maximal value 5-6 h later. To analyze cell cycle parameters and potential targets for α-tocopherol growth inhibition, cells were synchronized at the G_1/S boundary by using a combination of serum deprivation and hydroxyurea (HU) treatment. Upon removal of HU, cells entered synchronously into the S-phase (30).

RESULTS AND DISCUSSION

Effect of α- and β-tocopherol on VSMC proliferation: The concentration dependence of the inhibition of α- and β-tocopherol on the proliferation induced by fetal calf serum is shown in Fig. 2. At a concentration of 10 μM, α-tocopherol almost completely inhibited the FCS-induced growth of VSMC; no effect of β-tocopherol was observed. [³H]thymidine incorporation into DNA (17) was equally inhibited by α-tocopherol. Cell viability was always found to be greater than 95% as assessed by the trypan blue dye exclusion method.

248

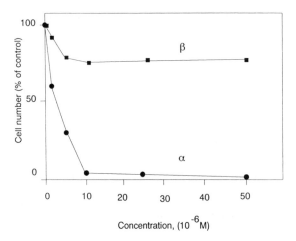

Fig. 2: The effect of α- and β-tocopherol on smooth muscle cell proliferation

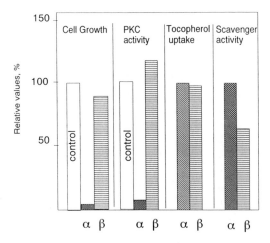

Fig. 3: The effect of α- and β-tocopherol on smooth muscle cell proliferation and protein kinase C activity. Cells were incubated with 100 μM of the tocopherol isomers as indicated and PKC activity was measured as described in Methods.

Previous data, showing inhibition of smooth muscle cell growth at somewhat higher tocopherol concentrations were obtained with ethanolic solutions of tocopherol. It appears that addition after adsorption on serum represents a more efficient way of administration.

To establish if the differential effects of the tocopherol isomers on smooth muscle cells proliferation were due to their unequal uptake, the amount of these compounds present in 24h-treated VSMC was measured. As shown the content of added α- and β-tocopherols was similar in smooth muscle cells. The amount of endogenous tocopherols present in untreated cells was found to be less than 10 pmol/10^6cells.

On the other hand the lack of effect of β-tocopherol compared with the growth inhibition obtained with α-tocopherol could not be attributed to a lack of antioxidant activity. In fact, although slightly less potent than α-, β-tocopherol is an excellent radical scavenger.

Thus, it appears that the inhibitory effect of α- and β-tocopherol on cell proliferation had precise structural requirements and did not depend on either their scavenging capacity or their transport efficiency into cells.

Effect of α- and β-tocopherol on PKC activity in permeabilized VSMC: PKC activity was previously shown to be inhibited by α-tocopherol (17). In order to study this reaction at a cellular level, streptolysin-O permeabilized VSMC and a PKC peptide substrate (29) were employed.

As can be seen in Fig. 2 and 3, α-tocopherol inhibited more than 95% PKC activity whereas β-tocopherol was ineffective.

The inhibition was observed regardless of whether the cells were preincubated with α-tocopherol or whether the compound was added just before carrying out the kinase C assay. This result suggests a direct specific interaction of PKC with α-tocopherol.

250

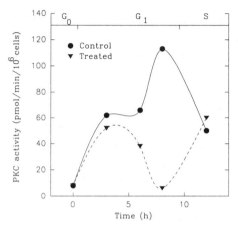

<u>Fig. 4:</u> Effect of α-tocopherol on PKC activity during the
cell cycle. Control (●) and α-tocopherol-treated (▼)
cultures were taken at different stages of the cell cycle
and activity of protein kinase C in permeabilized VSMC was
determined as described in Methods. Results are the mean ±
SD of three experiments.

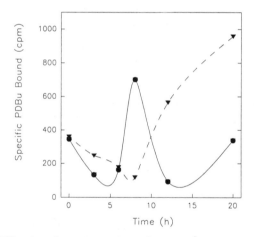

<u>Fig. 5:</u>. Effect of α-tocopherol on [³H]PDBu binding during
the cell cycle. Phorbol binding was measured as described in
Methods in cells at different stages of the cycle. Non-
specific binding was determined by using an excess of cold
PDBu. (●) Control, (▼) 100 μM α-tocopherol-treated
cells.

α-Tocopherol inhibition of growth occurs in late G_1 phase of the cell cycle: The inhibitory effect of α-tocopherol on smooth muscle cells-DNA synthesis was studied in greater detail by adding the compound at different times after stimulation of proliferation. The inhibitory response was observed when α-tocopherol was added either at the time of stimulation or at any time during the first 8 h in the G_1 phase. This result indicates that targets for α-tocopherol-mediated inhibition are not present in the early G_o/G_1 transition. However, when α-tocopherol was added 9 h after the initial stimulation, no inhibitory effect on DNA synthesis was seen, indicating that once the cells had progressed beyond commitment to DNA synthesis (around 8 h after stimulation), they became insensitive to the α-tocopherol effects.

Late G_1 phase inhibition of PKC activity by α-tocopherol in permeabilized VSMC: It is not known how kinase C activity is regulated during the smooth muscle cell cycle although it is speculated that both phosphorylation and dephosphorylation reactions may occur (31). Protein kinase C activity was assayed at various stages of the cell cycle in permeabilized smooth muscle cells using a 12-aminoacid peptide as a substrate, specific for kinase C. Maximal activity was observed during late G_1 phase of the cycle (Fig 4). Maximal inhibition by 50 μM α-tocopherol was also observed in the late G_1 phase.

Effect of α-tocopherol on [³H]PDBu binding to VSMC during the cell cycle: Protein kinase C is the major if not the only cellular target for phorbol esters as shown by radioactive phorbol 12,13-dibutyrate binding to a high-affinity specific receptors in a variety of tissues (32).In smooth muscle cells, [³H]PDBu binds to a homogeneous class of binding sites as determined by Scatchard analysis. We analyzed the binding of tritiated PDBu to protein kinase C during the cell cycle and an increase in binding was found during the last part of the G_1 phase concomitantly with the peak of kinase activity. This increase in phorbol binding was cycloheximide sensitive, indicating de novo synthesis of protein kinase C during the cycle. The increase in [^{3}H]PDBu binding was strongly inhibited by α-tocopherol in late G_1 phase but it was stimulated during the S and G_2. α-tocopherol may thus be involved in the regulation of the biosynthesis of protein kinase C. (Fig. 5).

The role of smooth muscle cell proliferation in the onset of arteriosclerosis and the protecting effect of α-tocopherol: Proliferation of smooth muscle cells, has a central role in the arteriosclerosis process. We have described above and previously that proliferation of smooth muscle cells in vitro is under the control of α-tocopherol. It is also established that the extent of arteriosclerosis lesions responsible for a clinical manifestation (ischemic heart disease mortality) inversely correlate with the plasma α-tocopherol concentration (33) (Fig. 6). α-Tocopherol may be thus placed in a central position in the onset of this degenerative disease (Fig. 7). As a major regulator of smooth muscle cell proliferation all events leading to a decrease of this compound may result in smooth muscle cell growth. The possible causes of a low blood α-tocopherol may be dietetic in nature or may be caused by the destructive effect of radicals on the compound itself. Thus radicals may have a damaging role at a cellular level both in activating protein kinase C and diminishing α-tocopherol level, both events resulting in a stimulation of smooth muscle cell proliferation and occlusive arterial phenomena.

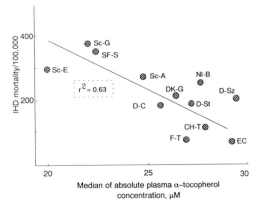

<u>Fig. 6:</u> Inverse correlation between age specific ischemic heart disease mortality and α-tocopherol concentration as detected in different populations. CH-T, Switzerland: Thun; D-C,-St and -Sz, Germany: Cottbus, Schwedt and Schleiz; DK-G Denmark: Glostrup; NI-B, Northern Ireland: Belfast; SC-A, - E, -G , Scotland: Aberdeen, Edinburgh and Glasgow; SF-S , Finland: Southwest region; E-C, Spain: Catalunia; F-T, France: Toulouse (redrawn from ref.33)

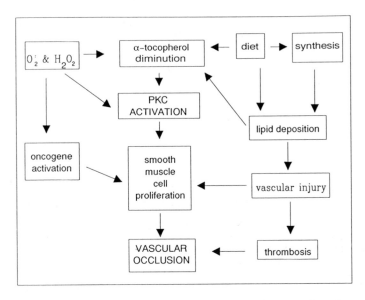

<u>Fig. 7:</u>. A unifying pathogenetic model of arteriosclerosis.

254

Acknowledgments: This work was supported by F. Hoffmann-La Roche AG and the Palm Oil Research Institute of Malaysia (PORIM). We greatly appreciated the helpful assistance of Ms. Barbara Conti and Mr. Peter Schmutz on part of the experiments described.

REFERENCES

1. Ross, R. (1986) N. Engl. J. Med. 314: 488-500.
2. Munro, J. M. and Cochran, R. S. (1988) Lab Invest. 58: 249-261.
3. Owens, G. K. and Reidy, M. A. (1985) Circ. Res. 57: 695-705.
4. Schwartz, S. M., Campbell, G. R. and Campbell, J. H. (1986) Circ. Res. 58: 427-444.
5. Ross, R., Glomset, J. A., Kariya, B. and Harker, L. (1974) Proc. Natl. Acad. Sci. USA 71: 1207-1210.
6. Gospodarowicz, D., Moran, J. S. and Braun, D. L. (1977) J. Cell. Physiol. 91: 377-385.
7. Geisterfer, A. A., Peach, M. J. and Owens, G. K. (1988) Circ. Res. 62: 749-756.
8. Raines, E. W., Dower, S. K. and Ross, R. (1989) Science 243: 393.
9. Ku, G., Doherty, N. S., Wolos, J. A. and Jackson, R. L. (1988) Am. J. Cardiol. 62: 77B.
10. Bobik, A., Grooms, A., Miller, J. A., Mitchell, A. and Grinpukel, S. (1990) Am. J. Physiol. 258: C408-C415.
11. Rozengurt, E. (1986) Science 234: 161-166.
12. Castellot, J. J., Pukac, L. A., Caleb, B. L., Wright, T. C. and Karnovsky, M. J. (1989) J. Cell Biol. 109: 3147-3155.
13. Loesberg, C., van Wijk, R., Zandbergen, J., van Aken, W. G., van Mourik, J. A. and de Groot, Ph. G. (1985) Exp. Cell Res. 160: 117-125.
14. Matsumoto, H. and Sasaki, Y. (1989) Biochem.Biophys.Res.Commun. 158: 105-109.
15. Thyberg, J. and Palmberg, L. (1987) Biol Cell 60: 125-132.
16. Desgranges, C., Campan, M., Gadeau, A-P, Guerineau, N., Mollard, P. and Razaka, G. (1991) Biochem.Pharm. 41: 1045-1054.
17. Boscoboinik, D., Szewczyk, A., Hensey, C. and Azzi, A. (1991) J. Biol. Chem. 266: 6188-6194.
18. Packer, L. and Landvik, S. (1989) Ann. NY Acad. Sci 570: 1-6.
19. Tappel, A. L. (1972) NY Acad. Sci. 203: 12-27.
20. Urano, S., Yano, K. and Matsuo, M. (1988) Biochem.Biophys.Res.Commun. 150: 469-475.
21. Kline, K., Cochran, G. S. and Sanders, B. G. (1990) Nutr.Cancer 14: 27-41.
22. Prasad, K. N. and Edwards-Prasad, J. (1982) Cancer Res 42: 550-555.

23. Slack, R. and Proulx, P. (1989) Nutr.Cancer 12: 75-82.
24. Egan, R. W., Gale, P. H. and Kuehl, F. A. (1979) J. Biol. Chem. 254: 3295-3302.
25. Pardee, A. B. (1989) G_1 events and regulation of cell proliferation. Science 246: 603-608.
26. Campisi, J. E., Medrano, E., Morreo, G. and Pardee, A. B. (1982) Proc. Natl. Acad. Sci USA 79: 436-440.
27. Griendling, K. K., Tsuda, T. and Alexander, R. W. (1989) Biol. Chem. 264: 8237-8240.
28. Ran, W., Dean, M., Levine, R. A., Henckle, C. and Campisi, J. (1986) Proc. Natl. Acad. Sci. USA 83: 8216-8220.
29. Alexander, D. R., Hexham, J. M., Lucas, S. C., Graves, J. D., Cantrell, D. A. and Crumpton, M. J. (1989) Biochem. J. 260: 893-901.
30. Ashihara, T. and Baserga, R. (1979) Methods Enzym. 58: 248-262.
31. Parker, P. J., Kour, G., Marais, R. M., Mitchell, F., Pears, C., Schaap, D., Stabel, S. and Webster, C. (1989) Mol.Cell.Endocrinol. 65: 1-11.
32. Adams, J. C. and Gullick, W. J. (1989) Biochem. J. 257: 905-911.
33. Gey, F.K., Puska, P. Jordan, P. and Moser, U.K. (1991) Am. J. Clin. Nutr. 53: 326S-334S.

Adenine Nucleotides in Cellular Energy Transfer and Signal Transduction
S. Papa, A. Azzi & J.M. Tager (eds)
© 1992 Birkhäuser Verlag, Basel/Switzerland

PROTEIN KINASE C AND G PROTEINS

E. Chiosi, A. Spina, F. Valente* and G. Illiano

3rd Chair of Biological Chemistry, Department of Biochemistry and
Biophysics, I Medical School, University of Naples, Via
Costantinopoli 16, 80138 Napoli, Italy.

*Supported by a AIRC Post-Doctoral Fellowship.

SUMMARY: In crude preparations of human platelet membranes we
have shown a PKC-mediated phosphorylation both of a_i and a_s G
protein subunits. The experiments have been carried out in
conditions of carefully controlled phosphatase activities.
   The proteic targets of the PKC-mediated phosphorylation have
been identified by immunoblotting and immunoprecipitation
experiments.
   The PKC-mediated $[^{32}P]$-labeling has been monitored through
autoradiography both of the immunoblots and of the
immunoprecipitates.
   The phosphorylation time course both of a_i and a_s have been
investigated. The results indicate that they are differentiated
for a_i and a_s.

INTRODUCTION

One of the most widespread mechanisms for modulation of protein conformation and activity is by means of phosphorylation and dephosphorylation. The protein kinase C's (PKC) are among the most active protagonists of such processes. They are serine-threonine kinases, Ca^{2+} and phospholipides (PL) (diacylglycerol/phosphatidylserine) dependent, and play a central role in biologically important cellular processes, such as intracellular transmission and amplification of external signals, cellular differentiation and/or proliferation (Nishizuka 1986). In the presence of Ca^{2+}/PL the PKC switch from a basal state into an active one. The basal state may be represented as an inactive catalytic site unable to recognize the physiological substrate because it is masked by its interaction with the PKC regulatory domain that functions as a "pseudosubstrate". The interaction of the PKC with Ca^{2+}/PL anchors the protein to the membrane lipids and through its conformational change frees the catalytic site to react with the physiological substrates. At least two components make up the general design of PKC. One component is catalytic and able to interact with the ATP and the serine-threonine residues binding sequence. The other is regulatory and binds alternatively either to the catalytic site or to the membrane depending on the absence or the presence of Ca^{2+}/PL.

The PKC reversion from the active to the basal state can be achieved through many mechanisms. The proteolytic cleavage between the catalytic and regulatory subunits has been well described. The catalytic component solubilized in the cytoplasm loses its Ca^{2+}/PL dependence, and its functional role is the object of investigation. A less "traumatic" regulatory mechanism based on the diacylglycerol (DG) modification (i.e. cleavage-inactivation by lipases, phosphorylation by kinases,) has been hypothesized. This regulation is complicated by the expression of many isoenzyme forms, seven being known at present (Nishizuka 1988). These are selectively localized in different tissues and may be activated differentially by DG depending on the fatty acid composition (Huang, et al. 1988).

Our paper concerns the relationship in human platelets between the PKC(s) associated with membranes and its protein targets. Human platelets are very sensitive to external signals, some of which are selectively internalized through the PLC pathway (i.e. thrombin), others through the adenylate cyclase (AC) system (i.e. PGE 1). Both pathways consist of three molecular rings: the peripheral one is the external signal receptor, the internal one is the amplifying system (PLC or AC) and the central one, with a very critical functional role, is the interconnecting G-protein (Birnbaumer et al. 1990; Baldassarre and Fischer 1986). The human platelets have been proved to be a very usefull experimental tool to explore the regulation of the previously mentioned metabolic pathways, as well as the network interconnecting them at various levels (Katada et al. 1985; Crouch et al. 1988).

MATERIALS AND METHODS: Human platelet membrane preparation: Twenty-four hours after blood drawing, the platelets were added with EDTA (5mM final concenration), washed three times with PBS and suspended in the sucrose saline buffer (0.33 M sucrose, 20 mM Tris HCl, 2 mM EDTA, 0.5 mM EGTA, 2 mM PMSF, 5 µg/mL leupeptin, pH 7.5). The platelets were homogenized with a glass-glass Dounce homogenizer (30 strokes) and after centrifugation of the homogenate at 27000 x g x 12 min, the pellet was resuspended with 5 mL of the same buffer without sucrose, and used as enzymic and substrate source.

Membrane phosphorylation: The incubation mixture (in 5 mL final volume) contained: 20 mM Tris HCl pH 7.5, 10 mM magnesium acetate, 5 µg/mL leupeptin, 0.1 mM ATP and platelets membranes (about 1 mg protein/mL incubation mixture). $CaCl_2$ 0.75 mM and phosphatidylserine/diolein (PL) lipid mixture, and $[^{32}P-]$-ATP $(1-2x10^6$ cpm/sample, from New England Nuclear) were added as indicated (see kinase assay). The incubation at 30° C for 3 min, was started by adding the membranes and stopped with 25% TCA 1 mL, unless otherwise indicated. In the immunoprecipitation experiments, the reaction was stopped with 2 volumes of ice cold 50 mM Tris HCl buffer, pH 7.5 enriched with protease and phosphatase inhibitors mixture (10 µg/mL leupeptin, 10 µg/mL pepstatin A, 1 mM PMSF, 5 mM NaF, 0.1 mM vanadate, 1 mM Na-pyrophosphate).

Kinase assay: The incubation mixture (250 µL) was the same used for the membrane phosphorylation and with addition of $[^{32}P-]$-ATP. The kinase C activity was triggered by adding a lipid mixture of 1.6 µg of 1,2 diolein and 24 µg phosphatidylserine in 20 mM Tris HCl pH 7.5; (Spina, et al. 1988; Thomas, et al. 1987). Total protein Kinase C activity is expressed as nmoles of $[^{32}P]$ incorporated/min x mg protein.

Electrophoretic separation of proteins: Proteins from phosphorylated or control membranes were separated by SDS-PAGE on 10% gels, according to Laemmli, 1970.

Immunoblotting: This was performed with Trans-Blot apparatus (BioRad) on Immobilon membranes (Millipore). The anti-a G proteins antibodies were kindly provided by Dr. A.M. Spiegel (NIH - Bethesda) (Simonds et al. 1989; Goldsmith et al. 1987). Goat anti rabbit Ig antibodies, alkaline-phosphatase coniugated, was used as visualization system (in according to the manufacturer's instructions).

Immunoprecipitation: Platelet membranes phosphorylated, after centrifugation at 27000 x g x 20 min, were resuspended in 2 mL of solubilization buffer (150 mM NaCl, 10 mM $MgCl_2$, 20 mM Tris HCl, 1% NP-40, pH 7.5) with protease and phosphatase inhibitors mixture. To the supernatant obtained after 10000 x g centrifugation for 10 min, was added anti-a subunits antiserum (1:100 v/v). After 4 h of incubation at 4° C, Protein A Sepharose (1:10 v/v, Ab/P-A Sep.) was added. The Ag-Ab-P-A complex, after incubation for 1 h at 4° C, was obtained by centrifugation at 10000 x g x 10 min and proteins were separated by SDS-PAGE. The dried gel was exposed for 6 days at -80° C with Kodak X AR 5 films and intensifying screen.

Protein concentration assay: The protein concentration was determined according to Bradford ,1976.

RESULTS

The results reported in Fig.1 show the time course of phosphate incorporation by membranes incubated in the presence of $\begin{bmatrix} ^{32} P- \end{bmatrix}$ -ATP, with or without Ca^{2+}/PL. They indicate that, beside the difference in the amount of the phosphate incorporated already evident after 1 min incubation, the incorporation slope is also very different. In fact, without Ca^{2+}/PL the phosphorylation plateau is reached with 5 min while in their presence the incorporation is still increasing after 5 min.

Fig.1: Time course of $\begin{bmatrix} ^{32}P \end{bmatrix}$-phosphate incorporation in human platelet membranes in the absence and in the presence of Ca^{2+}/PL.
Platelet membranes were prepared and phosphorylated as indicated in Methods. The experiment was performed twice and each sample was carried out in triplicate.

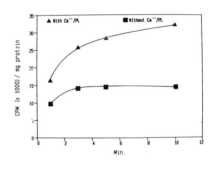

Fig.2: The effects of Ca^{2+} and PL added separately or in combination on the phosphate incorporation by human platelet membrane preparations. The membranes were incubated for 5 min in the presence of $\begin{bmatrix} -^{32}P \end{bmatrix}$-ATP.

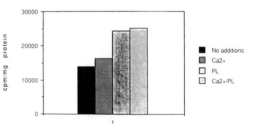

Fig.2 indicates the effects on the membrane phosphate incorporation when Ca^{2+} and PL are added separately or together. It is evident that a substantial increase in the phosphate incorporation is due to the presence of added PL while little if any contribution is due to the Ca^{2+}. This means that we are dealing with a peculiar PKC population that, although membrane-bound, show a marked PL dependence for optimal catalytic activity.

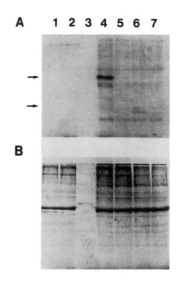

Fig.3: Phosphorylation of platelet membrane proteins by "endogenous" kinase activities. Autoradiograph (A) of the SDS-PAGE (B) of membrane proteins (for other details see Methods).
Lanes 1-2: membranes phosphorylated in the presence of Ca^{2+}/PL and then digested with alkaline or serine-threonine phosphatase respectively. Lane 3: Protein markers (LR BioRad). Lane 4: incubation in the presence of Ca^{2+}/PL. Lane 5: incubation in the presence of Ca^{2+}. Lane 6: incubation in the presence of 8-Br-cAMP. Lane 7: phosphorylated control membranes (only ATP).

1 = 4 + F.A.   5 = Ca^{++}
2 = 4 + S.F.   6 = 8Br cAMP
4 = Ca + PL   7 = ATP

The autoradiograph in Fig.3 shows some phosphorylated bands (A) of the membrane solubilized proteins resolved by SDS-PAGE (B). Lanes 1 and 2 indicate that the ^{32}P labeled bands of membranes phosphorylated as in lane 4 (with Ca^{2+}/PL added) are completely removed by a following digestion with alkaline phosphatase (which appears as a specific band in the panel B, lane 1), or with serine-threonine phosphatase type 1 (kindly provided by Dr. A.A. De Paoli-Roach Indiana Univ.). Lane 3 shows the protein markers (L.R.). In lane 4 the membranes phosphorylated with Ca^{2+}/PL have a specific labeled band that is absent in lane 5 (Ca^{2+} alone), in lane 6 (8-BrcAMP added) and in lane 7 (without Ca^{2+}/PL). A 25,000 MW band is specificly labeled in the cAMP presence, likely due to a kinase A activity.

The autoradiograph reported in Fig.4 shows the time course of membrane protein labeling without (upper panel) or with Ca^{2+}/PL added. The findings mirror the phosphate incorporation shown in Fig.1 and confirms that the extra amount of incorporated phosphates are selectively linked to specific membrane proteins.

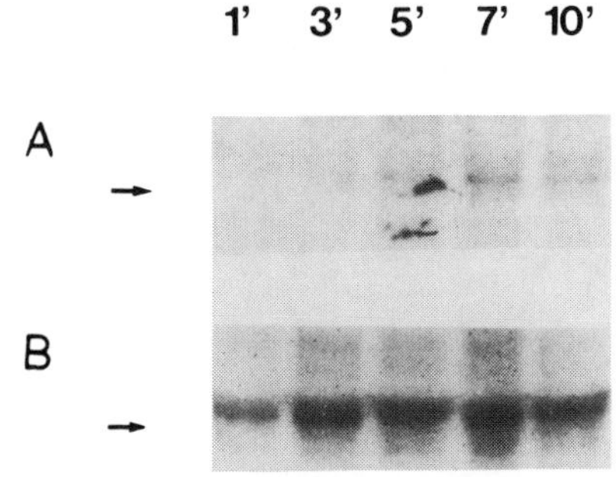

Fig.4:SDS-PAGE autoradiograph of membranes phosphorylated for different times.
The membranes were phosphorylated for the indicated times in the absence (A) or in the presence of Ca^{2+}/PL (B). The arrows indicate the position of the 40 Kd band.

The $\left[^{32}P\right]$-labeled proteins from the PKC-phosphorylated membranes have been identified by immunological tools. The first step was the immunoblotting of the G-protein a-subunits. The electrotransferred proteins after SDS-PAGE of the platelet membranes phosphorylated in presence and in absence of Ca^{2+}/PL, were probed with antibodies against a_i, a_s, and a_o and revealed by anti Ig conjugated with alkaline phosphatase. The results are shown in Fig.5: they indicate that this specific $\left[^{32}P\right]$-labeled band is mainly associated with a protein component immunoreacting with anti a subunit antibodies. The bands in lanes 2-4-6 and 8 of section B overlap both the anti-a_i, a_s, a_o immunoreacting bands of section A.

Further characterization was obtained by immunoprecipitation

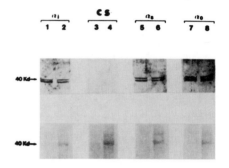

Fig.5: Autoradiographs and immunoblots of a_i, a_s and a_o subunits of human platelet membranes, phosphorylated without (1,3,5,7) or with (2,4,6,8) Ca^{2+}/PL added. The blots (upper panel)and autoradiograph (lower panel) were incubated with antiserum (a_i, a_s, a_o) and CS (control serum).

followed by electrophoretic and autoradiography analysis. After phosphorylation the proteins of platelet membrane were solubilized, clarified by Protein A Sepharose and incubated with the anti a_i serum. The immunocomplex, precipitated with Protein A Sepharose and collected by centrifugation, was submitted to

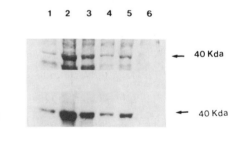

Fig.6: Immunoprecipitation of a-subunits.
A: SDS-PAGE.
B: autoradiographs.
Lane 1: solubilized membrane proteins; lane 2: P-A-Seph. clearing; lane 3: immunoprecipitation (IP) of a_i; lane 4: IP of a_s; lane 5: IP of a_o; lane 6: solubilized membrane proteins present in the supernatant after the immunoprecipitations.

SDS-PAGE and autoradiography, while the supernatant was submitted to the same procedure for immunoprecipitation with the anti a_s and the anti a_o serum and the sequential immunocomplexes processed as indicated. The results in Fig.6 indicate that part of the radioactivity is linked to proteins recognized by anti a_i antibodies, but also in part to proteins recognized by anti a_s and anti a_o-antibodies.

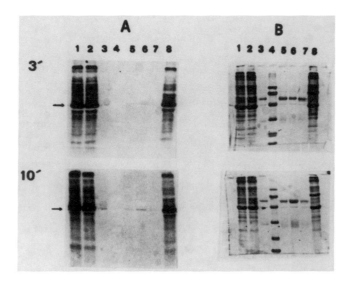

Fig.7: Autoradiography of immunoprecipitated a_i and a_s of platelet membranes phosphorylated for 3 and 10 min in the presence of Ca^{2+}/PL. A: autoradiographs; B: SDS-PAGE.

The immunoprecipitation experiments reported in Fig.7 refer to platelet membrane phosphorylation at different times with Ca^{2+}/PL. The experimental purpose was to verify the type of the

PKC-phosphorylation of a_i and a_s subunits. The results obtained indicate that both a_i and a_s are $[^{32}P]$-labeled after 3 min incubation, but after 10 min an increase of the a_s $[^{32}-P]$ incorporation appears, indicating that for a_s a dynamic equilibrium between kinase and phosphatase processes is reached later and at a higher level.

DISCUSSION

To date PKC activity has been shown to have many interactions with the AC signaling pathway through receptor, G-protein and effector system.

A PKC-mediated a_i phosphorylation and its functional inactivation in "in vivo" human platelets have been reported (Katada et al. 1985; Crouch et al. 1988). In crude membrane preparations of the same model, we confirm these data. We also demonstrate that a_s is PKC-phosphorylated. The PKC action on the two subunits overlaps in the first three min, but after 3 min the a_s phosphorylation prevails. The immunological tests have been performed with well characterized antibodies specific for a_i subtypes 1 and 2 (AS/7), for a_s (RM), while the anti a_o serum used is also known to crossreact with a_i subtype 3 (Simmonds et al. 1989; Goldsmith 1987). Since a_o is abundantly expressed in mammalian brain (Hescheler et al. 1987) but is poorly expressed, if at all, in human platelets, we think that the protein recognized by anti a_o serum is an a_i subtype 3 subunit.

Our results suggest that two or three a_i-subunit subtypes appear to be PKC-phosphorylated together with a_s. The first question to be answered is: why could other investigators not show a PKC-mediated phosphorylation of the a_s subunit (Bushfield et al. 1990)?

A critical point, in this type of experiment, may be the control of the phosphatase activities. We have monitored the activation of the phosphatase activities during the membrane protein solubilization procedure. During the two hours of antigen-antibody reaction this activity results in substantial loss of $[^{32}P]$-labeling unless an adequate phosphatase inhibitors cocktail is present. The reciprocal control of the phosphorylation/dephosphorylation processes on the G-proteins has not been described yet, but this a_s phosphorylation demonstrated under conditions of carefully controlled phosphatase activity, hints at a physiological role of the phosphatase control "in vivo".

A second important question is the functional role of both a_i and a_s phosphorylation. The a_i phosphorylation has been proved to weaken its inhibitory role, resulting in the exposure of the adenylate cyclase system to the a_s-mediated stimulating signals alone. The functional consequences for the a_s may be more complex in time; in fact, the a_s phosphorylation parallels a_i for the first minutes but lasts longer than a_i, resulting in a higher degree of phosphorylation.

The a-subunits have been shown to have more than one phosphorylation site. We hypothesize that the phosphorylation of more accessible serine residues might mean a conformational change followed by a functional change and, at the same time, a more favorable exposure of the other serine-threonine residues. The subsequent phosphorylations might modify and even revert the

268

previous conformational and functional states. The time course of the a_s phosphorylation (Fig. 5) implies a kinase cascade, with sequential conformational and functional changes.

REFERENCES

Baldassarre, J.J., Fischer, G.J. (1986) J. Biol. Chem. 261, 11942-11944.
Birnbaumer, L., Abramowitz, J. and Brown, A.M. (1990) Biochem. Biophys. Acta 1031, 163-224.
Bradford, M.M. (1976) Anal. Biochem. 72, 248-254.
Bushfield,M., Murphy, G.J., Lavan, B.E., Parker, P.J., Hruby, V.J., Milligan, G. and Houslay, M.D. (1990) Biochem. J. 268, 449-457.
Crouch, M.F. and Lapetina, E.G. (1988) J. Biol. Chem. 263, 3363-3371.
Goldsmith, P.K., Gierschik, P., Milligan, G., Unson, C.G., Vinisky, R., Malech, H.L. and Spiegel, A.M. (1987) J. Biol. Chem. 262, 14683-14688.
Hesheler, J., Rosenthal, W., Trautwein, W., Schultz, G. (1987) Nature 325, 445-447.
Huang, K., Huang, F.L., Nakabayashi, H. and Yoshida, Y. (1988) J. Biol. Chem. 263, 14839-14845.
Katada, T., Gilman, A.G., Watanabe, Y., Bauer, S. and Jakobs, K.H. (1985) Eur. J. Biochem. 151, 431-437.
Laenmli, U.K. (1970) Nature 227, 680-685.
Nishizuka, Y. (1986) Science 223, 305-312.
Nishizuka, Y. (1988) Nature 334, 661-665.
Simonds, W.F., Goldsmith, P.K., Wooward, C.J., Unson, C.G. and Spiegel, A.M. (1989) FEBS Lett. 249, 189-194.
Spina, A., Chiosi, E., Illiano, G., Berlingieri, M.T., Fusco, A. and Grieco, M. (1988) Biochem. Biophys. Res. Commun. 157, 1093-1103.
Thomas, P.T., Gopalakrishna, R. and Anderson W.B. (1987) Methods in Enzimology, 141, 397-411.

Adenine Nucleotides in Cellular Energy Transfer and Signal Transduction
S. Papa, A. Azzi & J.M. Tager (eds)
© 1992 Birkhäuser Verlag, Basel/Switzerland

CASEIN KINASES: AN ATYPICAL CLASS OF UBIQUITOUS AND PLEIOTROPIC PROTEIN KINASES

Lorenzo A. Pinna[1], Flavio Meggio[1], Oriano Marin[1], John W. Perich[2], Brigitte Boldyreff[3] and Olaf-G. Issinger[3]

[1]Dipartimento di Chimica Biologica and Centro per lo Studio della Fisiologia Mitocondriale del Consiglio Nazionale delle Ricerche, Università di Padova, via Trieste 75, 35121 Padova, Italy; [2]Biochemistry and Molecular Biology Unit, School of Dental Science, The University of Melbourne, Victoria, Australia and Centre CNRS-INSERM de Pharmacologie-Endocrinologie, Montpellier, France; [3]Institut für Humangenetik, Universität des Saarlandes, 6650 Homburg, Germany.

SUMMARY:

The term casein kinase applies to a small group of Ser/Thr specific protein kinases that by far prefer casein over histones as *in vitro* substrates. While however genuine casein kinases specifically expressed in the lactating mammary gland are actually responsible for the biosynthetic phosphorylation of casein fractions, two quite unrelated types of ubiquitous casein kinases, conventionally termed CK1 and CK2, are committed with the phosphorylation and regulation of a variety of protein targets involved in several cellular functions, with special reference to gene expression and signal transduction. CK2 is in particular a growth related protein kinase which, despite its heterotetrameric structure composed of two catalytic (α and/or α') and two non catalytic (β) subunits, is spontaneously active and apparently lacks any acute control mechanism. Its canonical consensus sequence is Ser(Thr)-Xaa-Xaa-Glu(Asp, SerP, TyrP) but it can be surrogated in peptide substrates by atypical motif(s) generated by previously phosphorylated serines, notably (SerP)-SerP-Ser-SerP. CK2 specificity moreover can be deeply altered by polycationic effectors, like polylysine, and by the subunit composition of the enzyme, as disclosed by using its recombinant catalytic subunit ($r\alpha$) and the reconstituted holoenzyme ($r\alpha_2 r\beta_2$). It is proposed that such a multifarious and flexible specificity could provide CK2 with a *sui generis* mechanism of regulation.

INTRODUCTION:

The term "casein kinase" denotes the members of a small group of Ser/Thr specific protein kinases sharing a marked preference for casein over histones as *in vitro* substrates. While however the "genuine" casein kinases (Moore et al., 1985; Bingham et al., 1988), specifically expressed in the lactating mammary gland, are committed with the biosynthetic phosphorylation of casein fractions, two other types of casein kinases, conventionally termed casein kinases-1 (CK1) and casein kinase-2 (CK2) are ubiquitous enzymes targeting a variety of proteins involved in different cellular functions (reviewed by Pinna (1990) and Tuazon & Traugh (1991)).

Despite their similar names CK1 and CK2 are quite unrelated enzymes, differing for their structure, specificity and responsiveness to a variety of compounds. The present interest for these ubiquitous enzymes, and in particular for CK2, has been prompted by a number of recent observations that suggest the involvement of CK2 in many biochemical processes, with special reference to signal transduction and gene expression. In several respects CK2 is an atypical protein kinase. Two distinctive features of CK2 will be considered here, namely its uncommon specificity and its apparent lack of any acute regulation, with the aim to show that the former might obviate the latter.

SPECIFICITY OF CK2.

CK2 uses GTP, besides ATP, as phosphate donor, a rare property among protein kinases, not shared by CK1. Two additional arguments support the view that the nucleotide binding site of

CK2 is atypical: firstly CK2 (similar to CK1) is insensitive to micromolar concentrations of staurosporin (unpublished data), a broad specificity, powerful inhibitor of protein kinases, whose IC_{50} values generally are in the nanomolar range (Meyer et al., 1989); secondly CK2 is readily inhibited by halogenated ribofuranosyl benzimidazole derivatives, that are nearly ineffective on most protein kinases tested so far (Meggio et al., 1990).

Specific inhibitors of CK2 could be theoretically designed also taking advantage of its peculiar peptide substrate specificity. While in fact most Ser/Thr specific protein kinases recognize phosphoacceptor sites specified by basic residues nearby, CK2 is acidophilic in nature, its specificity determinants consisting of acidic residues downstream from the target residue. Such a feature, recurrent in all known protein substrates of CK2, has beeen systematically analyzed with the aid of synthetic peptides (Meggio et al., 1984; Marin et al., 1986; Kuenzel et al., 1987; Marchiori et al., 1988; Meggio et al., 1989; Meggio et al., 1991a). These studies have shown that an individual acidic residue downtream of the phosphorylatable serine, at position +3, plays an especially crucial role, representing both a necessary and a sufficient condition for CK2 targeting. Such an acidic determinant can be either a carboxylic residue (Glu and Asp) or a phosphorylated side chain (phosphoserine and phosphotyrosine). In order to improve the otherwise low phosphorylation efficiency, however, additional positive determinants are required, consisting of either a cluster of multiple acidic residues (as it is found in most physiological targets of CK2) or a sequence prone to adopt a ß-turn conformation. On the other hand negative determinants, such as basic side chains, residues hindering β-bends and the motif Ser-Pro, may compromise the phosphorylation of otherwise suitable sites. The motif Ser-Pro is of special interest since it prevents phosphorylation by CK2 without diminishing, but rather increasing, the affinity of the peptide substrate for the enzyme.

It consequently provides a potential device for designing specific peptide inhibitors of CK-2.

The possibility that phosphorylated residues might generate atypical sites for CK2, specified by a consensus sequence different from the canonical one (Ser-Xaa-Xaa-Acidic) has been disclosed by showing that CK2 phosphorylates both the partially dephosphorylated triphosphopeptide AcSerP-SerP-SerP (Meggio et al., 1988) and Thr-4 in the phosphotyrosyl pentapeptide Asn-Glu-TyrP-Thr-Ala (Perich et al., 1990), both obviously lacking the expected consensus.

TABLE I: Phosphorylation by CK2 of phosphopeptides lacking the canonical consensus sequence.

Peptides	Phosphorylation rate (pmol/min)
AcSerP – SerP– Ser– SerP	35.6
AcGlu – SerP– Ser– SerP	4.5
AcAla – SerP– Ser– SerP	7.4
AcSerP – SerP– Ser – Glu	13.8
AcSerP – SerP– Ser – Ala	0.2
Glu – Glu – Ser – Glu	2.8

Experimental conditions as in Meggio et al. (1984). Peptide concentration was 1 mM.

A rationale for these unexpected findings has been recently provided using the triply phosphorylated tetrapeptide AcSerP-SerP-Ser-SerP and its derivatives in which phosphoseryl residues are variably replaced by either glutamic acid or alanine (Perich, J.W., Meggio, F., Reynolds, E.C., Marin, O. and Pinna, L.A., manuscript in preparation). As shown in Table I the motif SerP-SerP-Ser-SerP represents an excellent target for CK2 although it does not fulfil the canonical consensus sequence Ser-Xaa-Xaa-Acidic. Unlike the canonical consensus, moreover, that can be

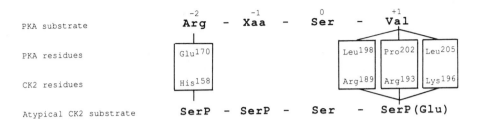

CK2-His[158]   SUBSTRATE - SerP(-2)

Figure 1:   CK2 residues homologous to PKA residues interacting
            with residues at position -2 and +1 of the peptide
            substrate.

PKA data are drawn from Knighton et al. (1991). The homologous
residues of CK2 have been calculated from Hanks & Quinn (1991).

determined equally well by either carboxylic aminoacids or
phosphoserine, the atypical consensus SerP-SerP-Ser-SerP is
crucially dependent on the phosphorylated aminoacids, as shown by
the very scanty phosphorylation efficiency of all the glutamyl
derivatives. Interestingly however, while a Glu substituted for
SerP-4 is still partially effective, as shown by the superiority
of SerP-SerP-Ser-Glu over SerP-SerP-Ser-Ala, a Glu substituted
for SerP-1 is totally ineffective, Glu-SerP-Ser-SerP actually
being a substrate even worse than Ala-SerP-Ser-SerP.

    An explanation for such a different effect of Glu for SerP
substitution could be provided by assuming that, in virtue of
their high degree of homology, the tertiary structures of the
catalytic domains of all Ser/Thr protein kinases are closely
reminiscent of that of PKA (Knighton et al., 1991), and
inspecting thereafter which residues in CK2 are expected to

interact with the aminoacids at positions -2 and +1 relative to the phosphorylatable serine, in its peptide substrates. As shown in fig. 1, Glu-170 of PKA, interacting with the arginyl residue at position -2 of the peptide substrates of this enzyme (Knighton et al., 1991) is replaced in CK2 by His-158. On the other hand three hydrophobic residues have been shown to interact with the hydrophobic residue which is usually adjacent to the C terminal side of serine in PKA substrates (Knighton et al., 1991); all of them are replaced in CK2 by either Arg or Lys (see fig.1). It seems conceivable therefore that while at neutral pH these latter strongly basic residues are fully protonated and ready to interact with negatively charged side chain(s) adjacent to the C terminal side of the target residue, His-158 is only partially protonated, its dissociation being possibly reinforced by the juxtaposition of a phosphate group, whose second dissociation, at neutral pH is far from complete (see inset of fig. 1). Such a proton transfer from the phosphoryl group of SerP to His-158 will give rise to a stable imidazolium ion capable to bind the negatively charged phosphate group of the peptide substrate through electrostatic interactions.

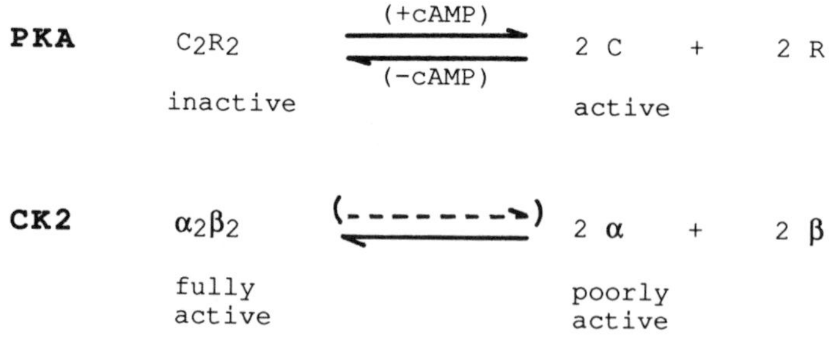

$$\textbf{PKA} \quad C_2R_2 \quad \underset{(-cAMP)}{\overset{(+cAMP)}{\rightleftharpoons}} \quad 2\ C \quad + \quad 2\ R$$

$$\text{inactive} \qquad\qquad\qquad \text{active}$$

$$\textbf{CK2} \quad \alpha_2\beta_2 \quad \overset{(------\rightarrow)}{\rightleftharpoons} \quad 2\ \alpha \quad + \quad 2\ \beta$$

$$\begin{array}{c}\text{fully}\\\text{active}\end{array} \qquad\qquad \begin{array}{c}\text{poorly}\\\text{active}\end{array}$$

Figure 2: Quaternary structure similarity and functional diversity between CK2 and PKA.

MODULATION OF CK2 ACTIVITY

Unlike most protein kinases CK2 is a spontaneously active enzyme, neither dependent on second messengers nor acutely affected by phosphorylation/dephosphorylation. Such an "independency" is especially strinking considering the unique similarities between the quaternary structures of CK2 and of PKA, the latter being the regulatable protein kinase *par excellence*. While however PKA holoenzyme is inactive and its activation occurs through dissociation into the catalytic and regulatory subunits, the opposite is true of CK2, whose heterotetrameric holoenzyme is fully active and it never dissociates unless under denaturing

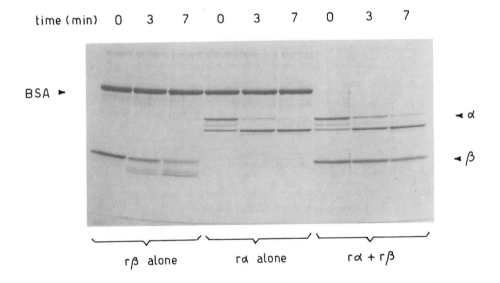

Figure 3: Susceptibility to trypsin treatment of recombinant α (rα) and β (rβ) subunits and of the reconstituted holoenzyme obtained by combining equimolar amounts of rα and rβ.

The trypsin to protein ratio was 1:40. The amount of protein was kept constant by replacing bovine serum albumin (BSA) for eiyher rα or rβ, where these subunits were omitted.

Table II: Effect of polylysine on the phosphorylation of calmodulin and CaM-peptide RKMKDTDSEEEIR by CK2 and by recombinant α subunit.

Substrate	Phosphorylation rate (pmol min^{-1}) by			
	CK2		rα	
	control	+polylysine	control	+polylysine
calmodulin	n.d.	2.9	2.1	1.0
CaM-peptide	7.9	42.5	0.7	1.9

Substrate concentration was 14 and 310 μM for calmodulin and CaM-peptide respectively. Polylysine was 2 μM. n.d.: not detectable.

conditions (see fig.2). Reconstitution experiments run by combining the recombinant α and β subunits in equimolar amounts, have shown that the β-subunit is required in order to confer to the catalytic α subunit its maximal activity (Grankowski et al., 1991) and increased stability under denaturing conditions (Meggio et al., 1991). The concept that the recombinant α and β subunits spontaneously give rise to $\alpha_2\beta_2$ tetramers is corroborated by the experiment shown in fig.3. Clearly the β-subunit, which is intrinsically quite susceptible to proteolysis by trypsin, becomes very resistant whenever it is associated with the α subunit, a behaviour which is reminiscent of the β subunit of the native CK2 holoenzyme (Meggio & Pinna, 1984). The susceptibility of the α subunit to tryptic proteolysis is conversely unaffected by the addition of the β subunit, consistent with the concept that the β-subunit might be relatively hidden within the quaternary structure of the enzyme, with the α subunit more exposed to external agents.

Altogether the available data support the view that the β-subunit, rather than a *sensu stricto* regulatory component is a structural element conferring to the catalytic subunit specific properties, including higher stability, different specificity and

enhanced susceptibility to a number of effectors. As shown in
Table II, in fact, calmodulin, which *in vitro* is not a substrate
for CK2 under basal conditions, is readily phosphorylated if
polylysine is added. Polylysine is not required however for the
phosphorylation of calmodulin by the recombinant α subunit: under
these circumstances polylysine actually becomes an inhibitor. If
calmodulin is replaced by a tridecapeptide reproducing its
phosphorylation site, phosphorylation occurs spontaneously in the
presence of either CK2 or the recombinant catalytic subunit; in
both instances polylysine stimulates the phosphorylation of the
peptide, though the responsiveness of the holoenzyme to
polylysine is much greater than that of the free catalytic
subunit. These data show that the specificity of CK2 is not

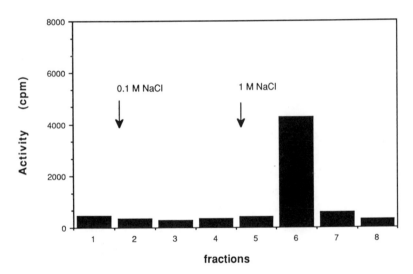

Figure 4: Polylysine-agarose column chromatography of
recombinant α subunit.

The column (1 x 0.5 cm) equilibrated with 50 mM Tris-HCl pH 7.5
containing 0.1M NaCl was washed with 5 ml of the same buffer
after application of the recombinant α subunit. Elution was
performed with the same buffer containing 1M NaCl. Casein kinase
activity was determined in the presence of a molar excess of
added recombinant β subunit.

rigidly pre-determined but can be modulated by several factors, including the tertiary structure of the protein substrate, the presence of polybasic modulators and the subunit composition of the enzyme itself.

As a general rule it can be concluded that the β subunit confers enhanced susceptibility to polybasic compounds: the residual stimulation observed with the recombinant α subunit alone, however, in conjunction with the findings that tha α subunit tightly binds to polylysine-agarose (fig.4) and that the β-subunit seems to be rather shielded inside the CK2 holoenzyme (see fig.3) support the view that polybasic effectors may also directly interact with the catalytic subunit, though the β-subunit is obviously required for rendering the α subunit highly responsive to them.

CONCLUSIONS

The resulting picture is that CK2, albeit spontaneously active, is endowed with a quite flexible specificity which can be modulated by a variety of factors impinging on either the substrate or the enzyme. Among the circumstances that have been show to have the potentiality to alter CK2 targeting are: previous phosphorylation of definite residues on the protein substrates by other kinase(s); small changes of pH that may influence the recognition of given sites by affecting the protonation of crucial histidyl residue(s); conditions altering substrate conformation; the presence of a variety of polybasic effectors, like histones, that are likely to play a relevant role in the nuclear compartment, where CK2 concentration is especially high (Filhol et al., 1990; Krek et al., 1991); and the subunit composition of CK2 itself. It is conceivable that these and other elements still to be detected may cooperate in providing CK2 with a subtle and multifarious control mechanism which could obviate the lack of any known specific effector capable to elicit an all-or-nothing response.

Acknowledgements: This work was supported by grants from Ministero per la Ricerca Scientifica e Tecnologica (Target Project on Biotechnology and Bioinstrumentation) and Consiglio Nazionale delle Ricerche.

REFERENCES

Bingham, E.W., Parris, N. and Farrel, H.M.Jr. (1988) J. Dairy Sci. 71, 324-336.
Filhol, O., Cochet, C. and Chambaz, E.M. (1990) Biochemistry 29, 9928-9936.
Grankowski, N., Boldyreff, B. and Issinger, O.-G. (1991) Eur. J. Biochem. 198, 25-30.
Hanks, S.K. and Quinn, A.M. (1991) Methods in Enzymology, vol.200, (T., Hunter, Ed.), Academic Press, New York, pp.38-62.
Knighton, D.R., Zheng, J., Ten Eyck, L.F., Ashford, V.A., Xuong, N.-H., Taylor, S.S. and Sowadski, J.M. (1991) Science 253, 407-414.
Krek, W., Maridor, G. and Nigg, E.A. (1991) J. Cell. Biol., in press.
Kuenzel, E.A., Mulligan, J.A., Sommercorn, J. and Krebs, E.G. (1987) J. Biol. Chem. 262, 9136-9140.
Marchiori, F., Meggio, F., Marin, O., Borin, G., Calderan, A., Ruzza, P. and Pinna, L.A. (1988) Biochim. Biophys. Acta 971, 332-338.
Marin, O., Meggio, F., Marchiori, F., Borin, G. and Pinna, L.A. (1986) Eur. J. Biochem. 160, 239-244.
Meggio, F., Boldyreff., Marin, O., Pinna, L.A. and Issinger, O.-G. (1991) Eur. J. Biochem., in press.
Meggio, F., Marchiori, F., Borin, G., Chessa, G. and Pinna, L.A. (1984) J. Biol. Chem. 259, 14576-14579.
Meggio, F., Perich, J.W., Johns, R.B. and Pinna, L.A. (1988) FEBS Lett. 237, 225-228.
Meggio, F., Perich, J.W., Meyer, H.E., Hoffmann-Posorske, E., Lennon, D.P.W., Johns, R.B. and Pinna, L.A. (1989) Eur. J. Biochem. 186, 459-464.
Meggio, F., Perich, J.W., Reynolds, E.C. and Pinna, L.A. (1991a) FEBS Lett. 279, 307-309.
Meggio, F. and Pinna, L.A. (1984) Eur. J. Biochem. 145, 593-599.
Meggio, F., Shugar, D. and Pinna, L.A. (1990) Eur. J. Biochem. 187, 89-94.
Meyer, T., Regenass, U., Fabbro, D., Altieri, E., Rösel, J., Müller, M., Caravatti, G. and Matter, A. (1989) Int. J. Cancer. 43, 851-856.
Moore, A., Boulton, A.P., Heid, H.W., Jarasch, E.-D. and Craig, R.K. (1985) Eur. J. Biochem. 152, 729-737.

Perich, J.W., Meggio, F., Valerio, R.M., Kitas, E.A., Johns, R.B. and Pinna, L.A. (1990) Biochem. Biophys. Res. Commun. 170, 635-642.

Pinna, L.A. (1990) Biochim. Biophys. Acta 1054, 267-284.

Tuazon, P.T. and Traugh, J.A. (1991) in Advances in Second Messenger and Phosphoprotein Research (Greengard, P. and Robison, G.A. eds.), Raven Press, New York, vol. 23, pp. 123-164.

Adenine Nucleotides in Cellular Energy Transfer and Signal Transduction
S. Papa, A. Azzi & J.M. Tager (eds)
© 1992 Birkhäuser Verlag, Basel/Switzerland

SPECTROPHOTOMETRIC ASSAY OF PROTEIN KINASES

Zuzana Technikova-Dobrova*,Anna Maria Sardanelli°and Sergio Papa°

Institute of Medical Biochemistry and Chemistry° and Centre for
the Study of Mitochondrial and Energy Metabolism° CNR,University
of Bari; Institute of Microbiology, Czechoslovak Academy of
Science*, Praha.

SUMMARY

Protein kinases are usually identified and characterized by
radioisotopic assay. The functional activity and substrates can,
however, be also analyzed by a coupled enzymatic assay. In this
method Pi transfer from ATP to the substrate protein is coupled
to the reaction of added pyruvate kinase and lactate dehydrogenase resulting in the final oxidation of NADH. This is monitored spectrophotometrically, thus allowing continuous recording
of the kinase activity.
The spectrophotometric assay is shown to be suitable for easy
and rapid analysis of fucntional activity of protein kinases.It
seems the method of choice for kinetic study of kinases as well
as for identification and functional analysis of their effectors.
The method can be used with any natural and synthetic polypeptide substrate of kinases, providing an estimate of the purity
and/or the molecular weight of protein substrates yet to be
characterized.
The performance of the method and some parameters so obtained
for protein kinase C and cAMP-dependent protein kinase are presented.

INTRODUCTION

Protein kinases are usually identified and characterized by

radioisotopic methods (Kitano et al.1986; Wylie et al.1988;

Roskowski et al.1991). The functional activity and substrates

of kinases can,however, be also analyzed spectrophotometri-

cally with coupled enzymatic assay. Spectrophotometric assay

of protein kinases was used by Cook and Roskoski (Cook et al.

1982; Roskoski, 1983) for kinetic study of protein kinase A.

In this paper a coupled enzymatic assay for protein kinases, developed in our laboratory (Technikova-Dobrova et al.1991)is presented. Its application in determining kinetic parameters of protein kinase C and the catalytic subunit of protein kinase A as well as its merits as compared to radioisotopic assay are examined. It appears that general application of spectrophotometric assay would facilitate the study of protein kinases.

RESULTS AND DISCUSSION

Reaction (1) catalyzed by protein kinase,in the presence of added pyruvate kinase (PK), phosphoenolpyruvate (PEP), lactate dehydrogenase (LDH) and NADH, is coupled with reaction 2 and 3. NADH oxidation with an excess of added anzymes gives directly and continuosly the rate of reaction (1).

$$(1) \quad \text{ATP + protein substrate} \xrightarrow{\text{protein kinase}} \text{ADP + phosphoprotein}$$

$$(2) \quad \text{ADP + PEP} \xrightarrow{\text{PK}} \text{ATP + pyruvate}$$

$$(3) \quad \text{pyruvate + NADH} \xrightarrow{\text{LDH}} \text{lactate + NAD}^+$$

$$(4) \quad \text{ATP + H}_2\text{O} \xrightarrow{\text{ATPase}} \text{ADP + Pi}$$

$$(5) \quad \text{phosphoprotein} \xrightarrow{\text{phosphatase}} \text{protein + Pi}$$

NADH oxidation is followed either with conventional or dual wavelength spectrophotometer at 340-374 nm.The latter provides higher sensitivity in the order of 20-30 µM and minimizes light scattering artefacts, which can arise when measuring the activity in crude extracts.Contaminating ATPase(reaction 4) and phosphatase activities (reaction 5) can be also detected and correction for them done.

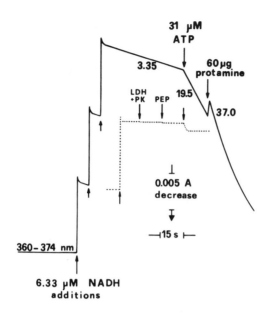

Fig.1. Determination of protein kinase activity in crude extract
of rat-brain
760 μg protein of the supernatant of rat-brain homogenate,were
suspended per ml of reaction mixture, containing 3.1 units/ml
(μM·min^{-1}) each of LDH and PK and 235 μM PEP. 6.33 μM addi
tions were made as indicated and served as internal sensitivity
standards. The other additions are specified on the illustration.
The figures on the traces refer to rates of NADH oxidation in
nmoles·min^{-1}·mg protein.The dotted line shows a control in the
absence of the brain extract. The small deflection caused by
ATP addition is due to the trace of contaminating ADP (0.2%).
Reproduced with permission from Technikova-Dobrova et al.1991.
For other details see this ref.

The Fig.1 shows determination of protein kinase activity in
a crude extract of rat-brain, which has a relatively high content of protein kinase C (Kikkawa et al.1982).

The assay was carried out at 360-374 nm to minimize interference from turbidity changes. Spectrophotometric traces showed

presence of NADH oxidase in the brain extract. ATP elicited an ATPase activity as indicated by the enhanced rate of NADH oxidation. The addition of protamine, general substrate of kinases, also in the absence of specific activators (Kikkawa,et al.1989) resulted in a sharp increase in the rate of NADH oxidation, which after correction for NADH oxidase and ATPase gave an activity for protein kinase C of 18 nmoles ATP utilized·min^{-1}·mg protein^{-1}.

In Fig.2 spectrophotometric determination of the activity of protein kinase C purified from rat brain is shown.
NADH oxidase and ATPase activity were very low in the sample

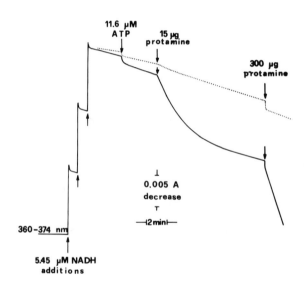

Fig.2. Determination of PKC purified from rat-brain
The 7.5 μg protein of purified rat-brain PKC were suspended per ml of reaction mixture containing LDH,PA, and PEP as specified in the legend of Fig.1. All the other additions are indicated in the illustration.The dotted trace refers to an experiment carried exactly under the same conditions of the measurement of the solid trace but in the presence of 85 μM staurosporine.(Reproduced with permission from Technikova-Dobrova et al. 1991).

of purified protein kinase. The addition of 15 g protamine
solid trace)resulted in rapid NADH oxidation. This continued
till all substrate was phosphorylated and the rate returned
to the value before its addition. An excess of protamine elicited maximal rate of NADH oxidation and a specific activity
of 385 nmol·min^{-1}·mg protein^{-1} can be calculated. The dotted trace
shows that in the presence of staurosporine, the known inhibitor of protein kinase C, protamine-induced NADH oxidation
was fully blocked.

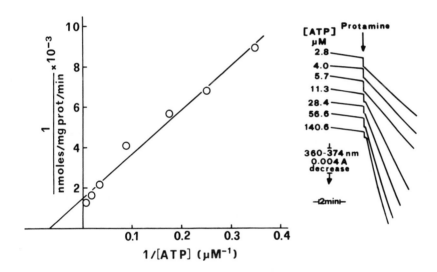

Fig.3. Spectrophotometric analysis of saturation of kinetics
     of rat brain PKC
5 µg protein of purified rat-brain PKC were incubated per ml
of reaction mixture containing LDH,PK and PEP as specified in
the legends to Fgs.1 and 2.The concentration of ATP was varied
in different samples as indicated in the illustration. The kinase reaction was started by the addition of 180 µg protamine/
ml (27.7 µM).For experimental conditions and other details see
legend to Fig.2. (Reproduced with permission from Technikova-
Dobrova et al.1991).

Spectrophotometric analysis of saturation kinetic of purified protein kinase C with respect to ATP is presented in Fig.3. From the initial rates of NADH oxidation, measured upon addition of increasing concentrations of ATP in the presence of an excess of protamine, a linear Lineweaven-Burk plot was obtained and K_m and V_{max} values were determined.

Table I summarizes kinetic parameters obtained with the spectrophotometric assay for protein kinase C and protein Kinase A. It can be noted that the values so measured were practically equal to those obtained on the same enzyme samples with the conventional radioisotopic method (Technikova-Dobrova et al.1991).

TABLE I. Kinetic parameters of protein kinases measured with the spectrophotometric assay

Protein kinase C (purified from rat-brain)

substrate protamine	Vmax	650 nmoles*
	K_m (ATP)	14 µM
	K_m (protamine)	1.8 µM
	pH optimum	8.1

Protein kinase A (catalytic subunit purified from bovine heart)

substrate protamine	Vmax	1500 nmoles*
	Km (ATP)	6 µM
	Km (protamine)	2.8 µM
	pH optimum	7.7

*nmoles·mg prot^{-1}·min^{-1}

DISCUSSION

The spectrophotometric enzyme-coupled assay appears to be suitable for easy and rapid analysis of functional activity of protein kinases. The assay can provide sensitivity around 250 nM with conventional spectrophotometers and around 30 nM with dual wavelength spectrophotometer. The latter instrument is particularly suitable to avoid light-scattering and turbidity artefacts which could arise in crude extracts or when phospholipids sus-

pensions or micelles (Bell et al.1986) have to be used to activate the kinase (PKC). The spectrophotometric assay allows to
measure on the same sample contaminating ATPase activity and
correct for it. The ATPase activity can, on the other hand,
represent, particularly in crude extracts, a serious compli-
cation for the radioisotopic assay.

The spectrophotometric assay, providing a direct continuous
recording of the enzymatic activity, represents the method of
choice for kinetic study of purified protein kinases as well
as for identification and functional analysis of their inhibitors and activators. An advantage of the spectrophotometric
assay derives from the fact that the concentration of ATP is
maintained constant in the course of the reaction by continuous regeneration by pyruvate kinase.

The assay can be used with any natural and synthetic polypeptide substrate of protein kinases. It is advantageous with
respect to the radioisotopic assay which requires separation
of ^{32}P-protein substrates from $|\gamma-^{32}P|$-nucleotides, a process
which needs rigorous procedures and is difficult to be achieved with certain polypeptide substrates (Roskosky,1983).

Some protein kinases display also protein phosphatase activity or can be contamined by phosphatases (Kitano et al.1986).
These will result in the hydrolytic removal of phosphate from
the phosphorylated protein substrate and unerstimate or prevent
measurement of the kinase activity with the radioisotopic assay.
The phosphatase activity will not prevent spectrophotometric
assay of the kinase activity. The phosphatase reaction being
coupled with reactions (1), (2) and (3) can also be detected in
the spectrophotometric assay. Finally the spectrophotometric
assay can provide an estimate of the purity and/or the molecular
weight of protein substrates yet to be characterized, or of the
number of phophoryl groups incorporated per mole of substrate
protein.

ACKNOWLEDGEMENTS

This work was in part supported by Grant 89.0037.75 of C.N.R. Italy. Dr.Z.Technikova-Dobrova was recipient of a travel grant from University of Bari.

REFERENCES

Bell,R.N., Hannum, Y. and Loomis, C. (1986), Methods in Enzymology 124, 353-354

Cook, P.F., Neville, M.E., Vrana, K.E., Hartl, F.T. and Roskoski R. Jr. (1982) Biochemistry 21, 5794-5799

Kikkawa, U., Takai, Y., Minakuchi,R., Inohara, S. and Nishizuka, Y. (1982) V.Biol.Chem.257, 13341-13348

Kikkawa, V., Kishimoto, A. and Nishizuka, Y. (1989) Annu.Rev. Biochem.59, 971-1005

Kitano,T., Masayoshi, G., Kikkawa, U. and Nishizuka, Y. (1986) Methods in Enzymology 124, 349-352

Roskosky, R.Jr. (1983) Methods in Enzymology 99 (Part.F),3.6

Roskosky, R.Jr. and Ritchie,P. (1991) Journal of Neurochemistry 56,1019-1023

Technikova-Dobrova, Z., Sardanelli, A. and Papa, S. (1991) FEBS Lett.292, 69-72

Wylie, D., Stock, A., Wong,C.Y., and Stock, J. (1988), Biochem. Biophys Res.Comm.151, 891-896

Adenine Nucleotides in Cellular Energy Transfer and Signal Transduction
S. Papa, A. Azzi & J.M. Tager (eds)
© 1992 Birkhäuser Verlag, Basel/Switzerland

KGF RECEPTOR: TRANSFORMING POTENTIAL ON FIBROBLASTS AND EPITHELIAL CELL-SPECIFIC EXPRESSION BY ALTERNATIVE SPLICING

Toru Miki, Donald P. Bottaro, Timothy P. Fleming, Cheryl L. Smith, Jeffrey S. Rubin, Andrew M.-L. Chan, and Stuart A. Aaronson

Laboratory of Cellular and Molecular Biology, National Cancer Institute, Bethesda, MD 20892, USA

SUMMARY: The mouse keratinocyte growth factor (KGF) receptor (KGFR) cDNA was isolated by an expression cDNA cloning strategy involving creation of a transforming autocrine loop. Characterization of the cloned 4.2 kb cDNA revealed a predicted membrane-spanning tyrosine kinase structurally related to the FGF receptor (FGFR). Structural analysis of the human KGFR cloned by the analogous procedure revealed identity with one of the fibroblast growth factor (FGF) receptors (bek/FGFR-2) except for a divergent stretch of 49 amino acids in their extracellular domains. Binding assays demonstrated that the KGFR was a high affinity receptor for both KGF and acidic FGF, while FGFR-2 showed high affinity for basic FGF and acidic FGF but no detectable binding by KGF. Analysis of the bek gene revealed two alternative exons responsible for the region of divergence between the two receptors. The KGFR transcript was specific to epithelial cells, and it appeared to be differentially regulated with respect to the alternative FGFR-2 transcript.

INTRODUCTION

Growth factor signalling pathways play critical roles in normal development and in the genetic alterations associated with the neoplastic process. Our laboratory has recently identified (Rubin et al., 1989) and molecularly cloned (Finch et al., 1989) a growth factor, designated keratinocyte growth factor (KGF), which has potent mitogenic activity for a wide variety of epithelial cells but lacks detectable activity on fibroblasts or endothelial cells. Its synthesis by stromal fibroblasts of a large number of epithelial tissues has suggested its likely role as an important paracrine mediator of normal epithelial cell proliferation. Recent studies have further indicated specific KGF binding to keratinocytes but not fibroblasts (Bottaro et al., 1990). We reasoned that the ectopic expression of the KGF receptor (KGFR) in the cells that secrete KGF might result in the transformed phenotype by creation of a transforming autocrine loop. We have recently developed a highly efficient expression cDNA cloning system with capability of plasmid rescue from stably-transfected cells (Miki et al., 1991b). Using this system, we prepared a cDNA expression library from epithelial cells that express the KGFR and introduced the library DNA into fibroblasts that express KGF but not its receptor. By this approach, it was possible to identify foci induced by expression of the KGFR cDNA and to molecularly clone the receptor.

MATERIALS AND METHODS

Construction of cDNA libraries: cDNA libraries were constructed in λpCEV27 (Miki et al., 1991b) by the automatic directional cloning method (Miki et al., 1989). Amplification of the library and preparation of the DNA were performed by standard procedures.

DNA transfection and cell culture: Library DNA (5 µg/plate) and carrier DNA (40 µg/plate) were introduced into cells using calcium phosphate transfection (Wigler et al., 1977). Cells were maintained in Dulbecco's modified Eagle's medium (DMEM) containing 5% calf serum.

PCR assays: The cDNA preparation (1 µl) synthesized from each RNA or human placenta genomic DNA (0.5 µg) was added to the reaction buffer (Boehringer Mannheim), 200 µM each of four dNTPs, 0.2 µM each of the upstream and downstream primers, and 0.025 unit/µl of Taq DNA polymerase (Boehringer Mannheim). The reaction was cycled 30 times at 94° C for 1 min, at 60° C for 3 min and at 72° C for 3 min.

Growth factor binding assays and crosslinking experiments: Recombinant KGF and bovine brain aFGF were purified and labeled with ^{125}I-Na as described (Bottaro et al., 1990). Bovine brain bFGF was obtained from R & D Systems. Bovine brain ^{125}I-bFGF was obtained from Amersham. Specific actvities of all three tracers were approximately 0.1 µCi/ng. Binding assays and covalent affinity crosslinking experiments were performed as described (Bottaro et al., 1990).

RESULTS AND DISCUSSION

Expression cDNA cloning of the mouse KGF receptor: We prepared a cDNA library from BALB/MK epidermal keratinocytes in λpCEV27 (Miki et al., 1991a). The scheme for expression cloning of the KGFR is shown in Fig. 1. Transfection of NIH/3T3 mouse embryo fibroblasts by the library DNA led to detection of several transformed foci. Each focus was tested and shown to be resistant to G418, indicating that it contained integrated vector sequences. Three representative transformants were chosen for more detailed characterization based upon differences in their morphologies. When we performed plasmid rescue, each transformant gave rise to at least 3 distinct cDNA clones as determined by physical mapping. To examine their biological activities, each clone was subjected to transfection analysis on NIH/3T3 cells. A single clone rescued from each transformant was found to possess high-titered transforming activity. To screen for cells expressing the KGFR, we performed binding studies with recombinant ^{125}I-KGF as the tracer molecule. One of the transformants demonstrated specific high affinity binding of ^{125}I-KGF, implying that the cDNA clone rescued from the focus encoded the KGFR, whose introduction into NIH/3T3 cells had completed an autocrine transforming loop.

292

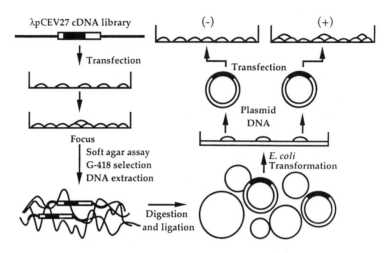

λpCEV27 cDNA library   (-)   (+)

Transfection

Transfection

Plasmid DNA

Focus

Soft agar assay
G-418 selection
DNA extraction

E. coli
Transformation

Digestion
and ligation

Fig. 1. Strategy for expression cDNA cloning of the KGFR. NIH/3T3 cells are transfected by λpCEV27-BALB/MK cDNA library DNA and scored at 14 to 17 days for transformed foci. Transformed cells are assayed for G-418 resistance to examine the presence of integrated vector sequence. Following expansion to mass culture, genomic DNA is isolated and subjected to plasmid rescue by digestion with either *Not*I, *Xho*I or *Mlu*I, followed by ligation at low DNA concentration and transformation to a suitable bacterial strain. Bacterial colonies resistant to both the ampicillin and kanamycin are isolated. Plasmid DNA extracted from each colony is tested by transfection analysis on NIH/3T3 cells to identify the transforming cDNA clone.

<u>KGF receptor is a membrane-spanning tyrosine kinase:</u> The rescued transforming plasmid contained a 4.2kb cDNA insert. We observed a single KGFR transcript of around 4.2 kb in BALB/MK cells (Miki et al., 1991a). Thus, our cDNA clone represented essentially the complete transcript. We next determined the nucleotide sequence of the 4.2kb cDNA insert. Analysis of the sequence revealed that it contained a long open reading frame for a deduced protein of 707 amino acids with a size of 82.5 kd. The amino acid sequence of the KGFR predicted a transmembrane tyrosine kinase closely related to the bFGF receptor (Fig. 2). The putative KGFR extracellular portion contained two immunoglobulin-like loops (Ig loops). The bFGF receptor contains a series of eight consecutive

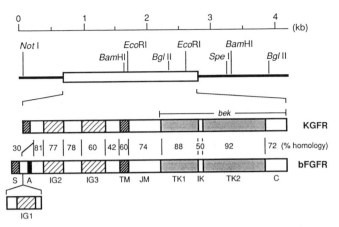

Fig. 2. Structural comparison of the predicted KGF and bFGF receptors. The region homologous to the published *bek* sequence is also shown. The schematic structure of the KGFR is shown below the restriction map of the cDNA clone. Amino acid sequence similarities with the smaller and larger bFGF receptor variants are indicated. S, signal peptide; IG1, IG2, and IG3, immunoglobulin-like domains; A, acidic region;, TM, transmembrane domain; JM, juxtamembrane domain; TK1 and TK2, tyrosine kinase domains; IK, interkinase domain; C, C-terminus domain.

acidic residues between the first and second IgG-like domains (Lee et al., 1989). However, the KGFR did not contain such an acidic domain. The intracellular portion of the KGFR was highly homologous to the bFGF receptor tyrosine kinase. A partial mouse cDNA, *bek*, isolated by bacterial expression cloning using phosphotyrosine antibodies (Kornbluth et al., 1988) was identical to the KGFR in the tyrosine kinase domain (Fig. 2), indicating that the KGFR is encoded by the *bek* gene.

Functional analysis of the cloned KGF receptor: Because of the existence of more than one receptor of the FGF family, we sought to characterize in detail the binding properties of the KGFR isolated by expression cloning. Scatchard analysis of ^{125}I-KGF binding to the NIH/3T3-mKGFR transfectant revealed expression of two similar high affinity receptor populations. Out of a total of ~3.8 x 10^5 sites/cell, 40% displayed a dissociation constant (Kd) of 180 pM, while the remaining 60% showed a Kd of 480 pM.

These values are comparable to the high affinity KGFRs displayed by BALB/MK cells (Bottaro et al., 1990). The pattern of KGF and FGF competition for ^{125}I-KGF binding to NIH/3T3-mKGFR cells was also very similar to that observed with BALB/MK cells. When ^{125}I-KGF crosslinking was performed with NIH/3T3-KGFR cells, we observed a single species of 137 kd, corresponding in size to the smaller species identified in BALB/MK cells. Detection of this band was specifically and efficiently blocked by unlabelled KGF. When glycosylation is considered, the size of the KGFR predicted by sequence analysis corresponds reasonably well with the corrected size (115kd) of the crosslinked KGFR in NIH/3T3-KGFR cells. Moreover, KGF stimulation of the transfectant rapidly induced tyrosine phosphorylation of several cellular proteins. These results suggested that the cloned KGFR was biologically active for signal transduction (Miki et al., 1991a).

Structural comparison of the KGF receptor with related molecules: Several human or avian cDNAs closely related to the KGFR have been reported. The external portions of human *bek* and TK14 and chicken *cek*3 proteins contain 3 Ig loops (Dionne et al., 1990; Houssaint et al., 1990; Pasquale, 1990). Computer analysis of these molecules (designated FGFR-2) showed that amino acid sequences are nearly identical. These FGFR-2 molecules also closely related to the KGFR but differ in that each contains an acidic region and is completely divergent in the carboxy terminal half of its third Ig-like domain from the KGFR. A gene, designated K-*sam*, was identified as an amplified sequence in a human stomach carcinoma (Hattori et al., 1990). A cDNA clone corresponding to one of the overexpressed K-*sam* transcripts predicts a 2-loop *bek* variant, whose Ig loops correspond to those of the KGFR. However, it differs in that it contains an acidic region and may be truncated at its carboxy terminus as well. To study the differences of these related molecules in detail, we cloned human KGFR and compared the structure with human K-*sam* and FGFR-2.

Human KGF receptor is identical to FGFR-2 except the third Ig loop region: We isolated the human KGFR cDNA from B5/589 mammary epithelial cell cDNA expression library by a similar strategy to the cloning of the mouse receptor (Miki et al., 1991c). Sequence analysis of the 4.5 kb cDNA insert revealed an open reading frame encoding a membrane-spanning tyrosine kinase closely-related to the mouse KGFR. Comparison of the predicted protein with FGFR-2 showed essentially complete identity with the exception of a strikingly divergent 49 amino acid stretch spanning the second half of the third loop into the stem region (Fig. 3).

```
mKGFR  ..--....................M............................
hKGFR  LK--HSGINSSNAEVLALFNVTEADAGEYICKVSNYIGQANQSAWLTVLPKQQAP
K-sam  ..--................................................
FGFR2  ..AAGVNTTDKEI...YIR...FE.....T.LAG.S..ISFH........---..
```

Fig. 3. Comparison of the amino acid sequences of the second halves of the third Ig loops of mouse KGFR (amino acids 196-250), human KGFR (311-365), human K-sam (222-276), and human FGFR-2 (311-365). The sequence of the human KGFR is shown in the second line. In the case of other molecules, only the amino acid residues different from the human KGFR are shown. The residues identical to the human KGFR are shown by dots. The dashes represent that the residues are not present in the molecule. The cysteine residue which is the start site of the third Ig loop is underlined.

Third Ig loop divergent region determines KGF-binding properties: The mouse KGFR has been shown to bind KGF and aFGF at similar high affinity and basic FGF at 20 fold lower affinity (Miki et al., 1991a). In contrast, FGFR-2 has been reported to bind both aFGF and bFGF at high affinity (Dionne et al., 1990), but there is no available evidence concerning its ability to interact with KGF. We isolated a human cDNA whose product closely resembled the mouse KGFR, and yet differed only by a stretch of 49 amino acids in its third Ig loop from FGFR-2 (Miki et al., 1991c). This allowed us to characterize its binding properties and compare them to those of FGFR-2. For those studies, we utilized NIH/3T3 transfectants overexpressing either protein and marker-selected NIH/3T3 cells transfected with the vector alone as a

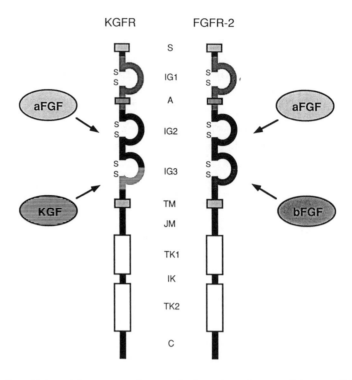

Fig. 4. Schematic comparison of the structure of human KGFR and FGFR-2 and domains responsible for binding of aFGF, bFGF and KGF.

control. The human KGFR transfectant demonstrated substantial ^{125}I-KGF binding, while neither FGFR-2 or the vector transfectant showed detectable binding (Miki et al., 1991c). These findings established that the cDNA encoded a human KGFR and suggested further that KGF lacked high affinity for FGFR-2. Evidence that FGFR-2 was indeed functional was derived from binding analyses with ^{125}I-aFGF and ^{125}I-bFGF. Both demonstrated a substantially greater number of binding sites on the FGFR-2 transfectant than on NIH 3T3 cells, which are known to express bFGFR (Ruta et al., 1989) and to show mitogenic response upon stimulation by either growth factor (unpublished results). Whereas the human KGFR transfectant also bound increased amounts of ^{125}I-aFGF, we observed no increase in ^{125}I–bFGF binding over the level observed with control NIH/3T3 cells (Miki et al., 1991c). All of these results demonstrated striking differences in the patterns of FGF and KGF binding by these two closely related human receptors (Fig. 4).

Determination of ligand-binding specificity by alternative splicing: The high degree of sequence identity of the FGFR-2 and KGFR strongly suggested that both the receptor species were encoded by the same gene. We reasoned that their divergent regions were encoded by different exons (exons K and B for KGF and bFGF, respectively) located between common upstream and downstream exons (U and D, respectively). To map these putative exons within the genomic sequence, we first compared nucleotide sequences of the divergent region. Two possible alternative locations of such exons could be postulated, and PCR analysis was performed to investigate the locations of these exons. The intron/exon map of this region was determined as shown in Fig. 5. Some of the PCR products were cloned and sequenced, and consensus sequences for intron/exon junctions were found in the expected positions. All of these results established that two receptors with different ligand-binding specificities were encoded by the same human gene (*bek*) and messages for the two receptors were generated by alternative splicing (Miki et al., 1991c).

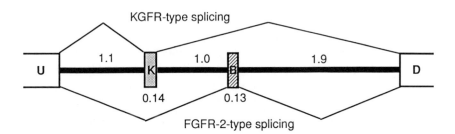

Fig. 5. The intron/exon structure of a part of the human *bek* gene which gives rise to the divergence. Exons and introns are shown by boxes and thick lines, respectively, with the approximate sizes in kb. The splicing events specific to generate the two receptors are also shown.

Tissue-specific expression of KGFR and FGFR-2 alternative
transcripts: To examine the level at which expression of the two
receptor species might be regulated in various cell types, we
used the PCR primers specific for the unique alternative exons of
the KGFR and FGFR-2. cDNA was synthesized from RNAs of different
human cell types using a primer specific to both the alternative
transcripts. The source of the RNAs include mammary epithelial
cells (B5/589), fibroblasts (M426), vesicular endothelial cells,
melanocytes, and monocytes, as well as several tumor cell lines.
The synthesized cDNA was then used to amplify KGFR or FGFR-2
specific sequences from the exon-specific primers. Fig. 6 shows
the striking contrast in patterns observed with only one of the
alternative transcripts demonstrated in each of the cells
analyzed. While the epithelial cells expressed transcripts

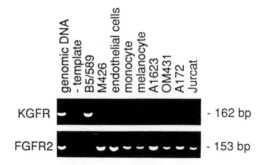

Fig. 6. Differential expression of the messages for KGFR and
FGFR-2 in various cell lines. Poly (A)+ RNAs extracted from the
cells shown at the top were reverse-transcribed by a primer which
can hybridize with the messages for either of the receptor
species. Segments specific to KGFR and FGFR-2 are amplified from
aliquote of the synthesized cDNAs by PCR using the primers
specific to KGFR or FGFR-2. Human placenta DNA was used as a
positive control for PCR to show equal amplification of the
segments of KGFR and FGFR-2 cDNAs (first lane). PCR assay was
also performed without template for a negative control (second
lane). In other lanes, cDNAs synthesized from human poly(A)+·
RNA were used as templates. Normal cells; B5/589 (mammary
epithelial cells), M426 (lung embryonic fibroblasts), umbilical
cord endothelial cells, monocytes from peripheral blood, and
melanocytes. Cell lines established from tumors; A1623
(anaplastic tumor groin node), OM431 (eye melanoma), A172
(glioblastoma), and Jurcat (lymphoma). PCR products were
separated by electrophoresis on a 3% NuSieve-GTG/1% SeaKem-GTG
agarose gel. Location of the expected PCR products are shown at
right in base pairs.

containing only the KGFR specific exon sequence, each of the other cell types expressed transcripts to corresponding to FGFR-2 specific exon sequence. These findings are in complete accordance with the tightly restricted specificity of KGF for cells of epithelial derivation (Rubin et al., 1989), suggesting that target cell-specificity of KGF was determined by cell type-specific alternative splicing of the *bek* gene transcript.

The evolution of increasingly complex multicellular organisms has been associated with significant expansion of gene families of growth factors and their receptors. In general, this has been associated with gene duplications. In our present studies, the KGFR alternative transcript was found to be specific to epithelial cells, while the FGFR-2 transcript was detected in cells of a variety of other tissue types. The strikingly different ligand-binding affinities of these two receptors encoded by a single gene combined with their different patterns of expression provides a new dimension to growth factor receptor diversity and may reflect a general mechanism for increasing the repertoire of these important cell surface molecules.

REFERENCES

Bottaro, D.P., Rubin, J.S., Ron, D., Finch, P.W., Florio, C., and Aaronson, S.A. (1990) J. Biol. Chem. 265, 12767-12770.
Dionne, C.A., Crumley, G., Bellot, F., Kaplow, J.M., Searfoss, G., Ruta, M., Burgess, W.H., Jaye, M., and Schlessinger, J. (1990) EMBO J. 9, 2685-2692.
Finch, P.W., Rubin, J.S., Miki, T., Ron, D. and Aaronson, S.A. (1989) Science 245, 752-755.
Hattori, Y., Odagiri, H., Nakatani, H., Miyagawa, K., Naito, K., Sakamoto, H., Katoh, O., Yoshida, T., Sugimura, T., and Terada, M. (1990) Proc. Natl. Acad. Sci. USA 87, 5983-5987.
Houssaint, E., Blanquet, P.R., Chanpion-Arnaud, P., Gesnel, M.C., Torriglia, A., Courtois, Y., and Breathnach, R. (1990) Proc. Natl. Acad. Sci. USA 87, 8180-8184.
Kornbluth, S., Paulson, K.E., and Hanafusa, H. (1988) Mol. Cell. Biol. 8, 5541-5544.
Lee, P.L., Johnson, D.E., Cousens, L.S., Fried, V.A., and Williams, L.T. (1989) Science 245, 57-60.
Miki, T., Matsui, T., Heidaran, M.A. and Aaronson, S.A. (1989) Gene 83, 137-146.
Miki, T., Fleming, T.P., Bottaro, D.P., Rubin, J.S., Ron, D. and Aaronson, S.A. (1991a) Science 251, 72-75.

Miki, T., Fleming, T.P., Crescenzi, M., Molloy, C.J., Blam, S.B., Reynolds, S.H., and Aaronson, S.A. (1991b) Proc. Natl. Acad. Sci. USA 88, 5167-5171.

Miki, T., Bottaro, D.P., Fleming, T.P., Smith, C.L., Burgess, W.H., Chan, A.M.-L. , and Aaronson, S.A. (1991c) Proc. Natl. Acad. Sci. USA (in press).

Pasquale, E.B. (1990). A distinctive family of embryonic protein-tyrosine kinase receptors. Proc. Natl. Acad. Sci. USA 87, 5812-5816.

Rubin, J.S., Osada, H., Finch, P.W., Taylor, W.G., Rudikoff, S., and Aaronson, S.A. (1989) Proc. Natl. Acad. Sci. USA 86, 802-806.

Ruta, M., Burgess, W., Givol, D., Epstein, J., Neiger, N., Kaplow, J., Crumley, G., Dionne, C., Jaye, M., and Schlessinger, J. (1990) Proc. Natl. Acad. Sci. USA 86, 8722-8726.

Wigler, M., Silverstein, S., Lee, L.-S., Pellicer, A., Cheng, Y.C., and Axel. R. (1977) Cell, 11, 223-232.

Adenine Nucleotides in Cellular Energy Transfer and Signal Transduction
S. Papa, A. Azzi & J.M. Tager (eds)
© 1992 Birkhäuser Verlag, Basel/Switzerland

301

THE RECEPTOR FOR THE HEPATOCYTE GROWTH FACTOR-SCATTER FACTOR: LIGAND-DEPENDENT AND PHOSPHORYLATION-DEPENDENT REGULATION OF KINASE ACTIVITY

Luigi Naldini, Elisa Vigna, Paola Longati, Lucia Gandino, Riccardo Ferracini, Andrea Graziani, and Paolo M. Comoglio

Department of Biomedical Sciences & Oncology, University of Torino, School of Medicine, Corso Massimo D'Azeglio 52, 10126 Torino, Italy.

SUMMARY: Hepatocyte Growth Factor (HGF) and Scatter Factor (SF) are identical molecules produced by the stromal fibroblats and non-parenchimal cells of many organs and also present in serum. HGF exerts an array of activities on epitelial cells, i.e. mitogenesis, dissociation of epithelial sheets, stimulation of cell motility, and promotion of matrix invasion. The receptor for HGF is the tyrosine kinase encoded by the MET proto-oncogene. It is widely expressed in normal epithelial tissues as a 190 kDa heterodimer of two disulfide-linked protein subunits. HGF binding triggers tyrosine autophosphorylation of the receptor β subunit in intact cultured cells. Autophosphorylation upregulates the kinase activity of the receptor, increasing the V_{max} of the phosphotransfer reaction. The major phosphorylation site has been mapped to Tyr 1235. Negative regulation of the receptor kinase activity occurs through distinguishable pathways involving protein kinase C activation or increase in the intracellular Ca^{2+} concentration. Both lead to the phosphorylation of serine residue(s) in a unique tryptic phosphopeptide of the receptor and to a decrease in its tyrosine phosphorylation and kinase activity. Receptor autophosphorylation also triggers the signal transduction pathways inside the target cells. The phosphorylated receptor associates phosphatidylinositol 3-kinase, indicating that the generation of the D-3 phosphorylated inositol lipids is involved in effecting the motility and/or growth response to HGF.

INTRODUCTION

Hepatocyte Growth Factor (HGF) is a powerful mitogen for
hepatocytes in primary cultures (Nakamura et al.,1986; Gohda et
al.,1988; Zarnegar and Michalopoulos,1989) and other epithelial
tissues, such as kidney tubular epithelium and keratinocytes (Kan
et al., 1991), endothelial cells and melanocytes (Rubin et al.,
1991). HGF is a disulfide-linked heterodimer of a heavy (α)
subunit of 55-65 kd and a light (β) subunit of 32 or 36 kD. The
α and β subunits of HGF originate from proteolytic cleavage of
a single 92 kd precursor, as indicated by the sequence of cloned
human cDNA (Miyazawa et al., 1989; Nakamura et al., 1989). HGF
is considered a major mediator of liver regeneration *in vivo*
(Michalopoulos, 1990). HGF was recently found (Naldini et
al.,1991c; Weidner et al.,1991) to be identical to Scatter Factor
(SF), a secretory product of fibroblasts which dissociates
epithelial cells increasing their motility and invasiveness
(Stoker et al.,1987; Weidner et al.,1990;).

Work from us (Naldini et al.,1991a; Naldini et al.,1991c)
and from another laboratory (Bottaro et al.,1991) has identified
the HGF receptor with the product of the *MET* protooncogene, a
transmembrane protein endowed with tyrosine kinase activity (Park
et al.,1987; Gonzatti et al,1988). The *MET* gene is expressed in
a variety of epithelial cells as well as in a number of tumors
of epithelial origin (Prat et al.,1991; Di Renzo et al.,1991).
The receptor is a 190 kD heterodimer made of a 50 kD α subunit
disulphide linked to a 145 kD β subunit synthesized as a single
chain precursor (Giordano et al.,1989a). Proteolytic cleavage
lead to the mature two-chains heterodimer (Giordano et al.,
1989b). The α chain and the N-terminal portion of the β chain are
exposed on the cell surface. The C-terminal portion of the β
chain is cytoplasmic and contains the kinase domain.

The role of the 190 kD *Met* heterodimer as HGF receptor was
proved by chemical cross-linking of the radiolabelled ligand to
the receptor β subunit and by reconstitution of a high-affinity
binding site for HGF into insect cells infected with a recombi-

nant baculovirus carrying the *MET* cDNA (Naldini et al.,1991c).

Here we discuss the regulation of the HGF receptor tyrosine kinase activity. Two levels of regulation were noted, one dependent on the ligand and one on the phosphorylation state of the receptor. The phosphorylation both of critical tyrosine and serine residues in the receptor molecule affected the kinase activity of the HGF receptor. The control of receptor phosphorylation on the signal transducing activity was also investigated.

RESULTS

Ligand-dependent activation of HGF receptor tyrosine kinase and tyrosine autophosphorylation

A549 human carcinoma cells respond to HGF with increased motility and matrix invasion. Monolayers of A549 cells were grown to confluence, serum starved and exposed to 1-2 nM HGF for 10 minutes at 37°C. The cells were then extracted and the HGF receptor was immunoprecipitated with antibodies directed against its C-terminal peptide. The immunoprecipitates were blotted and probed with phosphotyrosine antibodies. Stimulation with HGF induced the phosphorylation on tyrosine of the 145 kd β subunit of the HGF receptor (Fig.1). Similarly to other growth factor receptors, this indicates activation of the receptor kinase and autophosphorylation on tyrosine.

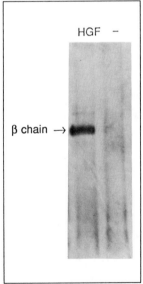

Regulation of the tyrosine kinase activity of HGF receptor by phosphorylation

The kinase activity of HGF receptor was positively regulated by tyrosine autophosphorylation. Immunocomplexes contain-

Figure 1 HGF-induced tyrosine phosphorylation of the β chain of its receptor.

ing the same amount of either unphosphorylated or phosphorylated HGF receptor were challenged with the exogenous substrate [Val⁵] angiotensin II and $^{32}P-\gamma-ATP$. The enhancement of activity involved an increase in the V_{max} of the enzyme catalyzed phospho-transfer reaction, without changes of the observed K_m for the substrates (Naldini et al.,1991b).

The tyrosine kinase activity of the HGF receptor was negatively modulated by protein kinase C activation. For these experiments we used the GTL-16 human gastric carcinoma cell line in which the *MET* gene is amplified and overexpressed (Giordano et al.,1989, Ponzetto et al.,1991). Overexpression of the receptor results in its activation and tyrosine phosphorylation even in the absence of ligand. Treatment of GTL-16 cells with the protein kinase C activator phorbol 12-myristate 13-acetate (TPA) for 1 hr at 37°C inhibited the tyrosine phosphorylation of the HGF receptor. The inactive analog α-phorbol 12,13-didecanoate had no effect. The inhibitory effect was already evident at concen-trations as low as 0.16 nM; maximum inhibition was reached at 16 nM (Gandino et al.,1990).

Increase of intracellular Ca^{2+} had a similar inhibitory effect via a PKC-independent mechanism. A rapid and reversible decrease of HGF receptor tyrosine phosphorylation was induced by a 5 min treatment of cells with 10 μM of the calcium ionophore A23187. When the cells were depleted of PKC by 20 hr treatment with 100 ng/ml of TPA, TPA or other phorbol esters were unable to exert any regulatory effect on the extent of tyrosine phos-phorylation of the HGF receptor. On the contrary, the calcium ionophore was still able to exert their inhibitory effect on HGF receptor tyrosine phosphorylation, showing that PKC and Ca^{2+} inhibit the receptor kinase by two independent mechanisms (Gandino et al.,1991).

Both inhibitory mechanisms involved phosphorylation of a critical serine residue in the HGF receptor. After TPA or A23187 treatment the HGF receptor was almost exclusively phosphorylated on serine residues. Phosphopeptide analysis showed that the increase in phosphoserine was due to phosphorylation of specific site(s) in a single phosphopeptide (Gandino et al.,1991).

Identification of the major tyrosine autophosphorylation site of HGF receptor

The HGF receptor was immunoprecipitated, allowed to phosphorylate in the presence of $[^{32}P]-\gamma$-ATP, and digested with trypsin. A major phosphopeptide was purified by reverse phase chromatography. The phosphorylated tyrosine was identified as residue 1235 (Tyr_{1235}) by Edman covalent radiosequencing. Metabolic labelling of cells with $[^{32}P]$-orthophosphate followed by immunoprecipitation and tryptic phosphopeptide mapping of the HGF receptor showed that Tyr_{1235} is a major site of tyrosine phosphorylation *in vivo* as well (Ferracini et al.,1991).

Association of the activated HGF receptor with phosphatidylinositol 3-kinase, a key enzyme involved in the transduction of signals generated by tyrosine kinase receptors

Stimulation of A549 target cells with purified HGF/SF resulted in the tyrosine phosphorylation of the HGF receptor and additional minor proteins. One of these putative receptor kinase substrates – of molecular weight 85 kDa – coprecipitated with the HGF receptor in immunocomplexes. 85 kDa is the molecular weight of the regulatory subunit of PI 3-kinase (Carpenter et al.,1990, Otsu et al.,1991, Escobedo et al.,1991). PI 3-kinase activity was present in immunocomplexes made with anti-receptor antibodies from HGF stimulated cells, as shown by TLC separation of the D-3 phosphorylated inositol lipids generated *in vitro* by incubation of the immunocomplexes with lipid substrates and $[^{32}P]-\gamma$-ATP. No PI 3-kinase activity was detectable in immunocomplexes precipitated from unstimulated cells. Furthermore, tyrosine phosphorylation of HGF receptor regulated its physical association with PI 3-kinase in intact cells and was required for binding of PI 3-kinase in vitro (Graziani et al.,1991).

DISCUSSION

The earliest detectable consequence of HGF stimulation of target cells is the tyrosine phosphorylation of its receptor, the kinase encoded by the *MET* protooncogene (Naldini et al.,1991a; Bottaro et al.,1991).

Autophosphorylation on tyrosine residues marks the activation of a tyrosine kinase receptor. It may have important functional consequences on the enzymatic activity of the molecule and it may affect the interaction of the receptor with other effectors or regulators of the signal transduction pathway.

Receptor autophosphorylation on tyrosine enhanced the kinase activity of HGF receptor. Thus, there are at least two independent mechanisms for upregulating the HGF receptor tyrosine kinase. Ligand-dependent activation would be the most powerful, recruiting more enzyme into an active oligomeric state. Autophosphorylation-dependent enhancement of activity would be an additional amplification mechanism. Tyr_{1235} is the major site of HGF receptor tyrosine autophosphorylation both *in vitro* and *in vivo*. The Tyr_{1235} residue is embedded in a consensus pattern for tyrosine phosphorylation (Pinna et al.,1990) located within the tyrosine kinase domain in a segment homologous to the major autophosphorylation sites of other receptor and non-receptor kinases (Hanks et al.,1988).

Serine phosphorylation of a critical site in the HGF receptor exerted a negative control on its kinase activity. This was effected by triggering the activity of protein kinase-C or by increasing the intracellular concentration of Ca^{2+} ions. It is known that PKC negatively modulates several tyrosine kinase receptors including EGF-receptor, Insulin-receptor and Insulin-like Growth Factor 1-receptor. In all these cases inhibition is mediated by threonine and/or serine phosphorylation of the receptor (Cochet et al., 1984; Hunter et al., 1984; Davis and Czech, 1985; Jacobs et al., 1983; Takayama et al., 1988). This should endow the HGF receptor with the capability of receiving a down-regulatory input from the signal transduction pathways activated by itself and other growth factors coupled to protein kinase C or Ca^{++} fluxes.

We have also investigated the effect of receptor phosphorylation on the signalling pathways triggered by HGF. The tyrosine phosphorylation of the HGF receptor was a switch for association of PI 3-kinase. This enzyme plays a key role in the signal transduction pathway by generating D-3 phosphoinositides, a novel class of putative intracellular second messengers (Cantley et al.,1991). PI 3-Kinase has been found to associate with and to be activated by a number of tyrosine kinase receptors, including the receptors for PDGF (Whitman et al.,1987), colony stimulating factor-1 (Varticovski et al.,1989), insulin (Ruderman et al.,1990), EGF (Bjorge et al.,1990) and stem cell growth factor (Lev et al.,1991). As with the HGF receptor, the association is ligand-dependent and mediated by tyrosine phosphorylation of the receptor cytoplasmic domain. These data strongly suggest that the signalling pathway activated by the HGF receptor includes generation of D-3 phosphorylated inositol phospholipids.

The elucidation of the mechanisms regulating the HGF receptor activity should help in understanding the unprecedented complexity of the HGF-*MET* ligand-receptor couple, which triggers such diverse biological responses as growth, motility and matrix invasion in its target cells.

Acknowledgements: This work was supported by grants from the Associazione Italiana Ricerche Cancro (AIRC) and the Italian National Research Council (CNR: PF ACRO) to PMC.

308

REFERENCES

Bjorge, J., Chan, T., Antczak, M., Kung, H., and Fujita, D. (1990) *Proc. Natl. Acad. Sci. U.S.A.* **87**, 3816-3820

Bottaro, D.P., Rubin, J.S., Faletto, D.L., Chan, A.M.-L., Kmiecick, T.E., Vande Woude, G.F., and Aaronson, S.A. (1991) *Science* **251**, 802-804.

Cantley, L.C., Auger, K., Carpenter, C., Duckworth, B., Graziani, A., Kapeller, R., and Soltoff, S.(1991) *Cell* **64**, 281-302.

Carpenter, C.L., Duckworth, B.C., Auger, K.R., Cohen, B., Schaffausen, B.S., and Cantley, L.C. (1990) *J. Biol. Chem.* **265**, 19704-19711

Cochet, C., Gill, G.N., Meisenhelder, J., Cooper., J.A., and Hunter, T. (1984). *J. Biol. Chem.*, **259**, 2553-2558.

Davis, R.J. and Czech, M.P. (1985). *Proc. Natl. Acad. Sci. USA*, **82**, 1974-1978.

Di Renzo, M.F., Narsimhan, R.P., Olivero, M., Bretti, S., Giordano, S., Medico, E., Gaglia, P., Zara, P., and Comoglio, P.M. (1991) *Oncogene*, **6**, 1997-2003.

Escobedo, J., Navankasattusa, S., Kavanaugh, M.,Milfay, D., Fried, V., and Williams, L. (1991) *Cell* **65**, 75-82

Ferracini, R., Longati,P., Naldini, L., Vigna, E., and Comoglio, P.M. (1991) *J. Biol. Chem.*, **266**, 19558-19654.

Gandino, L., M. F. DiRenzo, S. Giordano, F. Bussolino and P. M. Comoglio. (1990) *Oncogene* **5**, 721-725.

Gandino, L., L. Munaron, L. Naldini, M. Magni and P.M. Comoglio. (1991) *J. Biol. Chem.*, **266**, 16098-16104.

Giordano, S., Ponzetto, C., Di Renzo, M.F., Cooper, C.S. and Comoglio, P.M. (1989a). *Nature* **339**, 155-156.

Giordano, S., Di Renzo, M.F., Narsimhan, R., C.S. Cooper, C. Rosa, and Comoglio, P.M. (1989b) *Oncogene*, **4**, 1383-88.

Gohda, E., Tsubouchi, H., Nakayama, H., Hirono, S., Sakiyama, O., Takahashi, K., Miyazaki, H., Hashimoto, S., and Daikuhara, Y. (1988) *J.Clin. Invest.*, **81**, 414-419.

Gonzatti-Haces, M., Seth, A., Park, M., Copeland, T., Oroszlan, S., and Vande Woude, G.F. (1988) *Proc. Natl. Acad. Sci. USA*, **85**, 21-25.

Graziani, A., Gramaglia, D., Cantley, L.C., and Comoglio, P.M. (1991) *J. Biol. Chem.*, **266**, 22087-22090.

Hanks, S. K., A. M. Quinn, and T. Hunter. (1988) *Science* **241**, 42-51.

Hunter, T., Ling, N. and Cooper, J.A. (1984). *Nature*, **311**, 480-483.

Jacobs, S., Shayoun, N.E., Saltiel, A.R. and Cuatrecasas, P. (1983). *Proc. Natl. Acad. Sci. USA*, **80**, 6211-6213.

Kan, M., Zhang, G.H., Zarnegar, R., Michalopoulos, G., Myoken, Y., Mckeehan, W.L., and Stevens, J.L. (1991) *Biochem. Biophys. Res. Commun.* **174**, 331-337.

Miyazawa, K., Tsubouchi, H., Naka, D., Takahashi, K., Okigaki, M., Arakaki, N., Nakayama, H., Hirono, S., Sakiyama, O., Takahashi, K., Gohda, E., Daikuhara, Y., and Kitamura, N. (1989) *Biochem. Biophys. Res. Commun.* **163**, 967-973.

Nakamura, T., Teramoto, H., and Ichihara, A. (1986) *Proc. Natl. Acad. Sci. U.S.A.* **86**, 6489-6493.

Nakamura, T., Nishizawa, T., Hagiya, M., Seki, T., Shimonishi, M., Sugimura, A., Tashiro, K., and Shimizu, S. (1989) *Nature* **342**, 440-443.

Naldini L., Vigna, E., Narshiman, R., Gaudino G., Zarnegar R., Michalopoulos G., and Comoglio P.M. (1991a) *Oncogene* **6**, 501-504.

Naldini, L., E. Vigna, R. Ferracini, P. Longati, L. Gandino, M. Prat, and P. M. Comoglio. (1991b) *Mol. Cell. Biol.* **11**, 1793-1803.

Naldini, L., Weidner, M., Vigna, E., Gaudino G., Bardelli A., Ponzetto, C., Narsimhan, R., Hartmann, G., Zarnegar, R., Michalopoulos, G., Birchmeier, W., and Comoglio, P.M. (1991c) *EMBO J.*, **10**, 2867-2878.

Otsu, M., Hiles, I., Gout, I., Fry, M., Ruiz-Larrea, F., Panayotou, G., Thompson, A., Dhand, R., Hsuan, J., Totty, N., Smith, A., Morgan, S., Courtneidge, S., Parker, P., and Waterfield, M. (1991) *Cell* **65**, 91-104

Park. M., Dean, M., Kaul, K., Braun, M.J., Gonda, M.A., and Vande Woude, G. (1987) *Proc. Natl. Acad. Sci. USA*, **84**, 6379-6383.

Pinna, L. A. (1990) *Biochem. Biophys. Acta* **1054**, 267-284.
Lev, S., Givol, D., and Yarden, Y. (1991) *EMBO J.* **10**, 647-654

Ponzetto, C., Giordano, S., Peverali, F., Della Valle, G., Abate, M., Vaula, G. and Comoglio, P.M. (1991) *Oncogene*, **6**, 553-559.

Prat, M.P., Narsimhan, R.P., Crepaldi, T., Nicotra, M.R., Natali, P.G., and Comoglio, P.M. (1991) *Int. J. Cancer*, **49**, 323-328.

Rubin, J.S., Chan, A.M.-L., Bottaro, D.P., Burgess, W.H., Taylor, W.G., Cech, A.C., Hirschfield, D.W., Wong, J., Miki, T., Finch, P.W., and Aaronson, S.A. (1991) . *Proc. Natl. Acad. Sci. U.S.A.* **88**, 415-419

Ruderman, N.B., Kapeller, R., White, M.F., and Cantley, L.C. (1990) *Proc. Natl. Acad. Sci. U.S.A.* **87**, 1411-1415.

Stoker, M., Gherardi, E., Perryman, M., and Gray, J. (1987) *Nature* **327**, 239-242.

Takayama, S., White, M.F. and Kahn, C.R. (1988). *J. Biol. Chem.*, **263**, 3440-3447.

Varticovski, L., Drucker, B., Morrison, D., Cantley, L., and Roberts, T. (1989) *Nature* **342**, 699-702

Weidner, K.M., Arakaki, N., Vandekerchove, J., Weingart, S., Hartmann, G., Rieder, H., Fonatsch, C., Tsubouchi, H., Hishida, T., Daikuhara, Y., and Birchmeier, W. (1991) *Proc. Natl. Acad. Sci. U.S.A.*, **88**, 7001-7005.

Weidner, K.M., Behrens, J., Vandekerckove, J. and Birchmeier, W. (1990) *J. Cell Biol.* **111**, 2097-2108.

Whitman, M., Kaplan, D., Roberts, T., and Cantley, L. (1987) *Biochem. J.* **247**, 165-174

Zarnegar, R., and Michalopoulos, G. (1989) *Cancer Res.* **49**, 3314-3320.

Adenine Nucleotides in Cellular Energy Transfer and Signal Transduction
S. Papa, A. Azzi & J.M. Tager (eds)
© 1992 Birkhäuser Verlag, Basel/Switzerland

THE KIT RECEPTOR FOR THE STEM CELL FACTOR: GENETIC LESSONS IN SIGNAL TRANSDUCTION

Sima Lev[1], Janna Blechmann[1], Allain Berrebi[2], David Givol[1], and Yosef Yarden[1]

[1]Department of Chemical Immunology, The Weizmann Institute of Science, Rehovot 76100, Israel; [2]Hematology Institute, Kaplan Hospital, Rehovot 76100, Israel.

SUMMARY: Among the receptors for growth factors, Kit is uniquely characterized by the existence of multiple mutations, that are responsible for the phenotype of W mutant mice. The typical structure of this receptor consists of an extracellular ligand binding domain and a cytoplasmic part that carries a bisected tyrosine kinase sequence. The latter is the target for most structural mutations which lead to inactivation of the cat-alytic function. Phenotypically, the mutant mice display pleio-tropic defects in germ cells, erythroid progenitors and melanob-lasts.   However, different W mutants exhibit a variety of trait severity, and some heterozygotes display a more severe phenotype than expected from half gene dosage (negative dominance).
   To address the biochemical basis of these genetic phenomena we studied the mechanism of signal transduction by Kit. Using a chimeric Kit receptor we found that the tyrosine kinase is func-tionally coupled to a set of cytoplasmic signalling molecules that is related, yet distinct, from the sets utilized by the closest receptors for the macrophage- and the platelet-derived growth factors. In addition, constitutive ligand-mediated stim-ulation of the Kit kinase resulted in a transformed phenotype, analogous to the oncogenic retroviral form of the protein.
   The recent identification of the stem cell factor (SCF) as the ligand of Kit enabled direct biochemical analysis. We were able to localize the non-catalytic kinase-insert region as an effector domain of Kit, which mediates association with SH2-proteins. On the other hand, analysis of the interaction between SCF and Kit revealed that the receptor undergoes exten-sive dimerization subsequent to ligand binding. Moreover, the dimerization function appears to be mediated only by the extra-cellular domain as

was evident from experiments with a recombinant truncated receptor. These biochemical observations may have important implications to the genetics of Kit and W mutant mice. In addition, our preliminary work on Kit in human hematological disorders may link between the protooncogene and leukemia of stem cells.

INTRODUCTION

Increasingly complex growth regulatory mechanisms had to be evolved as evolution proceeded from unicellular to multicellular organisms. Mitosis in the latter is controlled by intercellular machinaries that include soluble factors as well as cell-surface attached structures such as adhesion molecules and neighboring tissues. A major boost to the research of the control of animal cell growth was the discovery that cancer causing agents (carcinogens) affect the rate of cell division by interaction with a single or a few proteins which are components of the control circuits. The corresponding genes, either oncogenes or anti-oncogenes, are thus activated by a variety of carcinogens including chemicals and viruses. For example, a group of RNA tumor viruses (retroviruses) direct the synthesis of mutant membrane receptors for soluble growth factors. Normally the receptors are transmembrane glycoproteins that bind the polypeptide factors to their extracellular portions. These all connected to cytoplasm-facing catalytic activity that phosphorylates tyrosine residues (Yarden and Ullrich, 1988). Aberrant viral forms of the receptor tyrosine kinases (RTKs) invariably carry the whole catalytic core, but are defective in the adjacent protein sequences. This reflects the ubiquitous requirement of the enzymatic function for the activity of both the normal or the transforming receptor proteins. Recent studies performed with various RTKs indicated that the ligand-activated receptors are functionally coupled to a cascade of signal-generating molecules that include protein kinases, phospholipases and nucleotide-binding proteins (Cantley et al., 1991; Ullrich and Schlessinger, 1990; Yarden and Kelman, 1991).

The present review will use the Kit/stem cell factor (SCF) receptor as an example of an oncogenic receptor for growth factor. Kit is a rather unique RTK as natural mutants of it exist in mice. The biological as well as the biochemical aspects of Kit and its mutants will be described with special emphasis on the mechanism of signal transduction.

The viral and cellular Kit genes

Hardy and Zuckerman reported in 1986 on the isolation of a new feline sarcoma virus from a domestic cat (Besmer et al., 1986). The genome of this virus, termed HZ4, contained a truncated gag gene fused to an oncogene that was denoted kit. The predicted v-Kit amino-acid sequence displayed partial homology with tyrosine kinases with the closest homolog being the viral fms oncogene. Unlike v-Fms, however, v-Kit displayed no sequence qualified to function as a transmembrane domain. Yet, both v-Kit and v-Fms displayed a bisected kinase domain. Later in 1986 a third tyrosine kinase that carried a split catalytic core was discovered. This was the murine receptor for the platelet-derived growth factor (PDGF, Yarden et al., 1986). The similarity in general architecture and sequence homology displayed by the receptors for the macrophage growth factor (CSF-1) and the PDGF-receptor therefore predicted that the cellular kit gene encodes a distinct receptor for a growth factor. This possibility was confirmed by molecular cloning of the human c-kit gene (Yarden et al., 1987). The latter was found to encode a 145 kilodalton glycoprotein that includes an extracellular domain with five immunoglobulin (Ig)-like domains. Thus, the receptors for CSF-1 and PDGF and the Kit protein shared structural features at their exoplasmic portions (five Ig-like domains) and at the cytoplasmic domains (split tyrosine kinase sequences). Despite overall structural similarities, the non-catalytic sequences of the three proteins displayed sequence heterogeneity. This is most remarkable for the ligand binding domains and the kinase inertions, implying receptor-specific functions for these protein regions. A fourth putative receptor of this class of RTKs has been recently isolated from fetal liver (Matthews et al., 1991). In addition, a differentially spliced RNA of c-kit, which results in an in-frame insertion of four amino-acids within the ectodomain, has been detected in murine mast cells (Hayashi et al., 1991; Reith et al., 1991)

The W locus and phenotype

Initial chromosomal localization studies mapped c-kit to human chromosome 4 (region q23 to q34), and mouse chromosome 5 (Yarden et al., 1987). Later and more precise mapping analysis revealed that c-kit was allelic with the W locus of the mouse (Chabot et al., 1988; Geissler et al., 1988). Mutations at the dominant white (W) locus of the laboratory mouse result in pleiotropic effects upon embryonic development and hematopoiesis in adult life, and they appear to affect proliferation and/or migration of progenitor cells (reviewed by Russell, 1979). Homozygous mice are usually sterile, have extensive white spotting and severe anemia. Experiments involving transplantation, in vitro co-culture or analysis of aggregation chimeras indicated that the defect in W mutant mice is intrinsic to the stem cells of the affected tissues, in contrast to the environmental defect of the phenotypically indistinguishable steel (Sl) mutation which affects chromosome 10. The identical phenotypes and the functional complementarity of the W and Sl mutations led, already more than a decade ago, to the hypothesis that the encoded proteins are a ligand and the cognate receptor (Russell, 1979). This hypothesis has been molecularily proven in the last year as will be described below.

Mapping of W mutations

Interestingly, a large number of W alleles exhibit independence of the pleiotropic effects. In other words, the severity of the macrocytic anemia, extent of coat spotting and whether or not the animal is fertile, vary in different W mutant mice (Russell, 1979; Silvers, 1979). A hint as to the basis of this phenomenon could be provided by molecular definition of the different mutations at the kit gene, and detailed analysis of c-kit expression in normal and mutant animals. In-situ hybridization and Northern blot analyses revealed that the gene is expressed in all tissues affected by W mutations including the blood islands of the yolk sac, mast cells and neural-crest-derived melanocytes (Orr et al., 1990; Nocka et al., 1989). Yet, high expression of c-kit was observed in the central nervous system with no known neurological defects in W mutant mice. Consistent with defective receptor functions, mast cells derived from W/W^V mutant mice displayed impaired tyrosine kinase activity of the Kit receptor (Nocka et al., 1989). This observation was later extended to other mutants including W homozygous mice and could be correlated with the severity of the physiological defects (Reith et al., 1990; Nocka et al., 1990). Point mutations that replace highly conserved amino-acids at the kinase domain were identified in the cases of W^{42} (Tan et al., 1990), W^{41}, and W^V (Reith et al., 1990; Nocka et al., 1990). The W^{37} mutation, however, similarily affects a residue at the juxtamembrane region outside of the catalytic core (Nocka et al., 1990; Reith et al., 1990). Besides point mutations, a deletion of an amino-acid sequence that includes the transmembrane domain and flanking regions was found in the W mutation (Nocka et al., 1989) and disruptions of sequences were located in W^{44} (deleted fragment: residues 240-342; Geissler et al., 1988) and W^x (residues 342-791; Geissler et al., 1988). In contrast with the mild phenotypes of W/+ and W^{19H}/+ heterozygous mice, the heterozygous phenotypes of W^{42}, W^{37}, W^V and W^{41} are more severe than would be expected by 50% gene dosage. This dominant negative phenomenon is probably due to the formation of receptor heterodimers between wild-type and mutant Kit proteins. Presumably the heterodimers have impaired signal transduction function.

The ligand molecule of Kit

As predicted by the similar phenotypes of W and Sl mutations, and the non-cell-autonomons effect of Sl mutations (Russell et al., 1979), the ligand of Kit was found to be encoded by the Sl locus (Copeland et al., 1990; Huang et al., 1990; Zsebo et al., 1990a). The ligand molecule was purified from the supernatants of either a buffalo rat liver cell line (Zsebo et al., 1990b), a murine bone marrow stromal cell line (Williams et al., 1990) or murine 3T3 fibroblasts (Huang et al., 1990). The factor, denoted stem cell factor (SCF), mast cell growth factor (MGF), or Kit ligand (KL) is a heavily glycosylated protein that forms non-covalent dimers. The ligand is synthesized as a precursor membrane protein which is present in two tissue-specific alternative splicing forms (Flanagan et al., 1991). It appears that both the membrane-bound and the soluble forms of SCF are capable of stimulating cell proliferation, but the surface-anchored form may be physiologically more important (Flanagan et al., 1991). SCF is by itself only a modest growth factor for hematopoietic progenitor cells, but it synergizes dramatically with other lymphokines such as GM-CSF and interleukin-7 (Zsebo et al., 1990b).

Signal transduction by Kit/SCF-receptor

The mechanism by which Kit transduces a mitogenic signal is interesting from the genetic point of view as it may explain the variable effect of W mutations on different tissues. It may also provide a molecular basis for the phenomenon of negative dominance. Another interesting aspect is the comparison of the

Kit signaling machinary with those of the closely related receptors for PDGFs and CSF-1. Apparently, despite extensive structural homology, each receptor evokes distinct cellular responses and activates specific gene programs.

To be able to study signal transduction by Kit when the endogeneous ligand was yet unknown we constructed a chimeric kit molecule (Lev et al., 1990). This molecule was composed of the extracellular portion of the receptor for the epidermal growth factor (EGF) and the cytoplasmic and transmembrane domains of p145^{kit}. The hybrid molecule was expressed in murine fibroblasts and was found to undergo activation in response to binding of the heterologous ligand. The latter not only stimulated the tyrosine kinase activity but also induced a potent mitogenic signal. Moreover, the ligand-stimulated Kit tyrosine kinase conferred a transformed phenotype to murine fibroblasts (Lev et al., 1990). This transforming potential of a chimeric Kit protein reflects the oncogenic function of the v-kit-encoded tyrosine kinase (Besmer et al., 1986). Employing the chimeric EGF-receptor/Kit, we undertook analysis of the Kit-specific signal transduction pathway (Lev et al., 1991). Biochemical studies revealed that upon stimulation the Kit kinase strongly associates with a phosphatidylinositol 3'-kinase (PI3K) activity and its regulatory 85 kilodalton subunit. In addition to PI3K, Kit leads to tyrosine phosphorylation of phospholipase C$_\gamma$ (PLC$_\gamma$) and serine and threonine phosphorylation of the Raf1 protein kinase. Nevertheless, the interaction with PLC$_\gamma$ is relatively limited and does not result in a significant change in the production of phosphatidylinositol (PI). When compared with the known signaling pathways of PDGF (Morrison et al., 1990) and CSF-1 (Baccarini et al., 1990), each of the three highly related proteins appears to have a specific combination of signal transfer molecules. Thus, PDGF which is the most mitogenic factor in the triad, couples to PI3K, PLC$_\gamma$, Raf1, the GTPase activating protein (GAP) of Ras and induces overall PI-turnover. On the other hand, CSF-1, which functions mostly as a survival factor (reviewed by Scherr, 1990), couples only to PI3K and Raf1. Our recent studies, as well as by others (Rottapel et al., 1991), indicated no interaction between Kit and ras-GAP. Based on the specific combination of signal-generating molecules that characterizes each receptor, we postulated that the specificity of hormonal signals may be encoded by a combinatorial code rather than by unique substrates. Accordingly, certain combinations are translated into specific gene transcription programs which are the basis for the long-term cellular response.

The effector domain of Kit

As the most efficiently coupled substrate of Kit was found to be PI3K, we undertook mapping of the receptor's domain that performs the interaction. For this analysis we employed the wild-type human Kit protein and a recombinant human SCF (a generous gift from K. Zsebo, Amgen). To localize the effector domain we separately deleted either the non-catalytic 68 amino acid long kinase insert (KI) domain, or the carboxy terminal portion distal to the catalytic sequences. Only the interkinase-deleted mutant receptor lost interaction with PI3K, indicating interactions of PI3K with the KI domain (Lev et al., 1991b). This was further supported by partial inhibition of the association by an anti-peptide antibody directed to the KI domain, and lack of effect of an antibody directed to the carboxy-tail of the SCF receptor. To examine the role of tyrosine phosphorylation of the KI region in establishment of association with PI3K we expressed the whole KI sequence in bacteria and tested its interaction with PI3K. Our analysis indicated that no interaction could be reconstituted unless the bacterially expressed protein was first phosphorylated on tyrosine residues. Based on these observations we concluded that the kinase insert domain of Kit, once tyrosine-phosphorylated, mediates interaction with PI3K. Apparently, no other receptor's domains or activities are involved in this process. This conclusion is consistent with the experiments done by Bernstein and his collaborators (Rottapel et al., 1991) that used a longer recombinant protein of 307 amino acid including the KI domain.

Dimerization of Kit/SCF receptor

Partial inactivation of the function of a wild type Kit receptor by a mutant receptor, as observed in certain W heterozygous mice, predicts the existence of receptor dimers. Indeed, using radiolabeled SCF and chemical crosslinking reagents we observed extensive dimerization of Kit (our manuscript in preparation). This was confirmed by using ^{35}S-methionine labeled Kit and also by Western blot analysis. When tested for tyrosine phosphorylation, either in vitro or in living cells, the dimerized receptors displayed much higher kinase activity as compared with the monomeric state. We therefore concluded that in the absence of

SCF the receptor exists in a monomeric form, and it undergoes an almost complete dimerization upon binding of the ligand. Apparently, the dimerization process is essential for activation of the tyrosine kinase. We currently attempt to study the dimerization mechanism. Our preliminary studies indicate lack of involvement of the transmembrane and cytoplasmic domains of Kit, as a recombinant protein that lacks these domains retained not only ligand binding but also SCF-induced dimerization of the isolated ectodomains. We hope that further studies along these lines will provide more information on the molecular basis of the genetic dominance of Kit mutants in W heterozygous mice.

Kit in human hematological disorders

In contrast with the wealth of information on the physiological role of Kit/SCF-receptor in mice, much less is known about Kit/SCF-receptor in human disorders. Following our initial observation of Kit overexpression in the A172 human glioblastoma cell line (Yarden et al., 1987), we found overexpression of the c-kit gene in 10-15% of human glioblastomas and cell lines derived from them. By using a Kit specific monoclonal antibody, expression of the protooncogene identified a sub-group of acute myelocytic leukemia (AML) patients with poor prognosis (Lerner et al., 1991). To address the relevance of c-Kit expression to hematogoligal diseases we have raised a panel of monoclonal antibodies directed to the human Kit/SCF-receptor. Using such a murine monoclonal antibody, denoted mAb94, we found that normal peripheral blood (PB) and bone-marrow (BM) samples contained 0-4% mAb94 positive cells. In contrast PB samples taken from patients in post-chemotherapy recovery showed 10-20% positive cells. Similarly, samples obtained from leukemic patients (CML and CMML) exhibited relatively high proportion of PB positive cells, up to 30%. In a single patient with myelodisplasia (MDS) evoluting rapidly to acute leukemia we observed that 60% of the blasts strongly expressed Kit together with other markers (HLA-DR, CD34, CD13 and CD33). Preliminary analysis revealed multiple chromosomal aberrations and a constant 4q$^-$ that most probably translocated to chromosome 11. Taken together our observations suggest that the MDS patient developed a stem cell leukemia. Currently we address this possibility by using additional markers and methods.

Acknowledgements: We thank Krisztina Zsebo for recombinant human SCF.   Our studies were supported by grants from The National Institutes of Health (Grant CA 51712), The Wolfson Foundation administered by The Israel Academy of Sciences and Humanities and The Israel Cancer Research Fund.   Y.Y. is a recipient of a Career Development Research Award from The Israel Cancer Research Fund.

REFERENCES

Baccarini, M., Sabatini, D.M., App, H., Rapp, U.R., and Stanley, E.R. (1990) EMBO J. 9, 3649-3657
Besmer, P., Murphy, J.E., George, P.C., Qiu, F., Bergold, P.J., Lederman, L., Synder, Jr, H.W., Brodeur, D., Zuckerman, E., and Hardy, W.D. (1986) Nature 320, 415-421
Cantley, L.C., Auger, K.R., Carpenter, C., Duckworth, B., Graziani, A., Kapeller, R., and Soltoff, S. (1991) Cell 64, 281-302
Chabot, B., Stephenson, D.A., Chapman, V.M., Besmer, P., and Bernstein, A. (1988) Nature 335, 88-89
Copeland, N.G., Gilbert, D.J., Cho, B.C., Donovan, P.J., Jenkins, N.A., Cosman, D., Anderson, D., Lyman, S.D., and Williams, D.E. (1990) Cell 63, 175-183

Flanagan, J.G., Chan, D.C., and Leder, P. (1991) Cell $\underline{64}$, 1025-1035

Geissler, E.W., Ryan, M.A., and Housman, D.E. (1988) Cell $\underline{55}$, 185-192

Hayashi, S.-I., Kunisada, T., Ogawa, M., Yamaguchi, K., and Nishikawa, S.-I. (1991) Nucleic Acids Res. $\underline{19}$, 1267-1271

Huang, E., Nocka, K., Beier, D.R., Chu,T.-Y., Buck, J., Lahm, H.W., Wellner, D., Leder, P., and Besmer, P. (1990) Cell $\underline{63}$, 225-233

Lerner, N.B., Nocka, K.H., Cole, S.R., Qin, F., Strife, A., Ashman, L.K., and Besmer, P. (1991) Blood $\underline{77}$, 1876-1883

Lev, S., Yarden, Y. & Givol, D. (1990) Mol. Cell. Biol. $\underline{10}$, 6064-6068

Lev, S., Givol, D., and Yarden, Y. (1991) EMBO J. $\underline{10}$, 647-654

Lev, S., Givol, D., and Yarden, Y. (1991b) Submitted for publication.

Matthews, W., Jordan, C.T., Wiegand, W.G., Pardoll, D., and Lemischka, R. (1991) Cell $\underline{65}$, 1143-1152

Morrison, D.K., Kaplan, D.R., Rhee, S.G., and Williams, L.T. (1990) Mol. Cell Biol. $\underline{10}$, 2359-2366

Nocka, K., Majumder, S., Chabot, B., Ray, P., Cervone, M., Bernstein, A., and Besmer, P. (1989) Genes Dev. $\underline{3}$, 816-826

Nocka, K., Tan, J.C., Chin, E., Chu, T.Y., Roy, P., Traktman, P., and Besmer, P. (1990) EMBO J. $\underline{9}$, 1805-1813

Orr-Urtreger, A., Avivi, A., Zimmer, Y., Givol, D., Yarden, Y., and Lonai, P. (1990) Development $\underline{109}$, 911-923

Qiu, F., Ray, P., Brown, K., Barker, P.E., Jhanwar, S., Ruddle, F., and Besmer, P. (1988) EMBO J. $\underline{7}$, 1003-1011

Reith, A., Rottapel, R., Giddens, E., Brady, C., Forrester, L., and Bernstein, A. (1990) Genes Dev. $\underline{4}$, 390-400

Reith, A.D., Ellis, C., Lyman, S.D., Anderson, D.M., Williams, D.E., Bernstein, A., and Pawson, T. (1991) EMBO J. $\underline{10}$, 2451-2459

Rottapel, R., Reedijk, M., Williams, D.E., Lyman, S.D., Anderson, D.M., Pawson, T., and Bernstein, A. (1991) Mol. Cell Biol. $\underline{11}$, 3043-3051

Russell, E.S. (1979) Adv. Genet. $\underline{20}$, 357-458

Scherr, C.J. (1990) Am. J. Hematol. $\underline{75}$, 1-12

Silvers, W.K. (1979) White Spotting patch and rump-white. In: The coat colors of mice: A model for gene action and interaction. pp. 206-241 Springer-Verlag, New York.

Tan, J.C., Nocka, K., Ray, P., Traktman, P., and Besmer, P. (1990) Science $\underline{247}$, 209-212

Ullrich, A. and Schlessinger, J. (1990) Cell $\underline{61}$, 203-212

Williams, D.E., Eisenman, J.,Baird, A, Ranch, C., Van Ness, K., March, C.J., Park, L.S., Martin, U., Mochizuki, D.Y., Boswell, H.S., Burgess, G.S., Cosman, D., and Lyman, S.D. (1990) Cell $\underline{63}$, 167-174

Yarden, Y., Escobedo, J.A., Kuang, W.-J., Yang-Feng, T.L., Daniel, T.O., Tremble, P.M., Chen, E.Y., Ando, M.E., Harkins, R.N., Francke, U., Fried, V.A., Ullrich, A., and Williams, L.T. (1986) Nature $\underline{323}$, 226-232

324

Yarden, Y. and Kelman, Z. (1991) Curr. Top. Struct. Biol. <u>1</u>, <u>582-589</u>

Yarden, Y., Kuang, W.-J., Yang-Feng, T., Coussens, L., Munemitsu, S., Dull, T.J., Chen, E., Schlessinger, J., Francke, U., and Ullrich, A. (1987) EMBO J. <u>6</u>, 3341-3351

Yarden, Y. and Ullrich, A. (1988) Ann. Rev. Biochem. <u>57</u>, 443-478

Zsebo, K.M., Williams, D.A., Geissler, E.N., Broudy, V.C., Martin, F.H., Atkins, H.L., Hsu, R.-Y., Birkett, N.C., Okino, K.H., Murdock, D.C., Jacobsen, F.W., Langley, K.E., Smith, K.A., Takeishi, T., Cattanach, B.M., Galli, S.J., and Suggs, S.J. (1990a) Cell <u>63</u>, 213-224

Zsebo, K.M., Wypych, J., McNiece, I.K., Lu, H.S., Smith, K.A., Karkare, S.B., Sachdev, R.J., Yuschenkoff, V.N., Birkett, N.C., Williams, L.R., Satyagal, V.N., Tung, W., Basselman, R.A., Mendiaz, E.A., and Langley, K.E. (1990b) Cell <u>63</u>, 195-201

Adenine Nucleotides in Cellular Energy Transfer and Signal Transduction
S. Papa, A. Azzi & J.M. Tager (eds)
© 1992 Birkhäuser Verlag, Basel/Switzerland

LIVER ADENYLYL CYCLASES: STRUCTURE AND REGULATION BY cAMP-DEPENDENT PHOSPHORYLATION

Richard T. Premont and Ravi Iyengar

Department of Pharmacology, Box 1215, Mount Sinai Medical Center, 1 Gustave Levy Place, New York, NY, 10029, United States of America

SUMMARY: Two distinct adenylyl cyclases have been cloned from a rat liver cDNA library.   These sequences share 70% identity, as compared to 30-35% shared identity among other cloned members of this enzyme family, indicating that these two liver forms comprise a distinct subfamily of adenylyl cyclases. The two liver adenylyl cyclase sequences have been identified in all tissues so far examined by polymerase chain reaction (PCR) amplification of first-strand cDNAs from specific oligonucleotide primers. Unlike other cloned forms of adenylyl cyclase with limited distributions, these two adenylyl cyclases have the wide distribution expected for the "ubiquitous" G_s-regulated adenylyl cyclase activity in mammalian tissues. The mouse homolog of one of these adenylyl cyclases has been identified in the S49 lymphoma cell line. The liver (and S49 cell) adenylyl cyclases share a single conserved protein kinase A consensus phosphorylation site, which may be involved in inhibitory regulation of adenylyl cyclase activity during heterologous desensitization.

Adenylyl cyclases are ubiquitous membrane-bound enzymes which respond to the presence of hormones by increasing the synthesis of cAMP within the cell (see review by Premont & Iyengar, 1990). The activity of adenylyl cyclase in most mammalian tissues is stimulated by the guanine nucleotide binding regulatory protein (G protein) called G_s (Pfeuffer, 1977; Ross & Gilman, 1977). The G_s protein becomes activated in response to agonist occupancy of cell surface receptor proteins, and the activated G_s binds to adenylyl cyclase and increases its rate of cAMP formation.

Functional evidence had indicated the existence of at least three distinct adenylyl cyclase activities over a decade ago (Neer, 1978). The most widely expressed form of adenylyl cyclase is the membrane bound G_s-stimulated enzyme which is insensitive to calmodulin (Sutherland et al, 1962). A membrane bound calmodulin-stimulated adenylyl cyclase is present mainly in neuronal tissues (Brostrom et al 1975; Cheung et al, 1975). A soluble calmodulin-sensitive, G_s-insensitive adenylyl cyclase activity has also been observed in testes (Braun & Dods, 1975).

Recent evidence from cDNA cloning indicates that the family of mammalian adenylyl cyclases is substantially larger than had been appreciated previously. A calmodulin- and G_s-sensitive adenylyl cyclase from bovine brain was cloned based on partial amino acid sequences of the purified protein (Krupinski et al, 1989). Using this sequence as a probe, cDNAs encoding several additional forms adenylyl cyclase have been isolated (Bakalyar & Reed, 1990; Parma et al, 1991; Krupinski, 1991; Feinstein et al, 1991). However, all of these forms of adenylyl cyclase appear to have limited distributions. We have cloned cDNAs encoding two adenylyl cyclases which have tissue distributions consistent with their being the "general" G_s-regulated adenylyl cyclases in mammalian tissues.

RESULTS AND DISCUSSION

A cDNA probe was prepared by PCR amplification and subcloning of the last 900 coding bases of the published bovine type 1 adenylyl cyclase sequence from a bovine brain cDNA library. This product was random-primed labeled and used to screen a rat liver cDNA library at low stringency. Four copies of a single virus were isolated, which contained a 600 bp EcoR1 fragment with high similarity to the probe sequence. Screening with this new probe led to the isolation of five independent clones, whose sequences segregated into two highly similar classes of adenylyl cyclases. Continued screening of three libraries has thus far failed to detect the 5' ends of either class of adenylyl cyclase sequence, or

RAT LIVER ADENYLYL CYCLASE, TYPE A CLONES

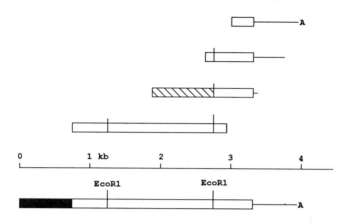

RAT LIVER ADENYLYL CYCLASE, TYPE B CLONES

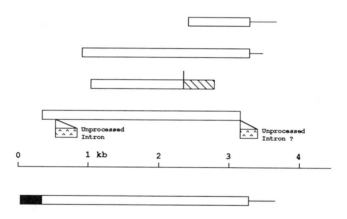

Figure 1: Schematic diagram of adenylyl cyclase clones obtained from the rat liver λZAP library. *Top*: Type A adenylyl cyclase clones. Open boxes are coding regions, and lines represent the 3' untranslated regions. EcoR1 restriction sites are indicated by vertical lines. The scale indicates approximate position in a full-length sequence, and the missing 5' portion of the sequence is highlighted. *Bottom*: Type B adenylyl cyclase clones.

any further adenylyl cyclase forms. Thus, rat liver appears to contain mRNAs for two distinct forms of adenylyl cyclase, which we have tentatively called A and B.

The currently known sequence of the rat A adenylyl cyclase spans 3200 bp, from the end of transmembrane domain 6 to the poly(A) tail. The rat type B sequence stretches 3200 bp, from transmembrane domain 3 to a point in the 3' untranslated region prior its poly(A) tail. Schematic diagrams representing the clones obtained and their relative positions in the sequences of the rat liver A and B adenylyl cyclases are shown in Figure 1.

The amino acid sequences deduced thus far appear to share the 12 membrane span repeated structure found in the other cloned adenylyl cyclases. The A adenylyl cyclase appears to have three N-linked glycosylation sites, while the B has only two. Both A and B adenylyl cyclases have one common protein kinase A consensus site, while A has seven and B eight protein kinase C consensus phosphorylation sites, only three of which are in common.

Comparison of the deduced amino acid sequences of the rat A and B adenylyl cyclases reveals that they share 71% identity and 91% similarity in 863 amino acids of overlap. Matrix comparison of the two sequences (Figure 2) demonstrates that most of the differences are concentrated in the membrane spanning regions 7-12 (amino acids 500-700 in B) and in one region toward the end of the first large intracellular domain (amino acids 350-450 in B).

We have also obtained and sequenced 2300 bp of adenylyl cyclase sequence from the murine S49 lymphoma cell line (UNC variant) by PCR amplification from a region in the first to a region in the second intracellular domain conserved among several adenylyl cyclase types. This S49 cell adenylyl cyclase is highly converved with the rat type B sequence (95% amino acid identity), and likely represents a mouse type B adenylyl cyclase.

A comparison of the rat A and B adenylyl cyclase sequences with all other available adenylyl cyclase sequences indicates 30-40% overall identity of both A and B forms with all the others. A dendrogram indicating the relationships among the cloned adenylyl cyclases is shown in Figure 3. It appears that the type A and B

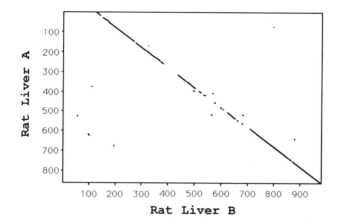

Figure 2: Matrix comparison of the deduced amino acid sequences of the rat liver A and B adenylyl cyclases. Dots indicate regions in the two sequences where 6 of 8 amino acids are identical.

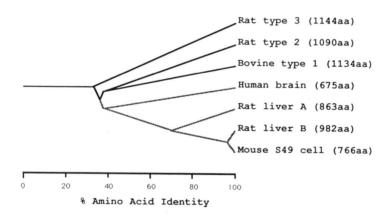

Figure 3: Dendrogram of relationships among cloned adenylyl cyclases. Adenylyl cyclase amino acid sequences were aligned with the PC/GENE CLUSTAL program, and a dendrogram constructed based on the number of identical residues found in each pair of sequences divided by the length of the shorter sequence. That portion of the tree based on partial sequences is indicated by the hashed lines.

adenylyl cyclases form their own sub-family of adenylyl cyclases, and indicates that these two forms of the enzyme may share common regulatory features.

Matrix comparison of the A and B adenylyl cyclases with any of the other adenylyl cyclase sequences reveals that regions of significant homology are confined to the two intracellular domains. Comparisons of the rat A and B adenylyl cyclases with the bovine type 1 sequence are shown in Figure 4. Alignment of the various sequences reveals extensive regions of conservation among the cloned mammalian adenylyl cyclases, and with the catalytic region of guanylyl cyclases (Koesling et al, 1991). These conserved regions are found throughout both intracellular loop regions, with several short stretches of near total identity. The distinct conservation of each of the two cytoplasmic domains despite their repeated nature hint that both loops may be intimately involved in ATP binding and catalysis, rather than only one.

In order to define the tissue distribution of the rat A and B adenylyl cyclase mRNAs, an 1100 bp fragment of the A and B

Table 1: Tissue distribution of rat A and B adenylyl cyclase mRNAs. Two PCR primers were prepared to sequences in the rat A and B adenylyl cyclases which are identical, but not conserved with other known adenylyl cyclases. These primers were used to amplify 1100 bp from the A and B adenylyl cyclase mRNAs using first-strand cDNAs from the indicated rat tissues. The amplified band was digested with Xho1, which cleaves the B sequence into 800 and 300 bp bands, but does not cleave the A sequence. Tissues were scored for the presence of 1100 or 800 + 300 bp bands. A "+" indicates the presence of a band in that tissue.

Tissue	A (1100bp)	B (800+300bp)
Liver	+	+
Brain	+	+
Kidney	+	+
Heart	+	+
Lung	+	+
Testes	+	+

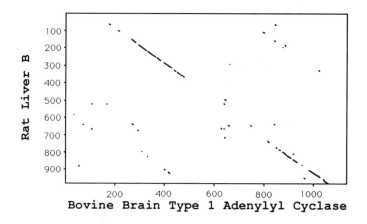

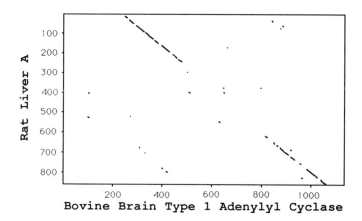

Figure 4: Matrix comparison of rat A and B adenylyl cyclases with bovine brain calmodulin-sensitive adenylyl cyclase (Krupinski *et al*, 1989). Dots indicate regions where 6 of 8 amino acids are identical. *Top panel*: Type B adenylyl cyclase partial sequence. *Bottom panel*: Type A adenylyl cyclase partial sequence.

sequences were co-amplified by PCR from first strand cDNAs prepared from poly(A) RNAs from various tissues, using primers which willnot match other known adenylyl cyclase sequences. The reaction product was then digested with XhoI, which cuts the B sequence only, to determine whether the 1100 bp band contained A or B type sequences.

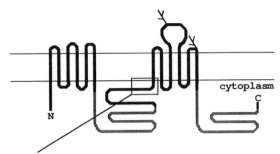

First intracellular domain:
protein kinase A phosphorylation site

Figure 5: Topographic model of adenylyl cyclase structure, with two repeats of six membrane spanning domains adjacent to a large cytoplasmic domain (Krupinski *et al*, 1989). Regions of highest homology among adenylyl (and guanylyl) cyclases in the two large cytoplasmic domains are indicated with the hashed lines, and a region in the rat A and B adenylyl cyclases containing a conserved protein kinase A phosphorylation site is indicated by the box.

These results are summarized in Table 1. Type A product bands, which remain uncut, are seen in all tissues examined. Type B products, which yield 800 and 300 bp fragments upon digestion, are also observed in these tissues. Additionally, 2.3 kb clones encoding A or B adenylyl cyclases have been obtained by PCR amplification from cDNAs prepared from several rat tissues. Thus, the rat A and B adenylyl cyclases appear to be widely distributed in several distinct tissue types.

We had previously observed the regulation of adenylyl cyclase by direct protein kinase A-mediated phosphorylation in heterologous desensitization of both cultured chick hepatocytes and murine S49 lymphoma cells (manuscript in preparation). Treatment of membrane fractions from hepatocytes or S49 cells with protein kinase A and ATP leads to the reduction of hormone and G_s-stimulated adenylyl cyclase activity. A similar loss of stimulation can be produced by hormone or 8-Br-cAMP treatment of intact cells, as well. Since protein kinase A is unable to significantly phosphorylate purified G_s or G_i proteins (Premont & Iyengar, 1989), and since G_s stimulation of adenylyl cyclase requires only the G_s and adenylyl cyclase proteins (Feder et al, 1986; May et al, 1985), it is likely that this effect is due to the direct phosphorylation of the adenylyl cyclase protein. We are currently attempting to demonstrate the hormone and cAMP-dependent phosphorylation of the adenylyl cyclase protein from S49 cells. Protein kinase C-mediated phosphorylation of a 135 kDa human platelet adenylyl cyclase protein has recently been reported (Simmoteit et al, 1991).

Interestingly, both the rat liver A and B and the S49 cell B adenylyl cyclases share a single conserved protein kinase A consensus phosphorylation site located adjacent to membrane span 7 in the first intracellular domain (Figure 5). This site is not present in any other adenylyl cyclases cloned to date. The role of this site in the regulation of adenylyl cyclase activity is an area of active interest in our laboratory.

The individual distributions and relative ratio of the A and B adenylyl cyclases within rat tissues are currently unknown, as are the actual sizes of the mature proteins. We have, however, partially purified a 135 kDa adenylyl cyclase protein from porcine liver membranes using forskolin affinity chromatography (unpublished). We are now preparing antisera specific for these two forms of adenylyl cyclase, and hope to individually identify the A and B adenylyl cyclase proteins. Since functional studies have given little indication that two distinct adenylyl cyclases are present in liver, the question remains whether both A and B adenylyl cyclases are present in hepatocytes. To address this will require mapping the cellular distributions of these two adenylyl cyclases within the rat liver.

Acknowledgements: This work was supported by NIH grants CA 44998 and DK 38761 to RI. We thank Dr. Richard Dixon for the bovine brain cDNA library, and Drs. Randall Reed and Jacques Hanoune for sharing adenylyl cyclase cDNA sequences prior to publication.

334

References

Bakalyar, H.A., and Reed, R.R. (1990) Science 250: 1403-1406.

Braun, T., and Dods, R.F. (1975) Proc. Natl. Acad. Sci. USA 72: 1097-1101.

Brostrom, C.O., Huang, Y., Breckenridge, B.McL., and Wolff, D.J. (1975) Proc. Natl. Acad. Sci. USA 72: 64-68.

Cheung, W.Y., Bradham, L.D., Lynch, T.J., Lin, Y.M., and Tallant, T.J. (1975) Biochem. Biophys. Res. Comm. 66: 1055-1062.

Feder, D., Im, M.-J., Klein, H.W., Helman, M., Holzhöfer, A., Dees, C., Levitzki, A., Helmreich, E.J.M., and Pfeuffer, T. (1986) EMBO J. 5: 1509-1514.

Feinstein, P.G., Schrader, K.A., Bakalyar, H.A., Tang, W.-J., Krupinski, J., Gilman, A.G., and Reed, R.R. (1991) Proc. Natl. Acad. Sci. USA 88: IN PRESS

Koesling, D., Schultz, G, and Böhme, E. (1991) FEBS Lett. 280: 301-306.

Krupinski, J. (1991) Mol. Cell. Biochem. 104: 73-79.

Krupinski, J., Coussen, F., Bakalyar, H.A., Tang, W.-J., Feinstein, P.G., Orth, K., Slaughter, C., Reed, R.R., and Gilman, A.G. (1989) Science 244: 1558-1564.

May, D.C., Ross, E.M., Gilman, A.G., and Smigel, M.D. (1985) J. Biol. Chem. 260: 15829-15833.

Neer, E.J. (1978) In: Advances in Cyclic Nucleotide Research (George, W.J., and Ignarro, L.J., eds.) Raven Press, New York 9: 69-83.

Parma, J., Stengal, D., Gannage, M.-H., Poyard, M., Barouki, R., and Hanoune, J. (1991) Biochem. Biophys. Res. Comm. 179: 455-462.

Pfeuffer, T. (1977) J. Biol. Chem. 252: 7224-7234.

Premont, R.T., and Iyengar, R. (1989) Endocrinol. 125: 1151-1160.

Premont, R.T., and Iyengar, R. (1990) In: G Proteins (Iyengar, R., and Birnbaumer, L., eds.) Academic Press, San Diego pp. 147-179.

Ross, E.M., and Gilman, A.G. (1977) J. Biol. Chem. 252: 6966-6969.

Sambrook, J., Fritsch, E.F., and Maniatis, T. (eds.) (1989) Molecular Cloning. A Laboratory Manual. Second edition, Cold Spring Harbor Press

Simmoteit, R., Schulzki, H.-D., Palm, D., Mollner, S., and Pfeuffer, T. (1991) FEBS Lett. 285: 99-103.

Sutherland, E.W., Rall, T.S., and Menon, T. (1962) J. Biol. Chem. 237: 1220-1227.

Adenine Nucleotides in Cellular Energy Transfer and Signal Transduction
S. Papa, A. Azzi & J.M. Tager (eds)
© 1992 Birkhäuser Verlag, Basel/Switzerland

STRUCTURAL AND FUNCTIONAL ORGANIZATION
OF THE CATALYTIC DOMAIN OF A BACTERIAL TOXIN:
BORDETELLA PERTUSSIS ADENYLATE CYCLASE

Hélène Munier[1], Evelyne Krin[2], Anne-Marie Gilles[1], Philippe Glaser[2],
Ahmed Bouhss[1], Antoine Danchin[2] and Octavian Bârzu[1]

[1]*Unité de Biochimie des Régulations Cellulaires, and*
[2]*Unité de Régulation de l'Expression Génétique,*
Institut Pasteur, 75015 Paris, France

Bacterial toxins frequently appear as multidomain proteins in which each domain corresponds to a well-defined function, such as recognition of cell surface receptors, translocation across membrane(s) and interaction with specific target(s).

Adenylate cyclase, secreted by virulent strains of *Bordetella pertussis*, is a single polypeptide of 1706 amino acid residues endowed with calmodulin (CaM)-dependent ATP-cyclizing activity and haemolysin activity (1, 2). Digestion of adenylate cyclase by trypsin (or trypsin-like bacterial enzymes) releases a 43-kDa form of enzyme (3, 4). This form corresponding to the first 399 amino acid residues has full CaM-dependent catalytic activity but is no more invasive or haemolytic. Further proteolysis of the 43-kDa form of adenylate cyclase released two complementary fragments corresponding to residues 1-224 (P_{1-224} or T25) and 225-399 ($P_{225-399}$ or T18) (3-5).

We attempted to identify different segments or amino acid residues of *B. pertussis* adenylate cyclase susceptible to interact with ligands or to affect indirectly binding or activation of enzyme by CaM, by using several approches : *(i)* isolation of peptides of *B. pertussis* adenylate cyclase able to interact with high affinity with CaM ; *(ii)* deletion of the adenylate cyclase gene and analysis of the CaM-binding and catalytic properties of the truncated proteins expressed in *Escherichia coli* ; *(iii)* site-directed mutagenesis of the residues (or clusters of residues) targeted as potentially involved in interaction with CaM or catalysis.

I/   Mapping the CaM-binding locus of *B. pertussis* adenylate cyclase.

Several pieces of evidence have accumulated over the last three years in our laboratories showing that the region situated around Trp_{242} in bacterial enzyme is essential for binding of CaM: *(i)* substitution by site-directed mutagenesis of Trp_{242} by Gly and Asp decreased the affinity of adenylate cyclase for CaM by a factor of 500 and 1700, respectively (6) ; *(ii)* a synthetic peptide of 20 amino acid residues corresponding to the sequence situated between Arg_{235} and Gly_{254} in *B. pertussis* adenylate cyclase ($P_{235-254}$) was shown to bind CaM ($K_d = 0.57$ μM) in a Ca^{2+}-dependent manner (4). Although highly flexible in solution, this peptide has a propensity to form helical structures as shown by circular dichroism and NMR spectroscopy (7) ; *(iii)* the C-terminal tryptic fragment of adenylate cyclase has a K_d for CaM of 20 nM which is 30 times smaller than that of $P_{235-254}$ but still 100 times higher than that of the uncleaved enzyme ; *(iv)* reassociation of T25 and T18 restores the high affinity CaM-binding properties of the intact protein (5).

To determine if the binding locus for CaM represents a single continuous sequence which spans the C-terminal end of T25 and the N-terminal end of T18 we constructed first a N-terminal deleted protein corresponding to residues 188-399 of *B. pertussis* adenylate cyclase ($P_{188-399}$). This protein interacted with CaM with a K_d of 2.5 nM, a 8-fold lower value than that exhibited by T18. Cleavage of $P_{188-399}$ by cyanogen bromide (CNBr) released several peptides. One of these peptides corresponding to the sequence situated between Pro_{196} and Met_{267} ($P_{196-267}$) was purified by CaM-agarose chromatography and HPLC. Competitive binding assays (8) indicated a K_d value of $P_{196-267}$ for CaM of 2.4 nM, identical with that exhibited by $P_{188-399}$. The fluorescence emission spectrum of $P_{196-267}$ after excitation at 295 nm (maximum at 347 nm), was close to that exhibited by $P_{235-254}$ or T18, respectively, and equimolar concentrations of CaM shifted this maximum to 332 nm with an increase of the quantum yield by a factor of 1.6.

We deduced, therefore, that: *(i)* the sequence downstream Met_{267} plays no role or is involved only indirectly in binding of CaM to adenylate cyclase ; *(ii)* the sequences situated within the first 188 amino acid residues and between Pro_{196} and Arg_{224} contribute each by about 10% to the binding energy of adenylate cyclase to CaM. To validate these estimations two experiments were considered. First, the positively charged cluster of four residues [221]Arg-Thr-Arg-Arg[224] was substituted by site-directed mutagenesis with Gln-Thr-Gln-Gln. The mutated protein did not generate, as expected, the two fragments T25 and T18, upon cleavage with trypsin. Its maximal activity represented 80% of that of parent adenylate cyclase, whereas the affinity for CaM (K_d of 0.76 nM) indicated a loss of only 6% of binding energy to activator protein. In the second approach the whole sequence situated between Ser_{207} and Ala_{220} was deleted by removing the corresponding nucleotides in the gene encoding adenylate cyclase. The internally truncated protein exhibited 2/3 of the activity of wild-type enzyme and a ten-fold lower affinity (K_d = 1.91 nM) for CaM which corresponds to a loss of about 10% of binding energy. It is difficult to say now if the sequence situated between Ser_{207} and Arg_{227} plays a purely structural role (contributing to a proper orientation of the CaM-binding sequence situated around Trp_{242} in its interaction with CaM) or interacting contacts between residues such as Arg_{221}, Arg_{223} or Arg_{224} and corresponding negatively charged residues in CaM increase by a direct effect the binding energy of interaction cyclase-CaM.

II/ Identification of amino acid residues involved in ATP binding and catalysis.

Adenylate cyclase from *B. pertussis* and *Bacillus anthracis* are enzymes whose activity is highly dependent on CaM (9, 10). Although they have different sizes and little sequence homology (11), the location of ATP- and CaM-binding sites and, probably, the mechanisms of activation of these two bacterial enzymes are similar. Therefore, we were interested to know if CaM-induced conformational change of cyclase molecules favored binding of ATP and/or activated catalysis. Kinetic analysis cannot discriminate between these possibilities. We looked therefore for probes such as ATP analogs possessing high affinity for bacterial adenylate cyclases and exhibiting sensitive signals which could differentiate between free and enzyme-bound nucleotides. In a search for ATP analogs which could act as potential substrates or inhibitors we found that an intact 6-amino group at the purine ring is necessary for enzyme-nucleotide recognition (12). Bulky substituents at the 3'OH of the ribose moiety in ATP or dATP substantially increased the affinity of nucleotide for bacterial enzymes with concomitant loss of ability to act as substrate (Table I).

TABLE I

Kinetic constants of *B. pertussis* adenylate cyclase for ATP and its analogs

Nucleotide		K_m or K_i (μM)
Substrate	Inhibitor	
ATP		600
o'ATP		3300
8-Bromo-ATP		2200
8-azido-ATP		3400
2'dATP		100
	3'dATP	350
	Bz$_2$ATP	10
	Ant-ATP	30
	Ant-dATP	9.1
	8-azido-Ant-dATP	74

Bz$_2$ATP : 3'-0-(4-benzoyl)benzoyladenosine 5'-triphosphate ; o'ATP : adenosine N$_1$-oxide 5'-triphosphate ; Ant-dATP : 3'-anthraniloyl-2'-deoxyadenosine 5'-triphosphate.

Binding of Ant-dATP to *B. pertussis* (and *B. anthracis*) adenylate cyclase as revealed by equilibrium dialysis or fluorescence analysis is the first experimental evidence showing that CaM is necessary not only for activating catalysis in a CaM-dependent enzyme, but also to ensure tight binding of the substrate to the active site (13). Whether conformational changes induced by CaM in these two bacterial enzymes result in reversible "closing" of the active site or in removal of an autoinhibitory sequence from the catalytic site remains to be established. 8-azido-Ant-dATP, a photoactivable and fluorescent analog of ATP, was found useful in detecting adenylate cyclase in SDS-polyacrylamide gel (14).

Despite little sequence similarity between *B. pertussis* and *B. anthracis* adenylate cyclases, three short regions of 13-24 amino acid residues in these proteins (residues 54 to 77 ; 184 to 196 ; and 294 to 314 in *B. pertussis* adenylate cyclase, versus residues 342 to 365 ; 487 to 499 ; and 573 to 594 in *B. anthracis* adenylate cyclase) exhibit between 66% and 80% identity suggesting that they might be implicated in some common and important catalytic or regulatory functions.

$$^{54}\text{GVATKGLGVHAKSSDWGLQAGYIP}^{77}$$
$$^{342}\text{GVATKGLNEHGKSSDWGPVAGYIP}^{365}$$

$$^{184}\text{PLTADIDMFAIMP}^{196}$$
$$^{487}\text{PLTADYDLFALAP}^{499}$$

$$^{294}\text{DVVQHGTEQNN PFPEADEKIF}^{314}$$
$$^{573}\text{DVVNHGTEQDNEEFPEKDNEIF}^{594}$$

Having in mind the seemingly "modular" construct of *B. pertussis* adenylate cyclase where the N-terminal fragment (P_{1-224}) corresponds mainly to the ATP-binding site, while the C-terminal fragment ($P_{225-399}$) belongs to the CaM-binding domain (5, 15), we hypothesized that the sequences 54-77 and 184-196 might contain amino acid residues involved in nucleotide binding or/and catalysis, whereas the sequence 294-314 should be involved in activation of adenylate cyclase by CaM. Our hypothesis is strengthened by the fact that the first sequence of homology between the two bacterial adenylate cyclases contains amino acids resembling the consensus sequence GXXXXGKT(S) present in many nucleotide-binding proteins (16, 17). On the other hand the sequence GIPLTADID in *B. pertussis* enzyme is highly similar to the GV(L)PGTIDND sequence in the phosphofructokinase from *Bacillus stearothermophilus* or *E. coli* (18). Using the latter enzyme to scan libraries we found that it can be also aligned with several other ATP-binding enzymes such as pyruvate kinase, thymidylate kinase or DNA polymerase.

Substitution by site-directed mutagenesis of Lys_{58}, His_{63}, Lys_{65}, Asp_{188} and Asp_{190} by several amino acids decreased catalysis in *B. pertussis* adenylate cyclase between two and four orders of magnitude (6, 8, 19). When Lys_{346} in *B. anthracis* adenylate cyclase (equivalent to Lys_{58} in *B. pertussis* enzyme) was replaced with Gln, the modified enzyme was completely inactive (20). From these data as well as from nucleotide-binding studies we deduced that Asp_{188} and Asp_{190} in *B. pertussis* adenylate cyclase interact with Mg^{2+}-

ATP (most probably as a β, γ-bidentate complex) as accepting a hydrogen bond from a water ligand of Mg^{2+} or/and directly coordinating the Mg^{2+} through the carboxylate group. Participation of Asp_{188} and Asp_{190} in the catalytic step would consist in stabilizing the transition state. It is significant in this respect that changes in charge (Asp $\to$ Asn substitution) altered catalysis much more than nucleotide-binding properties, and that replacement of Asp_{188} and Asp_{190} by unrelated amino acids (His or Tyr) abolished completely the activity. We assume that Lys_{65} (or Lys_{58}) interacts with the α-phosphate group of Mg^{2+}-ATP.

Stereochemical studies using phosphorothionate analogs of ATP showed that the reaction catalyzed by adenylate cyclase proceeds with inversion of the configuration at the α-phorphorus (21, 22). Gerlt *et al.* (21) proposed that cyclization occurs by a single nucleophilic displacement and that a general base catalysts is required to assist in the ionization of the 3'OH group of the pentose moiety. A candidate for such a role is the imidazole side chain of the histidine residue, due to its ability to substract protons at pH values close to the optimum pH of the reaction catalyzed by adenylate cyclase. Substitution of His_{63} with Arg, Glu, Gln or Val decreased the catalytic efficiency of adenylate cyclase from *B. pertussis* between two and three orders of magnitude, and altered the kinetic properties of the enzyme. These effects varied in relation to the nature of the substituting residue, pH and direction of the reaction, i.e. ATP cyclization (forward) or ATP synthesis (reverse). Arg was the best substituent of His_{63} as catalyst in the forward reaction, with shift of the optimum pH to the alkaline side, whereas Glu was the best substituent of His_{63} in the reverse reaction, with shift of the optimum pH to the acidic side. The marked increase in the hydroxyde dependence of ATP cyclization between pH 8 and 10 in His_{63} mutant, compared with the wild-type enzyme (Table II) favor the idea that another general acid/base catalyst that has a pKa value higher than 9 might be involved in the attack of the nucleotide substrate. Tyrosine seems to be such a candidate, the role of His_{63} being to shuttle protons from tyrosine.

TABLE II

Catalytic activity of His-modified forms of *Bordetella pertussis*
adenylate cyclase determined in the forward direction at two pH values[a]

Enzyme	Activity (μmol/min·mg of protein)[b]			
	pH 8		pH 10	
WT	1980	(100)	820	(100)
H63R	15	(0.76)	130	(15.8)
H63Q	8.3	(0.42)	90	(10.9)
H63V	13.0	(0.66)	47	(5.7)
H63E	10.3	(0.52)	18	(2.2)

[a]Wild-type and His-modified enzymes were purified by CaM-agarose chromatography.

[b]Activity was determined at 30°C, 5 mM ATP and 1 μM CaM concentrations in Tris-HCl buffer (pH 8) or glycine/NaOH buffer (pH 10). The numbers in parentheses are relative values, the cyclase activity of the wild-type protein being considered as 100%.

Another class of mutations yielding almost inactive forms of *B. pertussis* adenylate cyclase concerns His_{298} and Glu_{301}. These two residues belong to the third segment of identity of the two bacterial adenylate cyclases. Replacement of His_{298} with Arg, Leu or Pro as well as of Glu_{301} with Arg decreased the catalytic efficiency of *B. pertussis* adenylate cyclase by more than three orders of magnitude.

The residual cyclase activity of His_{298} and Glu_{301} mutants exhibited similar kinetic characteristics to that of wild-type adenylate cyclase. We suppose that these two residues are not involved directly in catalysis but participate in the activation of the bacterial enzyme by CaM. Interactive contacts, through hydrogen bondings, between these residues and segments of the polypeptide chain belonging to the active site of bacterial enzyme would optimize binding of the nucleotide substrate to the catalytic site and promote cyclization of ATP. Such "activating" contacts occur or are facilitated upon binding of CaM to adenylate cyclase.

III/ Reconstitution of catalytically active species of *B. pertussis* adenylate cyclase from fragments or inactive forms of enzyme.

The two complementary tryptic fragments of the catalytic domain of *B. pertussis* adenylate cyclase (T25 and T18) can reassociate in the presence of CaM to form species with similar catalytic and hydrodynamic properties to that of uncleaved enzyme (4, 5). The formation of the ternary active complex, involving two separated fragments of adenylate cyclase and CaM, is a unique phenomenon among CaM-activated enzymes and apparently has little analogy with any enzymatic system investigated so far. Reconstitution of active species of adenylate cyclase might also occur when largely overlapping inactive fragments are mixed in the presence of CaM. Thus, mixtures of a C-terminal truncated inactive form of adenylate cyclase (P_{1-343}) and T18 restored 10% of the original catalytic activity. On the other hand, mixtures of P_{1-343} and $P_{324-399}$ showed no increase in activity. These experiments showed that reassociation of adenylate cyclase fragments into active species occurs in a limited number of cases, even if the continuity of the polypeptide chain seems to be acquired.

The selectivity of reassociation was further confirmed by experiments where inactive enzymes resulted by site-directed mutagenesis were mixed with isolated tryptic fragments of adenylate cyclase. Modified proteins at Lys_{58} or Asp_{188} remained inactive in the presence of T18, but recovered between 1 and 3% of the original activity in the presence of T25. And conversely, modified protein at Glu_{301} recovered nearly 5% of the original activity in the presence of T18 but remained inactive in the presence of T25.

Even more surprising was the fact that mixtures of inactive 399 amino acids forms of adenylate cyclase produced active species. Combinations yielding active forms occurred also in well-defined pairs, for example, $K_{58}T + H_{298}R$ or $D_{188}Y + H_{298}R$. It seems that one inactive protein contributes by its intact T25 domain and the other one by its intact T18 domain. To explain this complementation we must assume that wild-type adenylate cyclase, as well as its inactive variants, form dimers or higher molecular weight aggregates. This hypothesis was tested by gel-permeation chromatography. About 90% of wild-type adenylate cyclase migrated as monomeric species and about 10% of the protein appeared under higher molecular weight forms. Aggregation of adenylate cyclase, mostly by dimerization, increased to 20% when CaM was complexed to the enzyme. When two modified inactive proteins are mixed, the only active species resulting were, as expected, the "heterodimer" produced by the association of two different mutants. From the experiments with wild-type protein and association of inactive forms, we can assume that the probability of formation of active dimers is 6.7%, obtained by dividing 20% by 3. In other words, the specific activity of an active dimer is not as low as it was supposed from the percentages given by complementation experiments, since it is estimated to be between 8 and 30% of the activity of wild-type enzyme.

In conclusion, the segment joining the ATP-binding domain with the C-terminal domain of adenylate cyclase is highly mobile. Binding of CaM ensures a proper mutual orientation of these two domains yielding catalytically active species. This argues in favor of further NMR experiments to confirm that the C-terminal tryptic domain of adenylate cyclase contains not only information concerning binding of CaM, but also residues involved in the process of activation. This concept, which makes adenylate cyclase a very particular CaM-activated protein, will need experimental proofs in the future.

Acknowledgments

We thank S. Michelson for helpful comments and M. Ferrand for excellent secretarial help. This work was supported by grants from the Institut Pasteur, the C.N.R.S. (URA D1129), and the I.N.S.E.R.M. (no. 910.615).

344

References

1. Glaser, P., Ladant, D., Sezer, O., Pichot, F., Ullmann, A., and Danchin, A.(1988) Molec. Microbiol. 2, 19-30
2. Glaser, P., Sakamoto, H., Bellalou, J., Ullmann, A., and Danchin, A. (1988) EMBO J. 7, 3997-4004
3. Ladant, D. (1988) J. Biol. Chem. 263, 2612-2618
4. Ladant, D., Michelson, S., Sarfati, S. R., Gilles, A.-M., Predeleanu, R., and Bârzu, O. (1989) J. Biol. Chem. 264, 4015-4020
5. Munier, H., Gilles, A.-M., Glaser, P., Krin, E., Danchin, A., Sarfati, S.R., and Bârzu, O. (1991) Eur. J. Biochem. 196, 469-474
6. Glaser, P., Elmaoglou-Lazaridou, A., Krin, E., Ladant, D. , Bârzu, O., and Danchin, A. (1989) EMBO J. 8, 967-972
7. Prêcheur, B., Siffert, O., Bârzu, O., and Craescu, C.T. (1991) Eur. J. Biochem. 196, 67-72
8. Glaser, P., Munier, H., Gilles, A.-M., Krin, E., Porumb, T., Bârzu, O., Sarfati, S.R., Pellecuer, C., and Danchin, A. (1991) EMBO J. 10, 1683-1688
9. Wolff, J., Hopecook, G., Goldhammer, A.R., and Berkowitz, S.A. (1980) Proc. Natl. Acad. Sci. U.S.A. 77, 3840-3844
10. Leppla, S.H. (1982) Proc. Natl. Acad. Sci. U.S.A. 79, 3162-3166
11. Mock, M., Labruyère, E., Glaser, P., Danchin, A., and Ullmann, A. (1988) Gene 64, 277-284
12. Bellalou, J., Sarfati, S.R., Predeleanu, R., Ladant, D., and Bârzu, O. (1988) Enzyme and Microbial Technology 10, 293-296
13. Sarfati, S.R., Kansal, V.K., Munier, H., Glaser, P., Gilles, A.-M., Labruyère, E., Mock, M., Danchin, A., and Bârzu, O. (1990) J. Biol. Chem. 265, 18902-18906
14. Sarfati, S.R., Namane, A., Munier, H., and Bârzu, O. (1991) Tetrahedron Lett. 32, 4699-4702
15. Gilles, A.-M., Munier, H., Rose, T., Glaser, P., Krin, E., Danchin, A., Pellecuer, C., and Bârzu, O. (1990) Biochemistry 29, 8126-8130
16. Moller, W., and Amons, R. (1985) FEBS Lett. 186, 1-7
17. Fry, D.C., Kuby, J.A., and Mildvan, A.S. (1986) Proc. Natl. Acad. Sci. U.S.A. 83, 907-911
18. Shirakihara, Y., and Evans, P.R. (1988) J. Mol. Biol. 204, 973-994
19. Munier, H., Bouhss, A., Krin, E., Danchin, A., Gilles, A.-M., Glaser, P., and Bârzu, O. (1992) J. Biol. Chem. (submitted)
20. Labruyère, E., Mock, M., Surewicz, W.K., Mantsch, H.H., Rose, T., Munier, H., Sarfati, S.R., and Bârzu, O. (1991) Biochemistry 30, 2619-2624
21. Gerlt, J.A., Coderre, J.A., and Wolin, M.S. (1980) J. Biol. Chem. 255, 331-334
22. Eckstein, F., Romaniuk, P.J., Heideman, W., and Storm, D.R. (1981) J. Biol. Chem. 256, 9118-9120

Adenine Nucleotides in Cellular Energy Transfer and Signal Transduction
S. Papa, A. Azzi & J.M. Tager (eds)
© 1992 Birkhäuser Verlag, Basel/Switzerland

SIGNAL TRANSDUCTION AS STUDIED IN LIVING CELLS BY ^{31}P-NMR: ADENYLATE CYCLASE ACTIVATION AND A NOVEL PHOSPHOETHANOLAMINE SYNTHESIZING PATHWAY ARE STIMULATED BY MSH IN MELANOMA CELLS

Yoram Salomon[1], John DeJordy[1,2] and Hadassa Degani[2]

Departments of Hormone Research[1] and Chemical Physics[2], The Weizmann Institute of Science Rehovot, 76100 ISRAEL

SUMMARY: ^{31}P nuclear magnetic resonance (NMR) spectroscopy can now uniquely provide a real-time panoramic view of the major intracellular phosphate metabolites and their concentrations in living cells/tissues. Hormone regulated cascades in many instances influence intracellular phosphate metabolism. This influence is apparent mainly at two levels: (1) cellular energy, i.e., modulation of ATP synthesis and utilization, and (2) cellular control mechanisms where regulation of key enzymes is often mediated by second messengers, themselves phosphate metabolites, such as 3'5' cyclic adenosine monophosphate (cAMP), 3'5' cyclic guanosine monophosphate (cGMP), inositol tris phosphate (IP$_3$) and/or by protein phosphorylation/dephosphorylation reactions. Certain aspects of energy-related phosphate metabolism have been previously reported in the literature. We were thus prompted to apply this non-invasive technique, to examination of signal transducing processes and the responsive cascades regulated by the melanocyte stimulating hormone (MSH) in live cultured M2R mouse melanoma cells. We describe here two major observations that were made in the course of these studies. (1) The discovery of a novel phospholipid associated pathway stimulated by MSH that leads to phosphoethanolamine (PE) production, and (2) that MSH stimulated synthesis of cAMP and a concomitant adenylate cyclase (AC) dependent depletion of ATP in these cells can be simultaneously monitored by ^{31}P NMR spectroscopy.

Cellular cAMP levels are set by tight control of the relative rates of the adenylate cyclase (AC) and cAMP phosphodiesterase (PDE) reactions (Sutherland et al., 1965). These enzymes are part of a common mechanism for hormonal control of a large number of cellular processes (Greengard & Robison, eds., 1984). It has been estimated that cAMP levels in cells fluctuate in the submicromolar to micromolar range to afford control upon cAMP dependent protein kinase (PKA), the key control point of the relevant cell-specific enzymic cascades (Rosen & Krebs, eds., 1981). Up until now a true comprehensive characterization of the reactions that are influenced by the hormone and their time course could not be continuously studied *in vivo*, mainly due to methodological limitations. Consequently, the study of these and other enzymic cascades was mainly accomplished by *in vitro* examination and cell-free reconstitution of partial reactions and of individual steps. This experimental approach relied heavily on simulation of the best physiological conditions that could be predicted and on an arbitrary set of assumptions which could never be accurately tested or proved. Therefore, the results obtained for each individual reaction, and even more so, the extrapolations made to the *in vivo* situation with respect to entire enzymatic cascades obviously left a relatively wide margin for interpretation.

In comparison to the classical biochemical approach for analysis of intermediary metabolism, the main advantages of *in vivo* NMR spectroscopy can be summarized as follows: (i) The method is non-invasive i.e. no perturbation of the system is necessary. (ii) A large number of metabolites can be identified and quantitated simultaneously. (iii) Continuous monitoring of metabolite-fluxes and their steady state levels can be obtained essentially under real-time conditions. (iv) Much of the information on control and stimulated states of the experimental system can be obtained from a single sample. It should be noted however that currently NMR spectroscopy has a relatively low sensitivity (>sub mM range).

MSH has long been thought to control melanin synthesis in melanocytes and melanoma cells *via* a cAMP mediated process. cAMP regulates the levels and activity of tyrosinase the first enzyme in the synthetic pathway leading from tyrosine to the pigment melanin (Lerner, 1959; Bitensky et al., 1973; Eberle, 1988). Using the [3H]adenine prelabelling technique (Humes et al., 1969) which labels the cell's adenine nucleotide pool, we determined the response to MSH of monolayers of M2R mouse melanoma cells. cAMP production (Gerst et al., 1986), was determined by conventional techniques which involve extraction and chromatographic procedures (Salomon, 1991). Under the conditions used cAMP-PDE activity was inhibited by inclusion of isobutyl methylxanthine (IBMX, 0.1mM) in the incubation medium and AC was stimulated by βMSH (1 μM) in the presence of a synergistic dose of forskolin (2 μM which, by itself, was barely stimulatory (Seaman & Daly, 1986). We found that cAMP accumulation was stimulated to a surprisingly high extent amounting, within 20 min, to approximately 45% of the total cell's adenine nucleotide pool, of which ATP is only one component (Gerst et al., 1986). Such a marked response of AC was considered to be amenable to analysis by ^{31}P-NMR spectroscopy possibly enabling the determination of cAMP synthesis under similar incubation conditions. Intracellular levels of cAMP were previously considered to be below detection limits by ^{31}P-NMR spectroscopy and changes in ATP levels due to AC stimulation were estimated to have an insignificant effect on cellular ATP levels (Schimizu et al., 1970; Krishna et al., 1970).

^{31}P NMR spectroscopy of melanoma cells in culture was performed in our laboratory (Degani et al., 1991) under conditions previously developed for cultured breast cancer cells (Neeman et al., 1988, Neeman & Degani 1989a; Neeman & Degani 1989b). In short, cells grown on agarose beads (2.5 ml) to a density of ~2x10^7 cells/ml were placed in a 10 ml NMR tube under continuous perfusion at a rate of 1.5 ml/min, using 60 ml of culture medium (DMEM:F12 1:1 10% horse serum) circulating in a closed sterile loop at 37°C under 95% O_2 + 5% CO_2 at atmospheric pressure. NMR experiments were performed with a Bruker AM-500 NMR spectrometer. ^{31}P spectra were recorded at 202.5 MHz. For each spectrum, 180 transients were accumulated, by applying 90° pulses, 10 sec repetition delay and composite pulse proton decoupling (Neeman & Degani, 1989a; Neeman & Degani, 1989b).

The ^{31}P-NMR spectrum of unstimulated cultured M2R cells, exhibited as major peaks the intracellular nucleotidetriphosphates (NTP) coinciding with α, β and γ ATP, UDPG, phosphocreatine (PCr), phosphodiesters (PDEs), phosphomonoesters (PMEs), and a sizeable peak of inorganic phosphate (Pi). The later representing primarily extracellular Pi present in the circulating medium (Degani et al., 1991). In order to maximally stimulate cAMP production in these cells we initially applied conditions for high stimulation of AC by MSH similar to those described above (Gerst et al., 1986) using the super potent αMSH analogue ([Nle^4,DPhe^7]αMSH (1 μM) (Sawyer et al., 1980), forskolin (1.7 μM) and IBMX (0.1 mM). In the spectrum accumulated for 30 min following the induction (Degani et al., 1991) a significant rise in intracellular cAMP (equivalent to ~40% of resting ATP levels) and an equal decline (~40 %) in ATP were observed while no changes in PCr levels were evident. Unexpectedly, we also noticed a significant increase (100-150%) of 3 peaks in the 4-5.5 ppm range of the ^{31}P NMR spectrum, a region that includes the PMEs. This unforeseen result will be further discussed below. No obvious changes in extracellular Pi levels were observed. It should be noted that water soluble molecules are visible in the spectrum and rarely phospholipids or polynucleic acids can be observed. Under the same conditions, we also studied the changes in these metabolites as a function of time over a period of 24 hrs. We found that the rise in cAMP and PMEs took place immediately and simultaneously within the time resolution of the experiment. cAMP levels reached peak values by 30 min and declined to basal levels by 60-90 min, apparently due to desen-sitization of AC. The levels of PME remained high initially but declined later over a period of 10 h in spite of the fact that the culture medium was replaced by 2 h with fresh medium devoid of stimulants. The decline and recovery in ATP levels coincided exactly in time and magnitude with the respective rise and fall in cAMP levels. Restimulation of the same cells once again, under the same conditions following 18 h of culture in the NMR spectrometer yielded similar results. Thus full viability and responsiveness of the cells were maintained.

The inverse relationship of the time related changes in cAMP and ATP (the product and substrate of AC) can be intuitively explained by the stimulation of the enzyme with MSH+forskolin in the presence of IBMX. However, the magnitude of the transient changes in both cAMP and ATP were quite surprising. On the basis of classical considerations, the AC reaction was not expected to withdraw significant amounts of ATP in the first place, but it was certainly not foreseen that AC stimulation can lead to a significant ,though transient, depletion of much of the cell's ATP. With the reservation that these stimulatory conditions are not physiological, we are presently unaware of any other pathway that would consume massive amounts of ATP concurrently with cellular stimulation by MSH. Most energy utilizing processes would be expected to give rise to equivalent amounts of ADP which would in turn be regenerated to ATP by the creatine kinase (CK) reaction. Creatine kinase and PCr are generally considered to buffer the high energy phosphates in cells and thus provide immediate replenishment of ATP. In muscle, for example, enhanced conversion of ATP to ADP is transiently reflected by a drop in PCr that is easily observed in the ^{31}P NMR spectrum (Shoubridge et al.,1984). It was therefore unexpected to find this MSH induced fall in cellular ATP under conditions were PCr levels remained unchanged. This unusual circumstance may be explained by the cAMP-"sink" mechanism (Fig.1). This mechanism suggests that under conditions where by the flux of ATP through the AC reaction (Fig.1, step 1) is stimulated by MSH+forskolin and the cAMP-PDE reaction (Fig.1, step 2) is efficiently inhibited by IBMX, considerable concentrations of cAMP accumulate in a sink in which much of the cell's adenine pool becomes temporarily stored. This mechanism is based on our earlier observation (Gerst et al., 1986) which prompted this study, in which MSH+forskolin/IBMX stimulation following [^{3}H]adenine prelabelling of the melanoma cells led to accumulation of nearly 45% of the cell's adenine pool in the form of cAMP. This is now reinforced by our ^{31}P NMR measurements which demonstrated that a sizeable portion (~40 %) of the ATP consumed after MSH+forskolin/IBMX stimulation accumulated as cAMP (Degani et al., 1991). However cAMP is neither a substrate of adenylate kinase (AK) nor of CK (Fig.1, steps 2 and 3, respectively) and can therefore not be readily converted into ATP. Moreover, recycling of cAMP for regeneration of ATP requires two ~P bonds. One ~P comes from of ATP at the AK reaction (Fig.1, step 3) to form ADP

350

and an additional ~P comes form of PCr in the CK reaction (Fig.1, step 4).

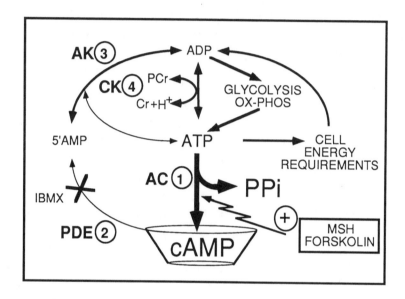

Figure 1. The "cAMP-Sink" mechanism: Stimulation of the adenylate cyclase (AC, reaction 1) with MSH+forskolin while the cAMP Phosphodiesterase (PDE) reaction 2. is blocked by IBMX leads to accumulation of cAMP in a "sink". Reconversion of cAMP to ATP requires one mole of ATP at the level of the adenylate kinase (AK, reaction 3) and one mole of PCr at the level of the creatine kinase (CK, reaction 4). Thus the cAMP-PDE reaction becomes the rate limiting step in the replenishment of the adenine moiety of ATP. At the same time ATP levels also limit the same pathway at the AK step tightening the blockade even more.

According to our ^{31}P NMR measurements PCr levels remained constant within the period of the experiment; therefore, it would be logical to conclude that the CK reaction was limited by ADP concentrations which are usually low and under stringent control.

The replenishment of ATP from cAMP was therefore limited by ATP concentrations for the ~P, at the AK step, and by the cAMP-PDE reaction for the adenine moiety. This doubly blocked pathway finally relaxes when AC activity desensitizes and ATP consumption slows down. This in turn will permit for a slow leak of cAMP through the PDE reaction to eventually take over and lead to the depletion of the cAMP-sink. The immediate product of this reaction, 5' AMP on the expense of one ATP molecule will then be converted to ADP in the AK step to allow for rapid regeneration of ATP by CK. The cAMP-PDE reaction which is rate limiting in this process, being inhibited by IBMX, appears to be much slower, in rate and capacity, then the generation of ATP/PCr by glycolysis and oxidative phosphorylation. Hence no sudden or large changes in the steady state level of PCr are observed. Cumulatively this process eventually resets ATP and cAMP to resting levels in the treated cells within approximately two hours. This observation may provide new insight into possibilities of cell specific manipulations of ATP levels. It may also stimulate new thoughts concerning the exact molecular basis of methylxanthine actions, widely used in medical treatment.

Simultaneously with the elevation of cAMP levels, PME concentration increased over two fold. The major peak within the PME area of the spectrum was assigned as PE by chemical shift and by comparison with authentic PE, in spectra obtained in perchloric acid cell extracts. Augmentation of the ^{13}C PE peaks when cells were grown in the presence of ^{13}C ethanolamine and proton/^{31}P two dimensional NMR spectral analysis further supported the identity of this peak in the ^{31}P spectrum as PE. Two additional smaller peaks within this region have not yet been identified. However, it should be noted that phosphocholine (PC) the major component of PMEs in the resting state clearly resolved in the ^{31}P spectrum from PE and the other two peaks. The level of PC was found to be unaffected by the experimental manipulations.

The major enzymatic reactions involved in PE metabolism are summarized in Fig.2. PE production according to this scheme can result from the degradation of phosphatidylethanolamine (PtdEtn) by a specific phospholipase C (PLC) (Fig.2, reaction 1), from phosphorylation of ethanolamine (EA) by ethanolamine kinase (EK) (Fig.2, reaction 2) or from degradation of sphingosine phosphate by sphingosine-1-phosphate aldolase (S-1-P ALDO) (Fig 2, reaction 3) (Ansel & Spanner, 1982).

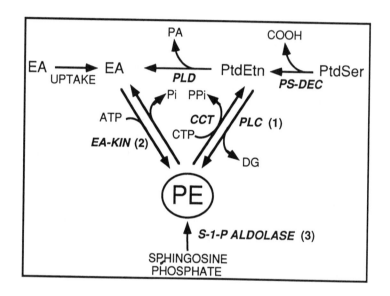

Figure 2. Major reactions in PE metabolism. Phosphoethanolamine (PE) is synthesized from phosphatidylethanolamine (PtdEtn) by a phospholipase C (PLC) (reaction 1), by phosphorylation of ethanolamine (EA) *via* EA kinase (EK, reaction 2) or by degradation of sphingosine-1-phosphate (S-1-P, reaction 3) catalyzed by S-1-P aldolase (ALDO). PE can be degraded by phosphatases or be utilized in the CTP Cytidyl transferase (CCT) reaction to produce PtdEtn.

Among the three possible reactions that lead to PE production the putative MSH-dependent activation of a PtdEtn specific PLC seems at the moment to be the most exciting one. MSH stimulated phosphorylation of EA by EK seems unlikely since the consumption of

ATP in this case and consequent generation of ADP would have probably been noticed in the ^{31}P NMR spectrum by a certain decline in PCr. It should be noted that peak PE levels were of a magnitude that even exceeded basal ATP levels. Furthermore, uptake of EA could be ruled out since the culture medium used in these experiments was determined to be devoid of EA. In addition, the generation of EA from PtdEtn *via* phospholipase D (PLD) would also be an unlikely step to be regulated by MSH. To this end we found that PLD activity as directly determined in M2R melanoma cells is MSH insensitive (Salomon & Liscovitch, unpublished). The inhibition of PE breakdown by phosphatase or enhanced utilization of PE in the CTP-Cytidyltransferase (CCT) reaction could also lead to transient elevations in PE levels. The involvement of MSH regulation in these reactions cannot be ruled out at this time and the final conclusion regarding the exact nature of the MSH responsive pathway must await further studies.

The question of whether PE represents a second messenger by itself or possibly is a downstream phospholipid-derived intermediate of an MSH regulated transduction pathway requires additional information. However, it is tempting to speculate that PE represents a true water soluble second messenger produced by an PtdEtn specific PLC. In this case, this MSH dependent pathway could be considered analogous to the hormone sensitive phosphatidylinositol (PtdIns) PLC (Berridge & Irvine.,1989). Yet, the fact that stimulated PE concentrations accumulated to mM levels, much higher then other known second messengers, may speak against such a possibility. Alternatively, the generation of PE could result from a secondary activation of a PLC by a protein kinase C (PKC) and Ca^{++} dependent mechanism for instance. This, in turn, would be analogous to the PKC dependent stimulation of phosphatidylcholine PLC that generates PC and diacylglycerol (DG) (Exton,1991). In this case, the identity of the primary second messenger generating mechanism stimulated by MSH remains open. So far, to the best of our knowledge, stimulation of the PtdIns pathway by MSH has not been demonstrated (see also Eberle, 1988). However, the implication of a Ca^{++} dependent process in this case

354

is of great interest due to the unique calcium dependence of MSH and melanocortin receptors studied in our laboratories (Gerst et al., 1987; Salomon, 1991). These peptide hormones depend on extracellular calcium for association with their respective receptors and therefore represent a special case in cellular regulation by hormones. Consequently, regulation of intracellular calcium dependent processes may have even more interesting implications.

The pathway that leads to PE production in melanoma cells represents a novel MSH regulated process that appears to be regulated simultaneously with melanin synthesis. The physiological role of PE in these cells and its mechanism of generation will have to be elucidated in the future.

In summary, we have demonstrated in this set of studies that NMR spectroscopy is a valuable tool in the study of signal transducing cascades and that the potential of this technique for *in vivo* studies has yet to be fully exploited. The non-invasive nature of this approach fulfills a dream of many of us, who have always been concerned about artifactual ⁄ influence caused by unavoidable manipulations, involved in the more classical biochemical tradition. Moreover, we demonstrated the ability of this technique to identify unexpected metabolites and reactions, such as hormone regulated PE synthesis in this case. We will now resort to the old faithful biochemical methods to complement and hammer out the molecular basis of our findings.

Acknowledgements: We wish to thank Josepha Schmidt-Sole and Anat Azrad for devoted technical help. This work was supported in part by a research grant to Y.S. from the Crown Endowment Fund for Immunological Research at the Weizmann Institute of Science, and by a research grant to H.D. from the German-Israeli Foundation for Scientific Research and Development. Y.S. is the Charles and Tillie Lubin Professor of Hormone Research.

REFERENCES

Ansel,G.B., and Spanner,S. (1982) In: Phospholipids, New Comprehensive Biochemistry (Hawthrone,J.N. & Ansel,G.B.,eds) Elsevier New York Vol.4,pp.1-49

Berridge, M.J. and Irvine, R.F. (1989) Nature 341, 197-205

Bitensky, M.W., Demopoulos, H.B. and Russell, V. (1973), In: Pigmentation, Its Genesis and Biologic Control, ed. Riley, V. (Appleton-Century-Crofts, N.Y.), pp. 247-255

Degani,H., DeJordy,O.J., and Salomon,Y., (1991) Proc. Natl. Acad. Sci. USA 88 ,1506-1510

Eberle, A.N. (1988) The Melanotropins (S. Karger, Basel)

Exton, J.H. (1990) J. Biol. Chem. 265, 1-4.

Gerst, J., Sole, J., Mather, J.P., and Salomon, Y. (1986) Mol. Cell. Endocrinol. 46, 137-147

Gerst, J., Sole, J., and Salomon (1987) J. Mol. Pharmacol. 31 81-88

Greengard, P., Robison, G.A., Paoletti, R., and Nicosia, S.,eds. In: Advances in Cyclic Nucleotide and Protein Phosphorylation Research (1984), (Raven Press, N.Y.), Vol. 18

Humes, J.L., Rounbehler, M., and Kuel, Jr., F.L. (1969) Anal. Biochem. 32, 210-217

Krishna, G., Forn, J., Voigt, K., Paul, M., and Gessa, G.L., In: Role of Cyclic AMP in Cell Function, eds. Greengard, P. and Costa, E., (Raven Press, N.Y.), Advances in Biochemical Psychopharmacology, Vol. 3, pp. 155-172

Lerner, A.B. (1959) Nature (London),184,674-677

Neeman, M., and Degani, H. (1989) Cancer Res. 49, 589-594

Neeman, M., and Degani, H. (1989) Proc. Natl. Acad. Sci. USA 86, 5585-5589

Neeman, M., Rushkin, E., Kadouri, A., and Degani, H. (1988) Mag. Reson. Med. 7, 236-242

Rosen, M.O., and Krebs,E.G.,eds.(1981) In: Protein Phosphorylation, Cold Spring Harbor Conferences on Cell Proliferation (Cold Spring Harbor Lab., Cold Spring Harbor, N.Y.Vol. 8, Books A and B.

Salomon,Y.(1990) Mol.Cell.Endocrinol,70,139-145

Salomon,Y.(1991) Methods in Enzymol.,195 , 22-29

Sawyer, T.K., Sanfilippo, P.J., Hruby, V.J., Engel, M.H., Heward, C.-B., Burnett, J.B., and Hadley, M.E. (1980), Proc. Natl. Acad. Sci. USA 77, 5754-5758

Seaman, K.B., and Daly, J.W. (1986), Adv. Cyclic Nucl. Res. 20, 1-150

Shoubridge, E.A., Bland, J., and Radda, G.K. (1984) Biochim. Biophys. Acta 805, 72-78

Shimizu, H., Creveling, C.R. and Daly, J.W. (1970) In: Role of Cyclic AMP in Cell Function, eds. Greengard, P. and Costa, E., (Raven Press, N.Y.), Advances in Biochemical Psychopharmacology, Vol. 3, pp. 135-154.

Sutherland, E.W., Øye, I., and Butcher, R.W. (1965) Recent Prog. Hormone Res. 21, 623-646

ADENYLATE CYCLASE-HAEMOLYSIN TOXIN OF
BORDETELLA PERTUSSIS :
REGULATION OF EXPRESSION AND ACTIVITY

Sophie Goyard, Peter Šebo and Agnes Ullmann
*(Unité de Biochimie des Régulations Cellulaires,
Institut Pasteur, 75015 Paris, France)*

In bacterial pathogens the molecular mechanisms governing expression of virulence factors and those responsible for intoxication are far from being completely understood. Adenylate cyclase toxin produced by *Bordetella pertussis*, the causative agent of whooping cough, appears a convenient model to study the mechanisms involved in the control of genetic expression, activation, secretion and penetration into mammalian cells.

Adenylate cyclase toxin, encoded by the *cyaA* gene, is a bifunctional protein of 200 kDa endowed with both calmodulin-activated catalytic activity-located in the 400-residue N-terminal domain, - and haemolytic activity, located in the 1300-residue C-terminal part of the CyaA protein (Glaser *et al.*, 1988a). The secretion of CyaA requires the products of three downstream genes (*cyaB,D,E*) (Glaser *et al.*, 1988b). To acquire cytotoxic activity the CyaA protein has to be modified post-translationally by the *cyaC* gene product (Barry *et al.*, 1991) before secretion. Once activated and secreted, the toxin is able to penetrate a range of eukaryotic cells whereupon it is activated by calmodulin, to produce high levels of cAMP which intoxicate the target cells (Weiss and Hewlett, 1986).

The expression of *B. pertussis* virulence factors is coordinately regulated in response to environmental signals by proteins encoded by the regulatory *bvg* locus which belongs to a family of bacterial regulators, known as "two-component systems" (Miller *et al.*, 1989). However, there is now increasing evidence that regulation of virulence factors may occur at two levels: a coordinate, central, regulation as well as specific control of individual factors.

I/   Adenylate cyclase toxin: relationship between structure and function.

We have previously shown that the *cyaA* gene product is synthesized and secreted as a 200-kDa protein and that the 400 amino acid residue

358

calmodulin-dependent catalytic activity could be released by mild proteolytic treatment (Bellalou *et al.*, 1990a). The structure-function relationship of the catalytic domain will be discussed elsewhere in this volume (see Bârzu *et al.* and Munier *et al.*). Here we will focus mainly on the haemolytic and invasive functions of the CyaA protein. By creating in-phase deletions in the *cyaA* gene, that removed 60 to 160 codons in the region encoding haemolytic activity we could show that the resulting mutant CyaA proteins had reduced haemolytic activities and lost their invasive activity, i.e. the ability to enter mammalian target cells and elevate intracellular cAMP levels (Bellalou *et al.*, 1990b). The finding that haemolytic activity could be separated from invasive toxin activity has been further corroborated by the following experiments:

We reconstructed in *Escherichia coli* the *cya* locus of *B. pertussis* by cloning the different genes under the control of strong promoters and *E. coli*-specific translational initiation signals (Sebo *et al.*, 1991). CyaA proteins were

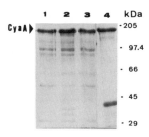

Fig. 1 - **SDS-PAGE analysis of the purified toxins.** Partially purified proteins were separated by SDS-PAGE (7.5%) as decribed in Sebo *et al.* (1991). Lane 1, CyaA expressed alone in *E. coli* ; lane 2, CyaA coexpressed with CyaC,B,D,E in *E. coli* ; lane 3, CyaA coexpressed with CyaC in *E. coli* ; lane 4, CyaA purified from *B. pertussis*.

synthesized in *E. coli* as 200-kDa polypeptides, in the absence as well as in the presence of other *cya* gene products (Fig. 1). As it can be seen in Fig. 2, CyaA produced alone in *E. coli* had no invasive or haemolytic activity and the *cyaC* gene product was sufficient to render the CyaA holotoxin fully invasive, but only partially haemolytic. The fact that in a reconstructed system in *E. coli* haemolytic and invasive activities could again be separated strongly suggests that distinct structural determinants within the CyaA protein are involved in the two functions. Besides, we have shown recently that the haemolysin part of CyaA could be expressed as a truncated protein, upon deletion of the adenylate cyclase moiety, and the protein still retained haemolytic activity (Sebo *et al.*, manuscript in preparation). Thus the lytic activity of CyaA is entirely provided by the C-terminal 1306 residues, and does not result from cAMP accumulation in the target cells. Therefore, on the basis of available experimental data, it seems reasonable to suppose that the role of the haemolytic moiety of the CyaA protein is to channel the N-terminal catalytic domain across the bacterial envelope and into eukaryotic target cells.

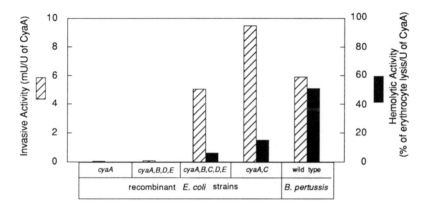

Fig. 2 - **Invasive and haemolytic activities of CyaA proteins.** Partially purified proteins, as depicted in Fig. 1 were tested for adenylate cyclase activity as described by Ladant *et al.* (1989) and haemolytic and invasive activites by incubating 1000 units of adenylate cyclase (one unit corresponding to 1 μmol cAMP formed per min at 30°C and pH 8) with sheep erythrocytes (5 x 10^8 cells/ml) at 37°C. Internalized adenylate cyclase activities were determined after 30 min incubation and haemolytic activities after 270 min incubation as described by Bellalou *et al.* (1990b).

II/   Regulation of the *cya* operon.

The *cya* operon of *B. pertussis* is composed of four genes, *cyaA,B,D* and *E* of ≈ 10 kb in length, encoding respectively adenylate cyclase-haemolysin (CyaA) and three secretion proteins (CyaB, D, E). The *cyaC* gene, located upstream from *cyaA* and in opposite orientation, encodes a protein required for modification of CyaA (Fig. 3). The analysis of the transcriptional organization of the *cya* operon and its genetic and environmental regulation showed that most transcripts, initiated at the *cyaA* promoter, terminate within the *cyaA-cyaB* intergenic region, but a low level of readthrough into the downstream genes can produce full length transcripts. Transcription from this promoter is only activated in virulent (*bvg*+) strains. There exists a second, weaker, start site of transcription in the intergenic *cyaA-cyaB* region which is activated in both virulent and avirulent strains (Laoide and Ullmann, 1990) (Fig. 3).

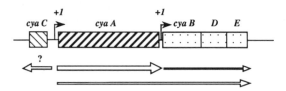

Fig. 3 - Schematic representation of the transcriptional organization of the *cya* locus. The arrows indicate direction of transcription ; the thickness of the arrows indicate relative abundance of transcripts.

Little is known about the molecular mechanisms involved in the regulation of individual virulence factors produced by *B. pertussis*. There is now converging evidence to indicate that the control locus, *bvg*, would regulate directly some virulence factors and indirectly some others: *(i)* the gene encoding filamentous hemagglutinin (*fha*), when cloned in *E. coli*, could be activated by the *bvg* locus introduced in *trans* ; under the same conditions the genes of pertussis toxin (*ptx*) were not activated (Miller *et al.*, 1989) ; *(ii)* when *B. pertussis* cultures are shifted from low temperature, where no virulence factors are expressed, to high temperature, which favors the expression of virulence factors, the *fha* promoter is activated immediately, whereas the activation of *ptx* and *cya* promoters occurs only after a long delay (Scarlato *et al.*, 1991).

To characterize the regulation of the *cyaA* promoter we constructed a *cyaA::lacZY* fusion encompassing the *cyaA* promoter, located 115 bp upstream from the translational start codon ATG, the first 336 bp of *cyaA* followed by *lacZY*; in addition it carries 569-bp sequences upstream from the transcription start site (Fig. 4). When the plasmid carrying the fusion was introduced into *E. coli* no expression of the *cya* operon, monitored by β-galactosidase synthesis, could be detected whether in the presence or in the absence of the *bvg* locus.

When the same *cyaA::lacZY* fusion was introduced into *B. pertussis* by homologous recombination, the recombinant strain produced high levels of β-galactosidase. In addition, both β-galactosidase and adenylate cyclase synthesis were under the control of virulence-dependent regulation, suggesting that 569-bp sequences upstream from the *cyaA* transcription start site are sufficient for *bvg*-dependent activation (Fig. 4) (Goyard and Ullmann, 1991). Taken together, these results strongly suggest that the activation of *cya* operon by *bvg* is indirect. It seems, therefore, that both *ptx* and *cya* operons require additional factors for their activation.

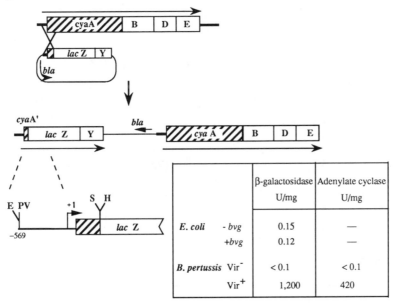

		β-galactosidase U/mg	Adenylate cyclase U/mg
E. coli	- bvg	0.15	—
	+bvg	0.12	—
B. pertussis Vir⁻		< 0.1	< 0.1
Vir⁺		1,200	420

Fig. 4 - Schematic representation of the integration of *cyaA::lacZY* fusion into the *B. pertussis* cya locus. The upper part represents the presumed recombination between *B. pertussis* DNA sequences contained in the fusion plasmid and their chromosomal homologs. The middle part represents the structure of *B. pertussis cya* locus after recombination and the lower part shows the enlargement of the *cyaA* promoter region. The expression of *cyaA::lacZ* fusion is shown in the table.

Searching for mutants of the *cya* regulatory region, that no longer require *bvg* and/or additional regulators for activation, we found two classes of mutants: the first class occurred by spontaneous integration of the insertion element IS2 in orientation II (Goyard *et al.*, 1991) and the second class could be obtained by chemical mutagenesis (Goyard and Ullmann, manuscript in preparation). The main feature of these up-promoter mutants is that the *cyaA* gene is no longer activated by the resident promoter, located 115 bp upstream from the translational start codon, but by newly created promoters. When reintroduced into *B. pertussis*, the mutants exhibited constitutive expression, i.e. they did not require *bvg* for activation. In the case of IS2 integration, transcription initiated from a promoter within IS2 (Table I).

TABLE I

cyaA::lacZY expression in *E. coli* and *B. pertussis*

Strain	Vir phenotype	Promoter	β-galactosidase U/mg
E. coli	—	*cyaA*	0.15
E. coli	—	IS2	220
B. pertussis	Vir[+]	*cyaA*	1,200
B. pertussis	Vir[-]	*cyaA*	0.01
B. pertussis	Vir[+]	IS2	280
B. pertussis	Vir[-]	IS2	185

The constitutive promoters obtained by chemical mutagenesis had the same characteristics as those obtained by IS2 integration, namely they were active in both *E. coli* and *B. pertussis*. Sequence analysis of the mutated promoter regions suggests that a new promoter, located 85 bp upstream from the translational start codon, was created which changed the spacing between the -35 and -10 region, bringing it closer to the *E. coli* promoter consensus sequences (Hawley and McClure, 1983) (Table II).

TABLE II

	-35		-10	Expression in		
				E. coli	*B. pertussis*	
					Vir[+]	Vir[-]
Wild type promoter	TTGTGC-	(N)$_{21}$-	AATAAT	—	+	—
cya up-promoter	TTGGAA-	(N)$_{19}$-	TATGAT	+	+	+
E. coli consensus	TTGACA-	(N)$_{15-19}$-	TATAAT			

It is noteworthy that no mutations, that could have changed the promoter strength, have been obtained in the conserved bases in the -10 and -35 regions of the promoter. The change in the spacing, from 21 to 19 bp thus rendered the promoter functional in *E. coli* as well as in *B. pertussis* and rendered it also independent of positive regulation. The 21-bp spacing between the -10 and -35 regions probably suffices to explain why wild type the *cya* promoter is not recognized by RNA polymerase in the absence of an activator.

The great structural diversity of the known promoters of the different virulence operons renders difficult to ascribe nucleotide sequences recognized by the *bvg*-encoded regulator. Besides, if an additional factor is involved in the activation of transcription initiation, important questions should be answered: How does each of them act, what is their respective role ? What is the role of local DNA secondary structure, negative supercoiling, for instance ? The understanding of the mechanisms underlying the specific regulation of the *cya* operon should provide an insight into the intricate mechanism of virulence control.

Acknowledgments

We thank M. Ferrand for page setting of the manuscript. This work was supported by the Institut Pasteur, the C.N.R.S. (URA D1129), I.N.S.E.R.M. (no. 91.06.15) and D.R.E.T. P. S. was financed by a fellowship from the Ministry of Research and Technology.

364

References

Barry, E.M., Weiss, A.A., Ehrmann, I.E., Gray, M.C., Hewlett, E.L., and St. Mary Goodwin, M. (1991) J. Bacteriol. 173, 720-726

Bellalou, J., Ladant, D., and Sakamoto, H. (1990a) Infect. Immun. 58, 1195-1200

Bellalou, J., Sakamoto, H., Ladant, D., Geoffroy, C., and Ullmann, A. (1990b) Infect. Immun. 58, 3242-3247

Glaser, P., Ladant, D., Sezer, O., Pichot, F., Ullmann, A., and Danchin, A.(1988a) Molec. Microbiol. 2, 19-30

Glaser, P., Sakamoto, H., Bellalou, J., Ullmann, A. and Danchin, A. (1988b) EMBO J. 7, 3997-4004

Goyard, S., and Ullmann, A. (1991) FEMS Microbiol. Letters 77, 251-256

Goyard, S., Pidoux, J., and Ullmann, A. (1991) Res. Microbiol. 142, 633-641

Hawley, D.K., and McClure, W.R. (1983) Nucleic Acids Res. 11, 2237-2255

Ladant, D., Michelson, S., Sarfati, R.S., Gilles, A.-M., Predeleanu, R., and Bârzu, O. (1989) J. Biol. Chem. 264, 4015-4020

Laoide, B.M., and Ullmann, A. (1990) EMBO J. 9, 999-1005

Miller, J.F., Mekalanos, J.J., and Falkow, S. (1989) Science 243, 916-922

Scarlato, V., Arico, B., Prugnola, A., and Rappuoli, R. (1991) EMBO J. 10, in press

Sebo, P., Glaser, P.,Sakamoto, H., and Ullmann, A. (1991) Gene 104, 19-24

Weiss, A.A., and Hewlett, E.L. (1986) Ann. Rev. Microbiol. 40, 661-686

Adenine Nucleotides in Cellular Energy Transfer and Signal Transduction
S. Papa, A. Azzi & J.M. Tager (eds)
© 1992 Birkhäuser Verlag, Basel/Switzerland

ARE PURINE NUCLEOSIDE TRIPHOSPHATE CYCLASES AN EXAMPLE OF CONVERGENT EVOLUTION ?

Antoine Danchin
Régulation de l'Expression Génétique, Institut Pasteur
28 rue du Docteur Roux, 75724 PARIS CEDEX 15, FRANCE

SUMMARY Adenylate cyclase genes from widely distant organisms have now been isolated. This allowed comparison of the primary sequence of their catalytic centers. Three classes stand out from the analysis of similarities. A first class comprises enzymes isolated from enterobacteria and related bacteria. A second class comprises the procaryotic toxic adenylate cyclases, activated by the host calmodulin, synthesised by two very distant organisms, *Bordetella pertussis* and *Bacillus anthracis*. Finally, a third class comprises adenylate cyclases isolated from organisms ranging from procaryotes to higher eucaryotes. It also contains the set of guanylate cyclases isolated from eucaryotes. This indicates that cAMP synthesis predated separation between eucaryotes and procaryotes, and suggests new hypotheses about the origin of 3'5'cyclic purine nucleotides. In parallel, the extreme diversity displayed between the different classes raises the question of the origin of present day cAMP : are cyclases a case of phylogenic convergence ? A tree can be constructed that links all enzymes to each other, but, even if one takes this tree for granted it appears that cAMP synthesis has probably been reinvented at least once.

cAMP was discovered by Sutherland as the product of an enzyme, adenylate cyclase, that generated cAMP and PPi from ATP, when eucaryotic cells were activated by adrenaline. Since this pioneering work, the study of cAMP-mediated effects permitted identification of the structure, function and regulation of many adenylate cyclases. At the beginning of the corresponding work, it was often admitted that "what is true for *Escherichia coli* is true for the elephant", so that the discovery, in 1968, of a specific effect of cAMP on catabolite repression in *E. coli* raised hopes that the study of *bacterial* adenylate cyclases would help understand what happened in eucaryotes. This hope was not substantiated, for it appeared that the study of adenylate cyclases was as difficult in procaryotes as in eucaryotes. However the recent discovery that several procaryotic adenylate cyclase genes synthesised proteins related to their eucaryotic counterparts has revived this hope (Beuve et al., 1990). But it now appears that adenylate cyclases do not constitute a homogeneous class of enzymes, but, rather, comprise at least three well defined types of enzymes, the enterobacterial type, the toxin type (corresponding to cyclases secreted by pathogens such as *Bacillus anthracis* and *Bordetella pertussis*) and a class, grouping the vast majority of cyclases, that comprises both procaryotic and eucaryotic enzymes.

I. THE ENTEROBACTERIAL CLASS (CLASS I)

The first adenylate cyclase gene that has been isolated and sequenced is the *E. coli cya* gene. It was then possible to clone *cya* genes from other bacterial types by direct complementation of *cya*-deficient *E. coli* mutants. Work performed on *Erwinia chrysanthemi* demonstrated that both the genes environment and the proteins were similar in size and overall organization to the *E. coli* genes and proteins at the *cya* locus (Hedegaard and Danchin, 1985; Danchin and Lenzen, 1988). A similar type of work was subsequently performed by several authors on a truncated *Salmonella typhimurium* gene, but the complete sequence was not obtained (Leib and Gerlt, 1983; Thorner et al., 1990; Fandl et al., 1990; Holland et al., 1988). Finally, the *cya* gene from two other enterobacteria, *Yersinia intermedia* and *Y. pestis* (the agent of plague), showed that all major features found in the *E. coli* gene and protein organisation were conserved (Glaser, Sismeiro and Danchin, unpublished). Knowledge of the catalytic domain sequence from five genes permits to construct a phylogenetic tree where *S. typhimurium* and *E. coli* are clustered together, as are the two *Yersinia* species (Figure 1).

In all cases the enzyme has the same structure : two domains constitute a 95 kDa polypeptide chain. The catalytic domain is amino-terminal, while the glucose-sensitive regulatory domain is carboxy-terminal. No significant similarity with other known proteins was found, indicating that these enzymes form a specific class. Finally, the gene from *Pasteurella multocida* (the causal agent of fatal septicemia in mammals) was sequenced by Mock et al., who demonstrated that it was very similar to the enterobacterial type, but that the contiguous genes were different (Mock et al., 1991). This indicated that the regulation of the protein activity (including glucose-mediated regulation) was preserved, while the expression of the gene may be different in Pasteurellaceae from the expression in Enterobacteriaceae.

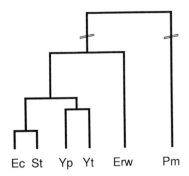

Figure 1. Phylogenesis of class I bacterial adenylate cyclases. Ec : *Escherichia coli*; Erw : *Erwinia chrysanthemi*; Pm : *Pasteurella multocida*; St : *Salmonella typhimurium*; Yi : *Yersinia intermedia*; Yp : *Yersinia pestis*.

B. THE CALMODULIN-ACTIVATED TOXIC CLASS (CLASS II)

B. pertussis, a Gram negative procaryote, causes whooping cough by secreting in the host many proteins having toxic activities. Among these proteins an adenylate cyclase activity was discovered (Hewlett and Wolff, 1976). In 1980 a further observation was made : the enzyme is activated by a *host* protein, calmodulin, which is not known to occur in bacteria (Wolff et al., 1980; Goldhammer and Wolff, 1982). Two years later Leppla demonstrated that another toxic adenylate cyclase, secreted by a Gram positive organism, *B. anthracis*, the etiological agent of anthrax, was also dependent on the presence of host calmodulin (Leppla, 1982). Several attempts to isolate the gene failed until a simple idea, introduction of a plasmid directing synthesis of calmodulin into an *E. coli* strain deficient in adenylate cyclase activity, permitted the cloning, sequencing, and expression of adenylate cyclase genes coding for the calmodulin-dependent cyclases (Glaser et al., 1988a; Escuyer et al., 1989).

It was thus discovered that *B. pertussis* adenylate cyclase is synthesized as a large bifunctional polypeptide of 1706 amino acid residues (Glaser et al., 1988). The N-terminal segment of the protein (400 residues) displays calmodulin-activated ATP-cyclizing activity, while the rest of the molecule was shown to be responsible for secretion as well as for the hemolytic activity of the pathogen (Ehrmann et al., 1991; Glaser et al., 1988a,1989;

Roger et al., 1991). Sequence analysis indicated that the cyclase domain was fused to a polypeptide similar to the polypeptide chain of *E.coli* hemolysin toxin (Goebel and Hedgpeth, 1982). Accordingly, the name cyclolysin was coined for the toxic adenylate cyclase from *B. pertussis* (Glaser et al., 1988b). Glaser and coworkers analysed the adjacent regions for functionally related genes and discovered that the protein is secreted using a mechanism similar to that permitting release in the external medium of *E. coli* hemolysin or cognate proteins in related media. Three genes located downstream from the cyclolysin gene were identified as required for secretion (Glaser et al., 1988b).

A different organisation was found for *B. anthracis* adenylate cyclase. It is encoded in a plasmid, together with another toxin (lethal factor, LF), and a carrier protein (protective antigen, PA), which is necessary for internalisation of both adenylate cyclase and LF in host target cells (Thorne et al., 1960; Beall et al. 1962; Gordon et al., 1988; Leppla et al., 1984). The gene product reflects the corresponding constraints. It is comprised of four segments. The first one is similar to signal peptides of other bacteria, permitting secretion of the protein. The second segment corresponds to the domain permitting binding with PA. It displays regions of significant similarity with a cognate region in LF. The third domain is adenylate cyclase. It is finally followed by a domain of unknown function (Figure 2).

Bordetella pertussis

Bacillus anthracis

Figure 2. Modular organisation of adenylate cyclase toxins. HLY : hemolysin; -PA binding domain for protective antigen.

Comparison between the catalytic region of the *B. pertussis* and *B. anthracis* enzymes permitted identification of four conserved regions that are involved in catalysis, calmodulin binding and activation. The first region (see figure 3) comprises a peptide, G----(G/A)KS having a sequence similar to the nucleotide binding region found in many ATP or GTP binding proteins. It was

```
                                                        catalytic
MQQSHQAGYANAADRESGIPAAVLDGIKAVAKEKNATLMFRLVNPHSTSLIAEGVATKGLGVHAKSSDWGL
 +   +   ++ = + ==+ +    =+ = +=+= =+ ++== ==   += ==  ======= ==+======
EKDRIDVLKGEKALKASGLVPEHADAFKKIARELNTYILFRPVNKLATNLIKSGVATKGLNVHGKSSDWGP

QAGYIPVNPNLSKLFGRAPEVIARADNDVNSSLAH----GHTAVDL-TLSKERLD----YLRQA-----GL
===== + +=== =    =   ++  =    =+ ++ =   =   ==   =+        =
VAGYIPFDQDLSKKHGQQLAVEKGNLENKKSITEHEGEIGKIPLKLDHLRIEELKENGIILKGKKEIDNGK

                                                  catalytic
VTGMADGVV-ASNHAGYEQFEFRVKETSDGRYAVQYRRKGGDDFEAVKVIGNAAGIPLTADIDMFAIMPHL
 + ++     +   ++ +\  \ = +=+ +=  \+   + = +\  ++   ===== =+==+ = =
KYYLLESNNQVYEFRISDENNEVQYKTKEGKITVLGEKFNWRNIEVMAKNVEGVLKPLTADYDLFALAPSL

          t                            α-helix
SNFRDSARSSVTSGDSVTDYLARTRRAASEATGGLDRERIDLLWKIARAGARSAV-GTEARRQFRYDGDMN
++ +    +   = = +       + =  ==    +== = +    +  == + =        +=
TEIKK--QIPQKEWDKVVNTP-----NSLEKQKGL----TNLLIKYGIERKPDSTKGTLSNWQKQMLDRLN

                     catalytic
IGVITDFELEVRNALNRRAHAVGAQDVVQHGTEQNNP-FPEADEKIFVVSATGESQMLTR-GQL-KEYIGQ
 +=           + + ++ ===+=====+= === =+ ==++ + == ++-+ ++    +=
EAV-------------KYTGYTGGDVVNHGTEQDNEEFPEKDNEIFIINPEGEF-ILTKNWEMTGRFIEK

      catalytic ?
QR-GEGYVFYENRAYGVAGKSLFDDGLGAAPGVPSGRSKFSPDVLETVPASPGLRRPSLGAVER
 +  =  =++= ==+=  + +      + = ++ ++    =    = +   +++==+= += +
NITGKDYLYYFNRSYNKIAPGNKAYIEWTDP-ITKAKINTIPTSAEFIKNLSSIRRSSNVGVYK
```

Figure 3. Comparison between the catalytic domains of *B. pertussis* and *B. anthracis* cyclases.

therefore proposed to be part of the catalytic site (Gilles et al., 1990; Glaser et al., 1989). Experiments with enzymes modified after *in vitro* mutagenesis substantiated this interpretation (Glaser et al., 1991; Sarfati et al., 1990). A second region, PLTADID, was also demonstrated to be involved in several aspects of catalysis, and it was proposed that the aspartate residues present in this region were involved in ribose and magnesium phosphate binding (Glaser et al., 1991). Finally, the more distal part of the protein has been found to be involved in catalysis. An amphiphilic helix, centrally located and comprising residue tryptophane 242 is necessary for calmodulin dependent activation, thus suggesting that the protein must be somehow folded around calmodulin in order to be active.

C. THE "UNIVERSAL" CLASS (CLASS III)

In spite of intense efforts adenylate cyclases from higher eucaryotes have long remained elusive, because the purification of the corresponding catalytic subunit is of extreme difficulty. In addition the activity of the enzyme is subject to a complex regulatory pattern, so that the catalytic center will display poor activity when expressed alone. Techniques as the one which has been used for the cloning of adenylate cyclase toxins are accordingly very difficult to implement, in particular because they may require the simultaneous presence of several regulatory subunits. The tedious but efficient method of Southern hybridization on cDNA libraries using oligonucleotides designed after the sequence of known peptides of the protein, has therefore been used. The availability of affinity chromatography using forskolin (Pfeuffer and Metzger, 1982) permitted purification of small amounts of the protein (Pfeuffer et al., 1985a; Coussen et al., 1985; Pfeuffer et al., 1985b,1989), that finally resulted in the cloning of one adenylate cyclase cDNA from bovine brain by Gilman and coworkers in 1989 (Krupinski et al., 1989).

The difficulty encountered with the cloning of the higher eucaryotes adenylate cyclase genes was, luckily, not so crucial in organisms such as *Saccharomyces cerevisiae*. Cloning was attempted in two laboratories. Masson et al. first cloned the wild type allele (Masson et al., 1984), and Wigler et al. then succeeded in determining the complete gene sequence after direct complementation of a *S. cerevisiae* mutant (Kataoka et al., 1985). In parallel, Masson et al. cloned the gene directly in a *cya*-deficient *E.coli* strain (Masson et al., 1986). These experiments demonstrated that this eucaryotic cyclase was completely different from the enterobacterial class, not only because of the sequence difference in the catalytic center, but also because the organisation of the gene was different : the catalytic domain is COOH-terminal in *S. cerevisiae* cyclase, whereas it is NH_2-terminal in *E. coli*. This suggested that the new protein constituted the first member of a third adenylate cyclase class. The yeast enzyme remained the only instance of its class until Garbers and coworkers recognized that the guanylate cyclase genes that had been cloned from several metazoans, were clearly derived from a common ancestor (Garbers, 1990; Lowe et al., 1989; Schulz et al., 1989; Singh et al., 1988). Another member of the class was subsequently discovered by Young *et al.* (1989) who cloned the adenylate cyclase gene from *Schizosaccharomyces pombe* by hybridization using the catalytic domain gene sequence from *S. cerevisiae* as a probe (Young et al., 1989). Finally, the first higher eucaryote adenylate cyclase gene, isolated in Gilman's laboratory from bovine brain, displayed features clearly reminiscent of this class (Krupinski et al., 1989). Later on Beuve *et al.* (1990) described an adenylate cyclase gene, isolated from the Gram negative procaryote *Rhizobium meliloti*, which was also a member of the same class (Beuve et al., 1990). This was remarkable because procaryotes are supposed to have separated from eucaryotes at least 2.5 billions years ago. Since then many other genes or cDNAs for adenylate or guanylate cyclases belonging to this class have been isolated and sequenced, not only from eucaryotes (*S. pombe, Trypanosoma brucei* and *T. equiperdum, Plasmodium falciparum*, human olfactive cells (Alexandre et al., 1990; Bakalyar and Reed, 1990; Garbers, 1990; Parma et al., 1991 ; Read and Mikkelsen, 1991; Schulz et al., 1989; Thorpe and Garbers, 1989; Young et al., 1989)), but also from other procaryotes : a Gram positive organism, *Brevibacterium liquefaciens* (known to synthesize a very active adenylate cyclase, requiring the presence of pyruvate for activation (Takai et al., 1974)) was found to be a member of the same class (Peters et al., 1991). It was also recently discovered that the Gram negative sliding bacteria, *Stigmatella aurantiaca*, harboured two cyclase genes, each of them corresponding to the same adenylate cyclase class (B. Lubochinsky, V. de la Fuente, O. Sismeiro, A. Danchin, in preparation).

Comparison of the catalytic domains of class III proteins is revealing. As shown in figure 4, several amino acid motifs are specific of the proteins, such as F-DI--FT-L, QVV-LLND, VK(E)T-GD--M, R-GIH-G, GDTVN-ASR. They are probably involved in the cyclisation reaction rather than in the base recognition, since they are common both to adenylate and guanylate cyclases. Analysis of the upstream part of the domain suggests however that the third motif could be involved in base discrimination, because it is different in adenylate cyclases (VKT) and guanylate cyclases (VET) (Peters et al., 1991). Unfortunately the set of individual adenylate cyclase examples encompasses a series of organisms that is much wider, from the point of view of phylogenesis, than the guanylate cyclase set, the former containing organisms from widely separated organisms whereas the latter was obtained only from metazoans. The difference could therefore reflect a coincidence.

```
GIESGHFGGRAAWRQEVTAMFTDIYDFTTI----SEGRSPEEVVAMLSEYFDLFSEVVA---AHDGTIIQFHGDSVFAMWNAP-VADTRHA        A RHIME
TTLRRLQPEISPPTGNLAMVFTDIKSSTFL-----WELFPNAMRTAIKTHNDIMRRQLR---IYGGYEVKTEGDAFMVAFPTPTSGLTWCL        A SACCE
SSLARMNREVSPPKGCIAMVFTDIKNSTLL-----WERHPIAMRSAIKTHNTIMRRQLR---ATGGYEVKTEGDAFMVCFQTVPAALLWCF        A SCHPO
MSNPRNMDLYYQSYSQVGVMFASIPNFNDFYIELDGNNMGVECLRLLNEIIADFDELMDKDFYKDLEKIKTIGSTYMAAVGLAPTAGTKA-       A BOVINb
PPERIFHKIYIQRHDNVSILFADIVGFTGL----ASQCTAQELVKLLNELFGKFDELAT---ENHCRRIKILGDCYYCVSGLTQ-PKTDHA       A BOVINa
NQDRGEDDGGTPLPLARAVGFADLVSYTSL----SRRMNERTLAQLVQRFEAKCAEIIS---VGGGRLVKTLLGDEVLYVAETPQAGAQIAL       A BRELI
RLQDLDRSETSPELREVTLLFADIRDFTSL-----SERLRPEQVVTLLNEYYGRMVEVVF---RHGGTLDKFIGDALMVYFGAPIADP-AHA      A STIAU1
ILKSPDTVVLTGEKREVTVLFADIRNFTGL-----AESLPPEQVVGVLNQVLGRLSDAVL---TCGGTLDKFLGDGLMAVWGAPVHRTDDAL      A STIAU2
RNKRDNDNAPKELADPVTLIFTDIESSTAQ-----WATQPELMPDAVATHHSMVRSLIE---NDCYEVKTVGDSFMIACKSPFAAVQLA-        A TRYBR2
HNTRDNNLAPKELTDPVTLIFTDIESSTAL-----WAAHPELMPDAVATHHRLIRSLIG---RYGCYEVKTVGDSFMIASKSPFAAVQLA-       A TRYEQ
GSKKRDEELYSQSYDEIGVMFASLPNFADFYTEESINNGGIECLRFLNEIISDFDSLLDNPKFRVITKIKTIGSTYMAASGVTPDVNTNG-       A RAT
EQLKRGETVQAEAFDSVTIYFSDIVGFTAL----SAESTPMQVVTLLNDLYTCFDAIID---NFDVYKVETIGDAYMVVSGLPGRNGQRHA       G ANPBRAT
EQLKRGETVQAEAFDSVTIYFSDIVGFTAL----SAESTPMQVVTLLNDLYTCFDAIID---NFDVYKVETIGDAYMVVSGLPGRNGQRHA       G ANPBHUMAN
EQLKRGETVQAEAFDSVTIYFSDIVGFTAL----SAESTPMQVVTLLNDLYTCFDAVID---NFDVYKVETIGDAYMVVSGLPVRNGRLHA       G ANPAHUMAN
EQLKRGETVQAEAFDSVTIYFSDIVGFTAL----SAESTPMQVVTLLNDLYTCFDAVID---NFDVYKVETIGDAYMVVSGLPVRNGQLHA       G ANPAMOUSE
SQLIKGIAVLPETFEMVSIFFSDIVGFTAL----SAASTPIQVVNLLNDLYTLFDAIIS---NYDVYKVETIGDAYMLVSGLPLRNGDRHA       G STRPU
EQLKRGETVQAEAFDSVTIYFSDIVGFTAL----SAESTPMQVVTLLNDLYTCFDAVID---NFDVYKVETIGDAYMVVSGLPVRNGQLHA       G ANPARAT
NELRHKRPVPAKRYDNVTILFSGIVGFNAFCSKHASGEGAMKIVNLLNDLYTRFDTLTDSRKNPFVYKVETVGDKYMTVSGLPEPCI-HHA      G CYTRAT
NELRHKRPVPAKRYDNVTILFSGIVGFNAFCSKHASGEGAMKIVNLLNDLYTRFDTLTDSRKNPFVYKVETVGDKYMTVSGLPEPCI-HHA      G CYTBOVIN
KSLKEKGIVEPELYEEVTIYFSDIVGFTTIC-K---YSTPMEVVDMLNDIYKSFDQIVD---NFDVYKVETIGDAYVVASGLPMRNGNRHA       G ARBPU
RQLWQGHAVQAKRFGNVTMLFSDIVGFTAIC---SQCSPLQVITMLNALYTRFDRQCG---ELDVYKVETIGDAYCVAGGL-HKESDTHA       G CYTBOVIN2
```

```
         LRR    V  EAFD VTILFSDIVGFTAL      SA     PMQVVTLLNDLYT FDAVID    NFDVYKVETIGDAYMVVSGLP RNGQ HA        G
                P     VTLIFADI   ST L      S      PEEVV LLNEH  F EL       GGY VKT GDSFMVV     P            A
                      M  T               W      QM        Y  M            IF   AL A                      "
```

```
EHACRCALAVEERLEAFNSAQRAS--------------GLP------------------EFRTRFGIHTGTAVVGSVGA-KERLQ----YTAM        A RHIME
-SGQ---LKLLDAQWPEEIT-SVQD------------GCQ-------------VTDRNGNIIYQGLSVRMGIHWGCPVPELDLV-TQRM------DYL        A SACCE
-SVQ---LQLLSADWPNEIVE-SVQ------------GRL-------VLGSKNEVLYRGLSVRIGVNYGPVVAGVIG--ARRPQ----YDY        A SCHPO
KKCISSHLSTL-ADFAIEMFD--------------------------VLDEINYQSYNDFVLRVGINVGPVVAGVIG--ARRPQ----YDIW       A BOVINb
----HCCV--EMGLDMIDTIT-SV----------------------AEATEVDLNMRVGLHTGRVLCGVLG--LRKWQ----YDVW       A BOVINa
-SLS-------RELAK--DEL---------------------------------FPQTRGAVVWGRLLSRLG----------DIY        A BRELI
RRGVQCA----LDMV-QELET--VNALRSAR-------GEP----------------CLRIGVGVHTGPAVLGNIGSATRRLE----YTAI        A STIAU1
RALQ--AAKMMMTAMVELRQAAQAEWAANERL--------GEP----------------LVLELGIGNSGLAVAGNIGGSMR-TE----YTCI        A STIAU2
QELQLRFLR----DWGTTVFDEFYREFEERHAEEGDGKYKPPTARL-DPEVYRQLWNGLRVRVGIHTGLC--------DIRYDEVTKGYDYY       A TRYBR2
QELQLCFLH---HDWGTNAIDESYQQLEQQRAEE-DAKYTPPTARL-DLKVYSRLWNGLRVRVGIHTGLC--------DIRRDEVTKGYDYY       A TRYEQ
FTSSSKEEKSDKERWQH-LADLADFALAMKDTLTNI---------------NNQSFNNFMLRIGMNKGGVLAGVIG--ARKPH----YDIW       A RAT
PEIARMALALLDA----------VSSFRIRH--------RP--------------HDQLRLRIGVHTGPVCAGVVGLKMPR------YCLF       G ANPBRAT
PEIARMALALLDA----------VSSFRIRH--------RP--------------HDQLRLRIGVHTGPVCAGVVGLKMPR------YCLF       G ANPBHUMAN
CEVARMALALLDA----------VRSFRIRH--------RP--------------QEQLRLRIGVHTGPVCAGVVGLKMPR------YCLF       G ANPAHUMAN
REVARMALALLDA----------VRSFRIGH--------RP--------------QEQLRLRIGVHTGPVCAGVVGLKMPR------YCLF       G ANPAMOUSE
GQIASTAHHLLES----------VKGFIVPH--------KP--------------EVFLKLRIGIHSGSCVAGVVGLTMPR------YCLF       G STRPU
REVARMALALLDA----------VRSFRIRH--------RP--------------QEQLRLRIGVHTGPVCAGVVGLKMPR------YCLF       G ANPARAT
RSICHLALDMMEI----------AGQVQ-------------------------VDGESVQITIGIHTGEVVTGVIGQRMPR------YCLF       G CYTRAT
RSICHLALDMMEI----------AGQVQ-------------------------VDGESVQITIGIHTGEVVTGVIGQRMPR------YCLF       G CYTBOVIN
VDISKMALDILSF----------MGTFELEH--------LP--------------GLPVWIRIGVHSGPCAAGVVGIKMPR------YCLF       G ARBPU
VQIALMALKMMEL----------SHEVVSPH--------------------------GEPIKMRIGLHSGSVFAGVVGVKMPR------YCLF       G CYTBOVIN2
```

```
         EI  RMALALLDA        V  F IRH         RP              EQLRLRIGIHTGPVCAGVVGLKMPR------YCLF        G
              LQ    L LLDADW E                  P              YNGLRVRVGIHTG  V    G  TRR Q    YD Y       A
                                                                                  E            W           "
```

```
GDTVNVASRLEG-MNKDY-GTSV-LAS-GAVVAQ-CKDMVKFRP---L-GTAKAKGRSTALDIYEVVGVRAVNTTEAGTAA*        A RHIME
GPMVNKAARVQGVADGGQIAMSSDFYSEFNKI---MKYHER----------VKGKESLKEVYGEEIIGEVLEREIAMLESI        A SACCE
GPVVNRTSRVVSVADGGQIAVSAEVVSVLNQLDSETMSSEKTNVNEMEVRALKQIGYIIHNLGEFKLKGLDTTEMISLVYPVQ        A SCHPO
GNTNVASRMDS--T-GVQ-GRIQVTEEVHRLLRRGSYRF-------VCGKVSVKGKGEMLTYFLEGRTDGNGSQTRSLNSER        A BOVINb
SNDVTLANVMEA--A-GLP-GKVHITKTTLACLN-GDYEVEP----------GHGHERNSFLKTHNIETFFIVPSHRRKIFP        A BOVINa
GPTVNMAARLTSLAEPGTVLTDAITANT-LRNDARFVLTAQEITAVRGFGDIQPYELSAGEGAGLVID*        A BRELI
GDTVNLASRIESLTKTRDVPILASRATRE-QAGDTFLWNEMAP-------ASVPGKSQPVAIFTPRNRTPAQQAGAPAAA*        A STIAU1
GDAVNVAARLCALAGPGE--ILAGERTRELVSHREMPFEDLPP------VRLKGKQQPVPLYRVL*        A STIAU2
GQTANTAARTESVGNGGQV-LMTCETYHSLSTAERSQFD-VTP---L-GGVPLRGVSEYVEYQLNAVPGRSFAELRLDRVL        A TRYBR2
GRTSNMAARTESVGNGGQV-LMTTAAYMSLSAEEREQID-VTA---L-GDVPLRGVAKPVEMYQLNAVPGRTFAGLRLEHEL        A TRYEQ
GNTVNVASRMESTGVMGNI----QVVEETQVILREYGFRFVR------RGPIFVKGKGELLTTFLKGRDRPAAFPNGSSVTLP        A RAT
GDTVNTASRMES--N-GQ-ALKIHVSSTTKDALDELGCFQLE-----LRGDVEMKGKGKMRTYWLLGERKGPPGLL*        G ANPBRAT
GDTVNTASRMES--N-GQ-ALKIHVSSTTKDALDELGCFQLE-----LRGDVEMKGKGKMRTYWLLGERKGPPGLL*        G ANPBHUMAN
GDTVNTASRMES--N-GE-ALKIHLSSETKAVLEEFDGFELE-----LRGDVEMKGKGKVRTYWLLGER-GSSTRG*        G ANPAHUMAN
GDTVNTASRMES--N-GE-ALRIHLSSETKAVLEEFDGFELE-----LRGDVEMKGKGKVRSYWLLGDR-GCSSRA*        G ANPAMOUSE
GDTVNTASRMES--N-GL-ALRIHVSPWCKQVLDKLLGGYELE----DRGLVPMNGKGEIHTFWLLGQDPSYKITKVKPPPQK        G STRPU
GDTVNTASRMES--N-GE-ALKIHLSSETKAVLEEFDGFELE-----LRGDVEMKGKGKVRTYWLLGERGCSTRG*        G ANPARAT
GNTVNLTSRTET--T-GEKG-KINVSEYTYRCLMSPENSDPQ-FHLEHRGPVSMKGKKEPMQVWFLSRKN--TGTEETNQDEN        G CYTRAT
GNTVNLTSRTET--T-GEKG-KINVSEYTYRCLMTPENSDPQ-FHLEHRGPVSMKGKKEPMQVWFLSRKN--TGTEETEQDEN        G CYTBOVIN
GDTVNTASRMES--T-G-LPLRIHMSSSTIAILRRTDCQFL----YEVRGETYLKGRGTETTYWLTGMKDQEYNLPTPPTVEN        G ARBPU
GNNVTLANKFESCSVPRKINVSPTTYRLLKDCPGFVFTPRSREELPPNFPSDIPGICHFLEAYQQGTTSKPWFQKKDVEEANA        G CYTBOVIN2
```

```
GDTVNTASRMES--N-GE ALKIHVSS TK  L E  GF LE-     LRGDVEMKGKGK RTYWLLG        G
G TVN AARMESVGNGGQ                              L   GKGK    Y              A
       S L LA                                                             "
```

Figure 4. Alignment of class III cyclases catalytic domains. A; adenylate cyclases : *R. meliloti*, *S. cerevisiae*, *S. pombe*, bovine brain (domains a and b), *B. liquifaciens*, *S. aurantiaca* (isozymes 1 and 2), *T. brucei* (isozyme 2), *T. equiperdum*, rat olfactive cells; B; guanylate cyclases : rat atrial natriuretic peptide receptor (ANP) b, human ANP b, human ANP a, mouse ANP a, *S. purpuratus*, rat cytoplasmic 1, bovine cytoplasmic 1, *A. punctulata*, rat cytoplasmic 2.

IV. ARE CYCLASES THE RESULT OF CONVERGENT EVOLUTION ?

In view of the apparent antiquity of cAMP, the hypothesis that it was originally involved in energy scavenging or production rather than regulation seems worth considering. It is thus of interest to note that there is a weak, but definite, similarity between this general class of cyclases and ATP synthases. Phosphates (and polyphosphates) certainly played a significant role at the origin of life (Bernal, 1951; Cairns-Smith, 1982; Wächtershäuser, 1988; Danchin, 1989). Generation of a chemiosmotic function allowing synthesis of ATP (from ADP) requires numerous steps that are difficult to see fulfilled at once. It may therefore be considered that, originally, cAMP could have been a building block for ATP. This however requires a widely spread source for cAMP synthesis : the catalytic activities of self-splicing introns are involving processes of transesterification, and one may wonder whether cAMP or cGMP could not be by-products of such reactions, providing an early source for these molecules. A first class of enzymes predating class III adenylate cyclases would therefore be involved in ATP synthesis ; a second, very important enzyme, adenylate kinase, would then scavenge AMP, generating ADP. A general source of ADP thus available might have triggered evolution towards generation of ATP synthase. It seems worth noting in this respect that the class of toxic adenylate cyclases can be phylogenetically related to adenylate kinase, at least when considering the first block of conserved amino-acid residues in B. pertussis and B. anthracis enzymes.

Another observation may substantiate the validity of this hypothetical relationship. Antibodies raised against B. pertussis adenylate cyclase, or even against a peptide motif common to the B. pertussis and B. anthracis enzymes could recognize a human brain adenylate cyclase (Goyard et al., 1989; Monneron et al., 1988; Orlando et al., 1991). This suggested that higher eucaryotes enzymes might display similarities with the calmodulin activated toxic proteins. Identification of an adenylate cyclase gene from bovine brain by the group of Gilman revealed however, that the enzyme was clearly similar to the yeast enzyme and not to adenylate cyclase toxins (Krupinski et al., 1989). It is therefore difficult at the moment to know whether there exists a true similarity between class II and class III enzymes, or whether the class of brain cyclases is heterogeneous.

A tentative phylogenetic tree could be proposed, as seen in Figure 5. This is an integrated view, where all cyclases derive from an ancestral nucleotide binding structure. Alternatively we could be faced with a specific case of convergent evolution, as for instance in other ATP binding enzymes, the different adenylate cyclases having independently evolved, the oldest activity corresponding to class III enzymes, derived from an ancestral NTP synthesizing enzyme. Some similarity can be discovered with such enzymes (Takeyama et al., 1990), as well as with enzymes using ATP for positive ions translocation (Tamanoi et al., 1984). This makes clear that we are much in need of crystallographic data for all three classes, a challenge that will be taken up in the next few years. In parallel, determination of the sequence of adenylate cyclase genes from the many organisms that are sensitive to cAMP, should enhance our knowledge and permit more convincing phylogenetic comparisons (Ide, 1971; Kaul and Wenman, 1986; Khandelwal and Hamilton, 1972; Paveto et al., 1990; Verni and Rosati, 1987).

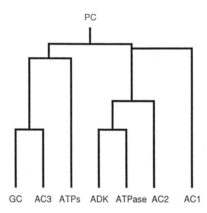

Figure 5. A tentative evolutionary tree linking the three classes of cyclases together. Similarities are so low that the structure of the tree is highly speculative.

REFERENCES

Alexandre, S., Paindavoine, P., Tebabi, P., Pays, A., Halleux, S., Steinert, M., and Pays E. (1990) Mol. Biochem. Parasitol. 43, 279-288

Bakalyar, H.A., and Reed, R.R. (1990) Science 250, 1403-1406

Beall, F.A., Taylor, M.J., and Thorne, C.B. (1962) J. Bacteriol. 83, 1274-1280

Bernal, J.D. (1951): The Physical Basis of Life, Routledge and Kegan Paul, London.

Beuve, A., Boesten, B., Crasnier, M., Danchin, A., and O'Gara, F. (1990) J. Bacteriol, 172, 2614-2621

Cairns-Smith, A.G. (1982): Genetic Takeover and the Mineral Origin of Life, Cambridge University Press, Cambridge

Coussen, F., Haiech, J., d'Alayer, J., and Monneron, A. (1985) Proc. Natl. Acad. Sci. USA 82, 6736-6740

Danchin, A. (1989) Prog. Biophys. Molec. Biol. 54, 81-86

Danchin, A., and Lenzen, G. (1988) Second Messengers and Phosphoproteins 12, 7-28

Ehrmann, I.E., Gray, M.C., Gordon, V.M., Gray, L.S., and Hewlett, E.L. (1991) FEBS Lett. 278, 79-83

Escuyer, V., Duflot, E., Sezer, O., Danchin, A., and Mock, M. (1988) Gene, 71, 293-298

Fandl, J.P., Thorner, L.K., and Artz, S.W. (1990) Genetics 125, 719-727

Garbers, D.L. (1990) The New Biologist 2, 499-504

Gilles, A-M., Munier, H., Rose, T., Glaser, P., Krin, E., Danchin, A., Pellecuer, C., and Barzu, O. (1990) Biochemistry 29, 8126-8130

Glaser, P., Ladant, D., Sezer, O., Pichot, F., Ullmann, A., and Danchin, A. (1988a) Mol. Microbiol. 2, 19-30

Glaser, P., Sakamoto, H., Bellalou, J., Ullmann, A., and Danchin, A. (1988b) EMBO J. 7, 3997-4004

Glaser, P., Elmaoglou-Lazaridou, A., Krin, E., Ladant, D., Barzu, O., and Danchin, A. (1989) EMBO J., 8, 967-972

Glaser, P., Munier, H., Gilles, A-M., Krin, E., Porumb, T. Barzu, O., Sarfati, R., Pellecuer, C., and Danchin, A. (1991) EMBO J. 10, 1683-1688

Goebel, W., and Hedgpeth, J. (1982) J. Bacteriol. 151, 1290-1298

Goldhammer, A., and Wolff, A. (1982) Anal. Biochem. 124, 45-52

376

Gordon, V.M., Leppla, S.H., and Hewlett, E.L. (1988) Infect. Immun. 56, 1066-1069

Goyard, S., Orlando, C., Sabatier, J-M., Labruyère, E., d'Alayer, J., Fontan, G., van Rietschoten, J., Mock, M., Danchin, A., Ullmann, A., and Monneron, A. (1989) Biochemistry 28, 1964-1967

Hedegaard, L., and Danchin, A. (1985) Mol. Gen. Genet. 201, 38-42

Hewlett, E.L., and Wolff, J. (1976) J. Bacteriol. 127, 890-898

Holland, M.M., Leib, T.K., and Gerlt, J.A. (1988) J. Biol. Chem. 263, 14661-14668

Kataoka, T., Broek, D., and Wigler, M. (1985) Cell 43, 493-505

Kaul, R., and Wenman, W.M. (1986) J. Bacteriol. 168, 722-727

Khandelwal, R.L., and Hamilton, I.R. (1972) Arch. Biochem. Biophys. 151, 75-84

Kiely, B., and O'Gara, F. (1983) Mol. Gen. Genet. 192, 230-234

Krupinski, J., Coussen, F., Bakalyar, H.A., Tang, W-J., Feinstein, P.G., Orth, K., Slaughter, C., Reed, R.R., and Gilman, A.G. (1989) Science 244, 1558-1564

Leib, T.K., and Gerlt, J.A. (1983) J. Biol. Chem. 258, 12982-12987

Leppla, S.H. (1982) Proc. Natl. Acad. Sci. USA 79, 3162-3166

Leppla, S.H. (1984) Adv. Cyclic Nucleotide Protein Phosphorylation Res. 17, 189-198

Lowe, D.G., Chang, M-S., Hellmiss, R., Chen, E., Singh, S., Garbers, D.L., and Goeddel, D.V. (1989) EMBO J. 8, 1377-1384

Masson, P., Jacquemin, J.M., Culot, M. (1984) Ann. Microbiol. 135, 343-351

Masson, P., Lenzen, G., Jacquemin, J.M., and Danchin, A. (1986) Curr. Genet. 10, 343-352

Mock, M., Crasnier, M., Duflot, E., Dumay, V., and Danchin, A.(1991) (in press) J. Bacteriol.

Monneron, A., Ladant, D., d'Alayer, J., Bellalou, J., Barzu, O., and Ullmann, A. (1988) Biochemistry 27, 536-539

Orlando, C., d'Alayer, J., Baillat, G., Castets, F., Jeannequin, O., Mazié, J-C., and Monneron, A. (1991) submitted

Parma, J., Stengel, D., Gannage, M-H., Poyard, M., Barouki, R., and Hanoune, J. (1991) Biochem. Biophys. Res. Commun., in press

Paveto, C., Egidy, G., Galvagno, M.A., and Passeron, S. (1990) Biochem. Biophys. Res. Commun. 167, 1177-1183

Peters, E.P., Wilderspin, A.F., Wood, S.P., Zwelebil, M.J.J.M., Sezer, O., and Danchin, A. (1991) Mol. Microbiol. 5, 1175-1181

Pfeuffer, T., and Metzger, H. (1982) FEBS Lett. 146, 369-375

Pfeuffer, E., Dreher, R-M., Metzger, H., and Pfeuffer, T. (1985a) Proc. Natl. Acad. Sci. USA 82, 3086-3090

Pfeuffer, E., Mollner, S., and Pfeuffer, T. (1985b) EMBO J. 4, 3675-3679

Pfeuffer, E., Mollner, S., Lancet, D., and Pfeuffer, T. (1989) J. Biol. Chem. 264, 18803-18807

Read, L.K., and Mikkelsen, R.B. (1991) Mol. Biochem. Parasitol. 45, 109-120

Rogel, A., Meller, R., and Hanski, E. (1991) J. Biol. Chem. 266, 3154-3161

Sarfati, R.S., Kansal, V.K., Munier, H., Glaser, P., Gilles, A-M., Labruyère, E., Mock, M., Danchin, A., and Barzu, O. (1990) J. Biol. Chem. 265, 18902-18906

Schulz, S., Chinkers, M., and Garbers, D.L. (1989) FASEB J. 3, 2026

Singh, S., Lowe, D.G., Thorpe, D.S., Rodriguez, H., Kuang, W-J., Dangott, L.J., Chinkers, M., Goeddel, D.V., and Garbers, D.L. (1988) Nature 334, 708-712

Takai, K., Kurashima, C., Suzuki-Hovi, Okamoto, H. and Hayaishi, O. (1974) J. Biol. Chem. 249, 1965-1972

Takeyama, M., Ihara, K., Moriyama, Y., Noumi, T., Ida, K., Tomioka, N., Itai, A., Maeda, M., and Futai, M. (1990) J. Biol. Chem. 265, 21279-21284

Tamanoi, F., Walsh, M., Kataoka, T., and Wigler, M. (1984) Proc. Natl. Acad. Sci. USA 81, 6924-6928

Thorne, C.B., Molnar, D.M., and Strange, R.E. (1960) J. Bacteriol. 79, 450-455

Thorner, L.K., Fandl, J.P., and Artz, S.W. (1990) Genetics 125, 709-717

Thorpe, D.S., and Garbers, D.L. (1989) J. Biol. Chem. 264, 6545-6549

Verni, F. and Rosati, G. (1987) J. Exp. Zoology 244, 289-298

Wächtershäuser, G. (1988) Microbiol. Rev. 52, 452-480

Wolff, J., Cook, G.H., Goldhammer, A.R., and Berkowitz, S.A. (1980) Proc. Natl. Acad. Sci. USA 77, 3840-3844

Young, D., Riggs, M., Field, J., Vojtek, A., Broek, D., and Wigler, M. (1989) Proc. Natl. Acad. Sci. USA 86, 7989-7993

Adenine Nucleotides in Cellular Energy Transfer and Signal Transduction
S. Papa, A. Azzi & J.M. Tager (eds)
© 1992 Birkhäuser Verlag, Basel/Switzerland

SENSORY TRANSDUCTION IN *DICTYOSTELIUM*

Peter J.M. Van Haastert

Department of Biochemistry, University of Groningen, Nijenborgh 16
9747 AG Groningen, The Netherlands

SUMMARY: In the micro-organism *Dictyostelium* extracellular cAMP
induces chemotaxis and cell differentiation by a complex sensory
transduction mechanism. cAMP binds to surface receptors which
activate multiple G-proteins that stimulate several second
messenger enzymes. This leads to the formation of intracellular
cAMP, cGMP, Ins(1,4,5)P$_3$ and Ca^{2+}. In this paper the biochemistry
and genetics of this sensory transduction process at the plasma
membrane of *Dictyostelium* are described and compared with that of
mammalian cells.

INTRODUCTION

Sensory transduction deals with information transfer and not
primarily with metabolism. Therefore, energy is not used to convert
molecules, but is mainly consumed to control the process. The
number of molecules involved in sensory transduction are generally
very low in the order of thousands to millions of molecules per
cell. A high turnover of these molecules or a rapid phosphory-
lation-dephosphorylation consumes a relatively small amount of
energy if compared to metabolism. It is likely that evolutionairy
stress has not optimized the efficient use of energy as in
metabolism, but may have resulted in accurate control mechanisms.

The organization of multicellular organisms depends on cell-
cell communication. The signals are often soluble molecules in the
extracellular fluid, but also include odors and light. A large
array of surface receptors is involved in the detection of these
signals. The receptors transduced the signals across the plasma
membrane so that enzymes at the inner face of the membrane are
activated, producing second messengers, which by a complex network
of interactions activate target proteins or genes.

Vertebrate cells have been used to study hormone and
neurotransmitter action, vision, the regulation of cell growth and
differentiation. By comparing sensory transduction in lower and

higher eukaryotes general principles may be recognized that are found in all organisms and deviations that are present in specialized systems. These comparative studies may also help to understand the differences between cell types within one organism and the importance of a particular sensory transduction pathway.

THE ORGANISM *DICTYOSTELIUM*

During the development of the cellular slime mold *Dictyostelium discoideum* an initially nearly homogeneous population of cells aggregate and differentiate into two final cell states (Loomis, 1985). About 80% of the cells form a ball of spores supported by the remaining 20% of the cells that form a dead, highly vacuolized, cellulose stalk. Triggeres by starvation, this differentiation program begins with the expression of genes that comprise a cAMP sensory transduction system (for reviews see Janssens and Van Haastert, 1987; Devreotes, 1989; Schaap, 1986; Van Haastert et al., 1991). This apparatus controls aggregation as well as early and late gene expression. Its components include surface cAMP receptors, G-proteins and adenylyl cyclase. During aggregation cAMP is secreted from central points at 5 min intervals. It diffuses to surounding cells which both move up the cAMP concentration gradient towards the aggregation center and synthesize and secrete additional cAMP. This results in the outward propagation of cAMP concentration waves, which direct periodic inward chemotactic movement steps, leading to the formation of a multicellular structure (see figure 1).

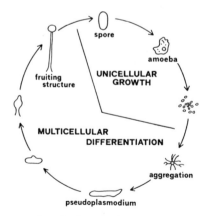

Figure 1. The life cycle of *Dictyostelium*

SECOND MESSENGER FORMATION

Stimulation of sensitive cells with cAMP leads to the formation of several second messengers in the cell, including cGMP, $Ins(1,4,5)P_3$, and Ca^{2+}. Maximal concentrations of cGMP and $Ins(1,4,5)P_3$ are obtained within 10 s after stimulation, and basal levels are recovered after about 30 s. Intracellular cAMP increases more slowly, reaching a maximal concentration after about 1-2 min. In contrast to cGMP and $Ins(1,4,5)P_3$ which are degraded intracellularly, cAMP is secreted in the medium (figure 2).

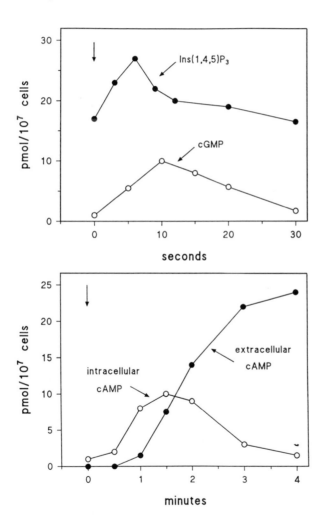

Figure 2. Second messenger formation in aggregation competent *Dictyostelium* cells after stimulation with 0.1 uM cAMP at t = 0s.

The rate of cAMP secretion is proportional to the intracellular concentration, suggesting that cAMP secretion is a first order proces and that the secretion mechanism is not activated by the receptor (see Janssens and Van Haastert, 1987). Several possible mechanisms of cAMP secretion may be present, including vescicular transport as in acelylcholine, a cAMP pump in the plasma membrane or passive diffusion. No conclusive experiments have been reported, although it has been demontrated that cAMP is immediately secreted upon permeabilization of the plasma membrane by electroporation, suggesting that cAMP accumulates in the cytosol before it is secreted (Schoen et al., 1989).

GENES IN SIGNAL TRANSDUCTION

Several genes have been cloned that encode proteins involved in signal transduction (see table I). The surface receptor gene cAR1 was cloned using an antiserum against the purified receptor (Klein et al., 1987; Klein et al, 1988). The deduced amino acid sequence of the protein predicted a topological structure now charasteristic of all receptors coupled to G-proteins: an extracellular N-terminal domain, seven putative transmembrane spanning domains and a cytoplasmic C-terminal domain. Three other genes, designated cAR2, cAR3, and cAR4, were cloned that are homologous with cAR1 in the transmembrane spanning domains and connecting loops (Johnson, Saxe, Kimmel and Devreotes, personal communication) however, the extracellular N-terminal domain and the cytoplasmic C-terminal domains differ markedly.

Table I Genes in *Dictyostelium* signal transduction

Protein	Number of genes	Main Investigators
cAMP receptor	4	Klein, Johnson, Saxe, Kimmel, Devreotes
Gα subunit	8	Kumagai, Pupillo, Firtel, Devreotes
G subunit	1	Lilly, Devreotes
Adenylyl cycl.	2	Pitt, Devreotes
PLC	1	Drayer, Van Haastert
protein kinase	>20	Many labs

Updated October, 1991; Most sequences have not been published yet.

Multiple genes encoding the α-subunit of G-proteins were cloned on the basis of the conserved GTP-binding domains of mammalian G-proteins (Kumagai et al., 1988; Puppilo, Firtel and Devreotes, personal communication). The genes show different developmental expression patterns, with Gα1 being expressed during growth, Gα2 and Gα3 during cell aggregation and Gα4 and Gα5 during multicellular development. Each Gα-protein sequence is unique, suggesting that each may fulfill a specific function in sensory transduction during development. Only one gene encoding a beta subunit hase been identified; this gene is expressed during the entire development (Lilly and Devreotes, personal communication).

The genes encoding some of the effector enzymes have been identified. Adenylyl cyclase is encoded by two genes that are expressed during cell aggregation and multicellular development, respectively (Pitt and Devreotes, personal communication). Phospholipase C (PLC) is expressed during the entire development with maxima during cell aggregation and fruiting body formation (Drayer and Van Haastert, unpublished results).

Many labs are involved in the identification of putative protein kinases by cloning relevant genes. Presently about 20 different protein kinases have been identified on a genetic level only. These genes were identified in about 7 different labs and only a few genes were identified twice. Statistics suggests that *Dictyostelium* may have more than 100 protein kinase genes.

DICTYOSTELIUM MUTANTS AND TRANSFORMANTS

As a micro-organism, *Dictyostelium* is very useful to exploit the power of classical and modern genetics to elucidate the role of sensory transduction for chemotaxis and development. Classical mutants are obtained after mutagenesis and phenotype-selection. Mutants have been isolated that are defective in the timing of differentiation, chemotaxis, locomotion, detecting cAMP signals, cell aggregation and proportioning and specificity of multicellular development. Here, only mutants are discussed that show defects in sensory transduction (table II).

Upon cloning of a gene it becomes possible to alter the expression of that gene by modern genetic techniques. The expression of a specific gene can be reduced by antisense expression or genedisruption (see Devreotes, 1989). Also wild-type or mutant genes can be expressed from a promotor that has a different activity than the endogeneous promotor. Finally, genes can be expressed in *Dictyostelium* that are from differnt organisms; these genes may encode for activities that are also present in Dictyostelium, or they may introduce completely new enzymes in this organism.

All signal transduction (Table II) depends on the presence of surface cAMP receptor as was established by studies with cells in which the expression of the receptor gene cAR1 was inactivated (Klein et al, 1988; Sun and Devreotes, 1991). Activation of a specific G-protein (Gα2) appears to be essential for most sensory transduction as was established using mutant *fgd*A which is defective in this G-protein (Coukell et al., 1983; Kesbeke et al., 1988; Kumagai et al., 1989). The effector enzyme of Gα2 is most likely PLC, because in this mutant GTPγS stimulation of PLC is absent, whereas GTPγS stimulation of adenylyl cyclase and guanylyl cyclase are unaltered (Bominaar and Van Haastert, unpublished observations). It is not known why in *fgd*A adenylyl and guanylyl cyclase are not stimulated by cAMP *in vivo*.

In mutant *synag*7 the the G-protein mediated activation of adenylyl cyclase is lost (Theibert & Devreotes, 1986; Van Haastert et al, 1987). The activation of guanylyl cyclase and PLC are essentially normal in these mutants, and cAMP induces normal chemotaxis and cell differentiation (Schaap et al, 1986). This suggests that the main function of the activation of adenylyl cyclase is to generate the extracellular cAMP relay signal.

Table II Sensory transduction in mutants and transformants

Mutant	Protein	Responses				Development
		Chem	cAMP	cGMP	IP3	
cAR1-GD	cAR1	-	-	-	-	-
*fgd*A	Gα2	-	-	-	-	-
*synag*7	?	+	-	+	+	+
*stm*F	cGMP-PDE	++	+	++	+	+
*fgd*C	?	±	+	+	-	-

Chem = chemotaxis; IP3 = Ins(1,4,5)P$_3$; +, present/normal; -, absent/reduced; ++, increased

The function of intracellular cGMP has been investigated using mutant *stm*F which appears to be defective in cGMP phosphodiesterase (Ross & Newell, 1981; Van Haastert et al, 1982). In this mutant extracellular cAMP induces elevated and prolonged cGMP accumulation, which is associated with prolonged chemotactic movement. The enhanced accumulation of cGMP in mutant *stm*F has no effect on the stimulation of adenylyl cyclase or PLC (unpublished observations). The importance of the cGMP pathway will be established more accurately when mutants lacking guanylyl cyclase are available.

Recently mutant *fgd*C isolated by Coukell et al (1983) was further characterized. Extracellular cAMP induces the normal activation of adenylyl and guanylyl cyclase, but no activation of phospholipase C. This defect is associated with the absence of cAMP-induced prespore and prestalk gene expression. These observations suggest that activation of phospholipase C may be required for cAMP-mediated gene expression, but is not required for the activation of adenylyl or guanylyl cyclase as was suggested previously (Bominaar et al., 1991). More mutants in the inositol phosphate/Ca^{2+} pathway are required to unravel the functions of the many components of the inositol cycle.

A COMPARISON BETWEEN *DICTYOSTELIUM* AND VERTEBRATES

The evolutionary distance between *Dictyostelium* and vertebrates is more than a bilion years, surface receptors, G-proteins, adenylyl cyclase, guanylyl cyclase and the inositol cycle are remarkably well conserved (see Van Haastert et al., 1991). The powerful genetic techniques that can be used in microorganisms are proving to be very useful to study sensory transduction in these organisms. In the yeast *S.cerevisiae* many components of the sensory transduction pathways were identified by a combination of biochemistry and molecular genetics. *Dictyostelium* has the additional advantage of a tightly programmed development of identical free moving amoeboid cells to a multicellular organism consisting of two or three cell types. It is expected that in the coming years we may learn in much more detail how simple extracellular signals regulate complex cellular functions such as chemotaxis and cell differentiation.

386

ACKNOWLEDGEMENTS
I thank Peter Devreotes and Rich Firtel for communication of unpublished results, and the members of the lab for fruitful discussions.

REFERENCES

Bominaar, A.A., Kesbeke, F., Snaar-Jagalska, B.E., Peters, D., Schaap, P., and Van Haastert, P.J.M. (1991) J. Cell Sci., in press.

Coukell, M.B., Lappano, S. and Cameron, A.M. (1983) Devel. Genet. 3, 283-297.

Devreotes, P.N. (1989) Science 245, 242-245.

Janssens, P.M.W. and Van Haastert, P.J.M. (1987). Microbiol. Rev. 51, 396-418.

Kesbeke, F., Snaar-Jagalska, B.E. and Van Haastert, P.J.M. (1988). J. Cell Biol. 107, 521-528.

Klein, P.S., Sun, T.J., Saxe III, C.L., Kimmel, A.R., Johnson, R.L. and Devreotes, P.N. (1988). Science 241, 1467-1472.

Klein, P., Vaughan, R., Borleis, J. and Devreotes, P. (1987). J. Biol. Chem. 262, 358-364.

Kumagai, A., Pupillo, M., Gundersen, R., Miake-Lye, R., Devreotes, P.N. and Firtel, R.A. (1989). Cell 57, 265-275.

Loomis, W.F. (1985) Dictyostelium discoideum a developmental system. New York, Acad. Press.

Schaap, P. (1986) Differentiation 33, 1-16.

Ross, F.M. and Newell, P.C. (1981). J.Gen. Microbiol. 127, 339-350.

Schaap, P., Van Lookeren Campagne, M.M., Van Driel, R., Spek, W., Van Haastert, P.J.M. and Pinas, J. (1986). Dev.Biol. 118, 52-63.

Schoen, C.D., Arents, J.C., Bruin, T., and Van Driel, R. (1989) Exp. Cell Res. 181, 51-62.

Sun, T.J., and Devreotes, P.N. (1991) Gene and Develop. 5, 572.

Theibert, A. and Devreotes, P. (1986). J.Biol.Chem. 261, 15121-15125.

Van Haastert, P.J.M., Van Lookeren Campagne, M.M. and Ross, F.M. (1982). FEBS Lett. 147, 149-152.

Van Haastert, P.J.M., Snaar-Jagalska, B.E. and Janssens, P.M.W. (1987). Eur. J. Biochem. 162, 251-258.

Van Haastert, P.J.M., Janssens, P.M.W. and Erneux, C. (1991) Eur. J. Biochem. 195, 289-303.

Adenine Nucleotides in Cellular Energy Transfer and Signal Transduction
S. Papa, A. Azzi & J.M. Tager (eds)
© 1992 Birkhäuser Verlag, Basel/Switzerland

STRUCTURE AND FUNCTION OF P-GLYCOPROTEINS

Alfred H. Schinkel, Carsten R. Lincke, Jaap J.M. Smit and Piet Borst

The Netherlands Cancer Institute, Plesmanlaan 121, 1066 CX Amsterdam, The Netherlands

SUMMARY: Overexpression of P-glycoprotein genes is one established cause of multidrug resistance in mammalian cells. We are studying P-glycoproteins and their genes in man, in the mouse and in the nematode worm *Caenorhabditis elegans* in order to understand the normal, physiological role of P-glycoproteins, and the mechanistics of P-glycoprotein function. Our data suggest that one common, and evolutionarily well-conserved function of P-glycoproteins is protection of the organism against naturally occurring xenotoxins present in ingested food. We consider it likely, however, that some P-glycoprotein variants perform a more specialized function in the organism.

INTRODUCTION

Chemotherapy of some human tumors can result in an initial favorable response leading to decrease of tumor size, or even an apparently complete disappearance of the tumor (remission). Unfortunately, after some time, more often than not, the tumor will increase in size again, now resistant to the drug(s) used in the initial chemotherapy regimen (acquired or secondary resistance). Frequently, these relapsed tumors are also cross-resistant to other drugs that were not used in the initial treatment, even when the structure and cellular target of these

388

drugs was entirely different. This phenomenon is called multidrug resistance (MDR). Other tumors turn out to be resistant to many different drugs from the very start of chemotherapy (intrinsic or primary resistance). It is likely that primary resistance can be caused by the same cell-biological mechanisms as acquired resistance, and that it depends on the biological make-up (differentiation) of the tumor cells. Obviously, both primary and acquired MDR form a major impediment for curative chemotherapy of cancer (for an overview, see DeVita et al., 1989).

One form of MDR has been unambiguously identified in cultured mammalian cells. It is caused by the overexpression of P-glycoprotein (P-gp) genes, which can cause resistance to a wide range of relatively large, hydrophobic cytotoxic drugs that do not share a common structure or cellular target. Most of these drugs are toxins from natural sources such as plants (*Vinca* alkaloids, epipodophyllotoxins, taxol, colchicine) or bacteria (actinomycin D, daunorubicin, gramicidin D), and several are regularly included in cancer chemotherapy protocols. Biochemical and genetic evidence indicates that P-gps are large, glycosylated cell membrane proteins, containing twelve transmembrane segments and two intracellular ATP binding sites. A schematic structure of P-gps is presented in Fig. 1. P-gp is thought to confer resistance by actively extruding drugs from the cell, thus lowering the drug concentration at the cellular target site (reviewed amongst others in Endicott & Ling, 1989 and Schinkel & Borst, 1991).

Although certainly other forms of MDR exist that are not caused by P-gps (see e.g. Baas et al., 1990), it is likely that P-gp-mediated MDR plays a role in the intrinsic or acquired drug resistance of at least some clinical tumors, such as ovarian cancer, myeloma, non-Hodgkin's lymphoma and childhood sarcoma (Chan et al., 1990; reviewed in amongst others Borst, 1991). The existence of so-called reversal agents, which are probably competitive inhibitors of the P-gp pump (see e.g. Yusa and Tsuruo, 1989), may provide a means for counteracting P-gp-mediated MDR in clinical tumors, and thus to improve chemotherapy

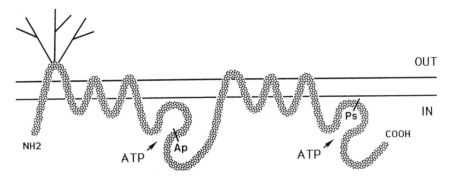

Fig. 1: Two-dimensional representation of P-glycoprotein as it spans the cell membrane. Some structural details are depicted as they occur in the human *MDR1* P-glycoprotein. In reality the transmembrane segments are probably closely associated with each other, possibly forming a three-dimensional barrel-like structure. The three extracellular N-linked glycosylation sites of *MDR1* P-glycoprotein are shown by branches sprouting from the polypeptide. The two putative ATP-binding sites are indicated with ATP. Ap and Ps indicate the equivalent positions in the protein of, respectively, the unique ApaI and PstI sites in the *MDR1* cDNA.

regimens. In order to be able to optimize methods to remedy P-gp-mediated MDR, we need to have better insight in the normal, physiological role of P-gps, and in the mechanistics of P-gp function. For this purpose, we are studying P-gps in man, in the mouse and in a simple metazoan organism, the nematode worm *Caenorhabditis elegans*.

RESULTS AND DISCUSSION

STRUCTURAL STUDIES ON THE HUMAN P-GLYCOPROTEINS: Man has two P-gp genes, *MDR1* and *MDR3* (also called *MDR2*). Although *MDR3* P-gp is highly similar to *MDR1* P-gp in amino acid composition (76% identity) and although it has the same overall structure (see Fig. 1), we have not been able to show that *MDR3* can confer drug resistance (Schinkel et al., 1991). When the N-terminal or C-terminal homologous halves of *MDR1* and *MDR3* P-gp were exchanged at the unique ApaI site in the highly conserved first ATP binding fold (see Fig. 1), the resulting chimeric *MDR1/MDR3* or *MDR3/MDR1* genes also failed to confer drug resistance. In contrast, replacement of the C-terminal end of *MDR1* from the PstI site onwards (Fig. 1) by the homologous *MDR3* sequence, resulted in a protein that could still effectively confer resistance to vincristine. Similar and more extensive results have been reported for the mouse *mdr1b* and *mdr2* genes (Buschman and Gros, 1991). Together, these results suggest that *MDR3/mdr2* P-gp has a substrate specificity different from that of *MDR1/mdr1* P-gp, and that the substrate specificity is mainly determined by the transmembrane regions of the protein. No substrates for *MDR3/mdr2* P-gps are known at this moment.

All mammalian P-gps analyzed to date carry two or more potential N-linked glycosylation sequences (N-X-S/T) in the first putative extracellular loop. At least one N-linked glycosylation site is also found at this position in all eukaryotic P-gps analyzed so far, including those of *C. elegans* (see below), *Leishmania tarentolae* (Ouellette et al., 1990) and *Saccharomyces cerevisiae* (Kuchler et al., 1989; McGrath and Varshavsky, 1989). The only exception is the *Plasmodium falciparum pfmdr1* P-gp, but this protein is mainly present in the membrane of the digestive vacuole, not in the cell membrane (Cowman et al., 1991). The strong conservation of these N-linked glycosylation sites suggested an important role in P-gp function, at least in mammals, so we mutated all three N-linked glycosylation sites present in the human *MDR1* P-gp (see Fig. 1). These mutations did not detectably affect the ability of the protein to confer drug resistance. Also, the mutant P-gps were still mainly found in the cell membrane, as tested by immunocytochemistry. This indicates that N-linked glycosylation is not necessary for the basic drug transport activity of *MDR1* P-gp, nor for its correct routing to the cell membrane. Yet, because of the strong conservation, we speculate that glycosylation of P-gp may be important in the living organism, although we don't have a clear idea of its possible role. Transplantation of transfected tumor cells containing these mutant P-gps in nude mice may give a clue to the role of glycosylation at the organismal level.

INACTIVATION OF MOUSE P-GLYCOPROTEIN GENES IN EMBRYONIC STEM CELLS: Experiments are in progress to inactivate each of the mouse P-gp genes (mdr1a and mdr1b, both homologues of human MDR1, and mdr2, homologue of human MDR3) in the mouse germline by gene targeting in embryonic stem cells. The phenotype of mice homozygously deficient for one (or more) of these genes will hopefully give clues to their normal function in the mouse and, by inference, to the normal function of their human homologues. So far, we have succeeded in disrupting the mouse mdr1a, mdr1b and mdr2 genes in embryonic stem cells. The generation of mice containing these mutated genes in the germline is in progress.

CHARACTERIZATION OF THE P-GLYCOPROTEIN GENES OF C. ELEGANS:
C. elegans is a small, free-living nematode worm, that feeds mainly on microbes and organic debris in the soil, which is its natural habitat. Its low complexity (the mature organism consists of only about 1000 cells) and highly programmed development make it a suitable subject for developmental studies and analyses of gene function and expression in a simple multicellular organism (Wood, 1988).

We have identified and cloned at least three genes encoding proteins highly similar to mammalian P-gps from C. elegans by cross-hybridization with human MDR1 cDNA. These genes, named cepgpA, cepgpB and cepgpC, were physically mapped by A. Coulson (Cambridge, U.K.) to respectively chromosome IV (near unc-31), I (near dpy-5) and X (near lin-14), and they were transcribed to relatively low abundant mRNAs of 5, 4.5 and 4 kb. Mutant phenotypes have not yet been described for these loci. Sequence analysis of genomic and cDNA clones showed that cepgpA covers 8kb and consists of 14 exons. cepgpC is 5kb long, and it is composed of 13 exons (see Fig. 2). cepgpB has not yet been analyzed in

392

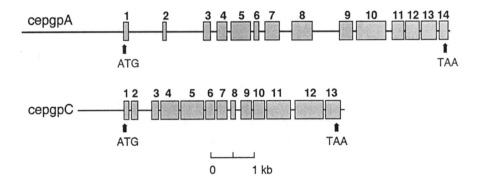

Fig. 2: Genomic structure of the *C. elegans* P-glycoprotein genes *cepgp*A and *cepgp*C. Numbered grey boxes indicate exons. The positions of translation start and termination codons are indicated by ATG and TAA respectively. The genes are depicted at the same scale (see scale bar), and lined up at the initiation codons. (Based on unpublished results of C.R.L.)

detail. The exon composition of the nematode genes does not resemble the highly conserved exon composition of the mammalian P-gp genes (27 coding exons in all human, mouse and hamster P-gp genes analyzed to date). A few intron positions are, however, absolutely conserved between nematodes and mammals (4 in *cepgp*A; 5 in *cepgp*C). In contrast, *cepgp*A and *cepgp*C share only a single conserved intron position. *cepgp*A and *cepgp*B pre-mRNA are trans-spliced to a trans-spliced leader (SL1) at splice acceptor sites only a few nucleotides upstream of the translation initiation codons.

The lengths of the predicted proteins are 1321 amino acids for *cepgp*A, and 1254 amino acids for *cepgp*C, similar to mammalian P-gps (about 1280 amino acids). Their hydrophobicity profiles and overall structures are virtually superimposable on those of mammalian P-gps, predicting two similar halves, each with 6 transmembrane domains, followed by cytoplasmic ATP-binding consensus motifs (Fig. 1). The overall similarity/identity with human and mouse P-gps is about 65/45%, but approaches 100% in the regions around the nucleotide binding domains. Surprisingly, the identity values for the *C. elegans* P-gps compared with each other are only about 50% (compared to >75% in human and mouse P-gps). Thus, the *intra*-species differences in the nematode P-gp genes are considerably larger than in mammals, possibly indicating a stronger divergence in function.

We have used RNase protection assays to quantitatively examine the expression of *cepgp* genes (relative to actin-IV mRNA) upon exposure of nematodes to various kinds of chemical and physical stress: sodium arsenite (reported to induce human *MDR1* transcription (Chin et al., 1989)) at sub-lethal concentrations had no effect on the mRNA levels of any of the P-gp genes. Heat shock led to an apparent five-fold increase in *cepgp*A RNA levels, but since heat shock also severely affected trans-splicing in this pre-mRNA, it is likely that the level of functional *cepgp*A mRNA decreased rather than increased. *cepgp*C RNA levels increased slightly under these conditions, whereas *cepgp*B RNA levels were not clearly changed. No induction of *cepgp* gene transcription was found upon exposure to actinomycin D or emetine (substrates of mammalian P-gps).

Initial studies with several MDR-related drugs showed that *C. elegans* was extremely resistant to most of them, generally 1000 times more than mammalian tumor cells. To investigate whether overexpression of *cepgp* genes can confer drug resistance to whole nematodes, we cultured EMS-mutagenized worms in the presence of emetine or actinomycin D for several months. Non-mutagenized animals exposed to either of the drugs did not respond with detectable changes in expression levels. However, prolonged selection of mutagenized worms with emetine yielded lines that had elevated *cepgp*A and/or *cepgp*C mRNA levels (2- to 10-fold). These lines had a low level (2- to 4-fold) of resistance to emetine compared to wild-type animals. These results suggest that *C. elegans* P-gps share both structural and functional properties with their mammalian counterparts.

TISSUE-SPECIFIC AND DEVELOPMENTALLY REGULATED EXPRESSION OF *C. ELEGANS* P-GLYCOPROTEIN GENES: Using RNase protection, we have analyzed the mRNA levels of *cepgp*A, B and C during development. All three genes were expressed throughout the life cycle of the nematodes. *cepgp*B mRNA levels were invariable. In contrast, *cepgp*A and *cepgp*C were found to be developmentally regulated with maximum steady state mRNA levels in early (*cepgp*A) and late larval development (*cepgp*C).

To analyze tissue-specific expression of *cepgp*A and *cepgp*C, we fused the promoter areas of these genes (sometimes including the first one or two exons) to a LacZ marker gene, containing a functional nuclear localization signal (see Fig. 3 for details). To create transgenic worms, these constructs were injected into the gonad together with a selectable marker (*rol-6*). *Rol-6* transmitting lines were established with each construct, and worms were fixed and stained with the chromogenic substrate X-gal. The nuclear localization of the LacZ marker makes identification of the cells where it is expressed relatively straightforward, since the position of all *C. elegans* cell nuclei is well-defined (see Wood et al., 1988).

In each established line, distinct nuclear staining was seen almost exclusively in all intestinal cells (18-24, depending on the developmental stage). Independent lines from all three constructs had similar staining patterns, although the intensity of staining was somewhat variable, even between lines transformed with the same construct. No staining was observed proximal of the pharyngeal-intestinal valve and distal of the anus. Interestingly, the staining was most intense in early (L1/L2) stage animals transformed with pPD26pgpA.X and pPD26pgpA.BP, whereas animals transformed with pPD26pgpC.XX also showed staining in later stages of development. This was in good agreement with the RNase protection analysis of stage specific expression (see above). Apparently, the LacZ-fusion protein is not very stable over longer periods of time.

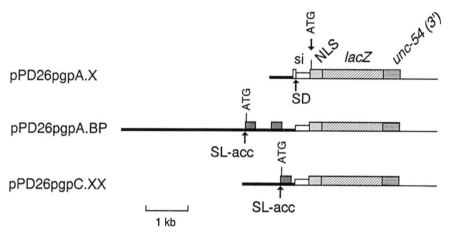

Fig. 3: Structure of *cepgp-lacZ* fusion constructs: We made three constructs starting from the pPD26.50 plasmid kindly donated by A. Fire (Baltimore, MD).
pPD26pgpA.X: A 460 bp fragment from *cepgpA*, containing the 5′ flanking sequence of the first exon but not its spliced leader acceptor, was inserted in a slightly modified pPD26.50.
pPD26pgpA.BP: The glp-1 promoter and part of the synthetic intron of pPD26.50 was removed and replaced by a 4 kb fragment containing 2.9 kb of 5′ flanking sequence of *cepgpA*, its first two exons and part of the second intron.
pPD26pgpC.XX: Analogous to pPD26pgpA.BP, but containing a 1200 bp fragment with 1 kb of 5′ flanking sequence of *cepgpC*, its first exon and part of the first intron.
The bold line indicates *cepgp* genomic sequence with exons shown as hatched boxes. Other domains of the constructs are designated in the upper construct. Only the genomic sequences are depicted approximately on scale (see scale bar). The given sizes and positions of the exons are approximate. The polyadenylation signal was derived from *unc*-54. Abbreviations: NLS: nuclear localization signal; SD: splice donor; si: synthetic intron; SL-acc: spliced leader acceptor.

CONCLUSION

The expression of the *C. elegans* P-gp genes *cepgpA* and *cepgpC* in intestinal cells, and the overexpression of these genes upon mutagenesis and drug selection of complete animals suggest that these P-gp genes protect *C. elegans* from ingested toxins. This pattern of expression is in good agreement with that of the mammalian *MDR1/mdr1* P-gp genes, which are most abundantly expressed in epithelia of the gut and of excretory organs (see *e.g.* Thiebaut et al., 1987). Together these data suggest that one of the common, and evolutionarily well-conserved functions of P-gps is the protection of the organism against naturally occurring xenotoxins present in ingested food (see also Ames et al., 1990). It is likely, however, that some P-gp variants (such as human *MDR3*) perform a more specialized function (compare the specific transport of the **a** mating factor peptide by yeast *STE6* P-gp (Kuchler et al., 1989)). Analysis of mice deficient in one or more of the *mdr* genes may provide a critical test of the nature and importance of these functions in mammals.

Acknowledgements: We thank A. Broeks, E. Wagenaar, L. van Deemter, I. The, M. van Groenigen and S. Kemp for help with experiments, and Dr. R. Plasterk for advice and support. This work was supported by the Netherlands Cancer Foundation, grant NKI 88-6.

REFERENCES

Ames, B.N., Profet, M., and Gold, L.S. (1990) Proc. Natl. Acad. Sci. USA 87, 7782-7786
Borst, P. (1991) Rev. Oncol. 4 (in Acta Oncol. Vol. 30, No. 1), 87-105
Buschman, E., and Gros, P. (1991) Mol. Cell. Biol. 11, 595-603
Baas, F., Jongsma, A.P.M., Broxterman, H.J., Arceci, R.J., Housman, D., Scheffer, G.L., Riethorst, A., van Groenigen, M., Nieuwint, A.W.M., and Joenje, H. (1990) Cancer Res. 50, 5392-5398
Chan, H.S.L., Thorner, P.S., Haddad, G., and Ling, V. (1990) J. Clin. Oncol. 5, 1452-1460
Chin, K.V., Tanaka, S., Darlington, G., Pastan, I, and Gottesman, M.M. (1990) J.Biol. Chem. 265, 221-226
Cowman, A.F., Karcz, S., Galatis, D., and Culvenor, J.G. (1991) J. Cell. Biol. 113, 1033-1042
DeVita, V.T. Jr., Hellman, S., and Rosenberg, S.A., Eds. (1989) Cancer: Principles and practice of oncology, J.B. Lipincott Co., Philadelphia
Endicott, J.A., and Ling, V. (1989) Annu. Rev. Biochem. 58, 137-171
Kuchler, K., Sterne, R.E., and Thorner, J. (1989) EMBO J. 8, 3973-3984
McGrath, J.P., and Varshavsky, A. (1989) Nature 340, 400-404
Ouellette, M., Fase-Fowler, F., and Borst, P. (1990) EMBO J. 9,

1027-1033

Schinkel, A.H., and Borst, P. (1991) Semin. Cancer Biol. $\underline{2}$, 213-226

Schinkel, A.H., Roelofs, M.E.M., and Borst, P. (1991) Cancer Res. $\underline{51}$, 2628-2635

Thiebaut, F., Tsuruo, T., Hamada, H., Gottesman, M.M., and Pastan, I. (1987) Proc. Natl. Acad. Sci. USA $\underline{84}$, 7735-7738

Wood, W.B., Ed. (1988) The nematode *Caenorhabditis elegans*, Cold Spring Harbor Laboratory, Cold Spring Harbor, New York

Yusa, K., and Tsuruo, T. (1989) Cancer Res. $\underline{49}$, 5002-5006

Adenine Nucleotides in Cellular Energy Transfer and Signal Transduction
S. Papa, A. Azzi & J.M. Tager (eds)
© 1992 Birkhäuser Verlag, Basel/Switzerland

FUNCTION AND REGULATION OF THE CYSTIC FIBROSIS TRANSMEMBRANE CONDUCTANCE REGULATOR

Matthew P. Anderson[1], Devra P. Rich[1], Richard J. Gregory[2], Seng Cheng[2], Alan E. Smith[2], and Michael J. Welsh[1]

[1]Howard Hughes Medical Institute, Departments of Internal Medicine and Physiology and Biophysics, University of Iowa College of Medicine, Iowa City, IA 52242; [2]Genzyme Corporation, Framingham, MA 01701

INTRODUCTION

Electrolyte transport across airway epithelia controls the quantity and composition of respiratory tract fluid. In patients with cystic fibrosis (CF), cAMP-stimulated Cl^- secretion by airway epithelia is defective. This defect results from a failure of cAMP to activate apical membrane Cl^- channels (Quinton, 1990).

Genetic mapping studies identified a single gene which is mutated in cystic fibrosis chromosomes (Riordan et al., 1989; Kerem et al., 1989). This gene encodes a protein called the cystic fibrosis transmembrane conductance regulator (CFTR). Amino acid sequence analysis and comparison with other proteins (Riordan et al., 1989) suggests that CFTR consists of two repeats of a unit containing six membrane spanning segments and a putative nucleotide binding domain (Fig. 1). The two repeats are separated by a large polar segment called the R domain, which contains multiple potential phosphorylation sites. The predicted topology of CFTR, with the exception of the R domain, resembles that of a number of other membrane proteins such as the multiple drug resistance P-glycoprotein, the yeast STE6 gene product, and several bacterial periplasmic permeases (Riordan et al., 1989; Ames et al., 1990; Hyde et al., 1990).

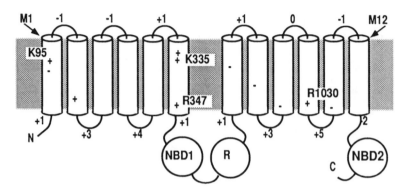

Figure 1. Model showing predicted topology of CFTR in the membrane. A. M1 to M12 indicate membrane spanning sequences, NBD-1 and NBD-2 refer to the nucleotide binding domains, R refers to the Regulatory domain. Sites of charged residues in M1-M12 are indicated by + and -. Predicted net charge of intracellular and extracellular loops are indicated; charge of first 5 amino acids only, before 11 and M7 and after M6 and M12, are indicated. Not drawn to scale. Amino acids mutated in this study are indicated.

Expression of CFTR, but not CFTR containing the most common mutation found in CF chromosomes (CFTRΔF508), corrects the Cl⁻ channel defect in epithelial cells from patients with CF (Rich et al., 1990; Drumm et al, 1990). This result demonstrated a causal relationship between mutations in CFTR and defective Cl⁻ transport, the hallmark of the disease. However those results did not identify the function of CFTR. As a first step in trying to understand its function, we expressed CFTR in cells which lack both CFTR and cAMP-activated Cl⁻ channels.

Heterologous CFTR Expression Generates cAMP-Stimulated Cl Channels:
We used HeLa (a human cervical carcinoma cell line), CHO (a Chinese hamster ovary cell line), NIH 3T3 (a mouse fibroblast cell line), and T84 cells (a human colon carcinoma epithelial cell line). We found that T84, but not HeLa, 3T3, or CHO cells,

contained CFTR by using antibodies that immunoprecipitate CFTR and the more sensitive reverse transcriptase/polymerase chain reaction to detect CFTR mRNA (Gregory et al., 1990; Anderson et al., 1991b).

To express CFTR, we used the transient vaccinia virus-T7 hybrid expression system developed by Moss and colleagues (Elroy-Stein et al., 1989). We previously used this system to express functional CFTR in CF airway epithelial cells (Rich et al., 1990; Gregory et al., 1990). We also used a stable retrovirus expression system developed by Mulligan and colleagues (Anderson et al., 1991a).

Using the standard whole-cell patch-clamp technique we found that cAMP did not increase current in HeLa or CHO cells expressing mutant CFTR (ΔF508) or in untransfected cells (Fig. 2). Similar results were obtained with 3T3 fibroblasts. In contrast, cAMP activated currents in all three cells when transfected with CFTR cDNA (Fig. 2). The fact that not every cell responded is

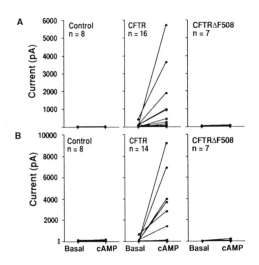

Figure 2. Comparison of currents measured at +80 mV before (Basal) and after (cAMP) addition of agents which increase cAMP in HeLa (A) and CHO (B) cells. Forskolin (20 μM) alone, forskolin with IBMX (100 μM), and CPT-cAMP (500 μM) gave similar results. cAMP significantly increased current in cells transfected with plasmid encoding CFTR (P < 0.004 for HeLa, P < 0.01 for CHO). cAMP did not significantly increase currents in either of the other two groups.

consistent with our previous studies with the transient vaccinia virus-T7 hybrid expression system in CF airway epithelial cells (Rich et al., 1990) and with the heterogeneous response found using the SPQ assay (Anderson et al., 1991b). Stable expression of CFTR in NIH 3T3 cells produced large cAMP-stimulated currents in all cells. Although we cannot completely exclude the possibility that CFTRΔF508 has some effect on Cl⁻ current, the data suggest that such an effect is at most 3% of that observed in cells transfected with CFTR cDNA.

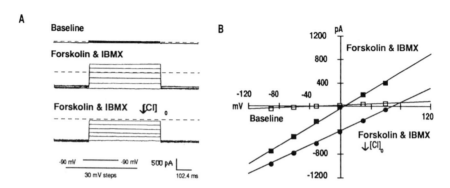

Figure 3A and B. Expression of CFTR induces cAMP-activated Cl⁻ currents in HeLa cells. A) Currents before (top) and after (bottom) addition of forskolin (10 μM) and IBMX (100 μM) in cells transfected with plasmid encoding CFTR. Insets show the voltage protocol for current tracings and the time and current scales. B) Current-voltage (I-V) relationship obtained before (open symbols) and after (closed symbols) addition of forskolin and IBMX. I-V relationships were obtained in symmetrical Cl⁻ solutions (squares) and during reduction in bath [Cl⁻] (circles) from 139.8 mM to 4.8 mM (Cl⁻ replaced by aspartic acid, expected reversal potential +84.8 mV).

The cAMP-activated currents are due to the activation of Cl⁻ channels: a relatively impermeant cation N-methyl-D-glucamine (NMDG) was used, and changes in Cl⁻ concentration shifted the current-voltage (I-V) relation as expected for a Cl⁻ current (Fig. 3A & B). The increase in current was reversible: when forskolin was removed current returned to basal values (Fig. 4). CFTR expression has also been shown to produce cAMP-stimulated Cl⁻ channels in Sf9 insect cells (Kartner et al., 1991).

Thus expression of CFTR, but not CFTRΔF508, generated cAMP-stimulated Cl⁻ currents not previously observed in these cells. This result suggested that CFTR may be a Cl⁻ channel.

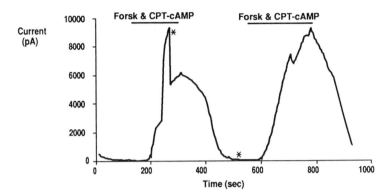

Figure 4. Expression of CFTR induces cAMP-activated Cl⁻ current in
CHO cells. Figure shows time course for activation and reversal
of currents. Because the current was so large, series resistance
compensation was temporarily turned off between the asterisks.

Recombinant CFTR and Native Apical Membrane Cl Channels Have
Similar Properties: If CFTR is a cAMP-activated Cl⁻ channel, then
currents generated by recombinant CFTR should have properties
similar to those of cAMP-activated Cl⁻ currents in the apical
membrane of epithelial cells which normally express CFTR. We
emphasize apical membrane, because that is where the CF defect is
manifest in secretory epithelia (Quinton, 1990), and for a channel
to govern Cl⁻ secretion, it must be located in the apical
membrane. To specifically investigate apical membrane Cl⁻
channels, we studied cells grown on permeable filter supports so
that they differentiated, formed tight junctions, and polarized,
segregating apical and basolateral proteins. Because
transepithelial current is determined by apical and basolateral
membranes in series, we functionally eliminated basolateral
membrane resistance by adding either nystatin or Staphylococcus
alpha toxin (Anderson et al., 1991a; Ostedgaard et al., 1991) to
the submucosal solution. Apical membrane Cl⁻ conductance was
measured as the current generated by a Cl⁻ concentration gradient
with transepithelial voltage clamped to 0 mV.

    We compared four properties of cAMP-activated Cl⁻ currents in
cells expressing recombinant CFTR with those of the apical

membrane of normal human airway epithelia and T84 intestinal epithelia. T84 cells are a Cl⁻ secreting epithelium (Madara et al., 1987) that contain CFTR located in the apical membrane (Gregory et al., 1990; Denning et al., 1991). First, we examined the effect of several Cl⁻ channel inhibitors: addition of diphenylamine-2-carboxylate (DPC, 0.5 mM) inhibited apical Cl⁻ current, but 4,4'-diisothiocyanostilbene-2,2'-disulfonic acid (DIDS, 0.5 mM), Zn^{2+} (0.1 mM) and an indanyloxyacetic acid (IAA-94, 40 µM) had minimal if any effect (Fig. 5). Similar responses were obtained with cAMP-activated Cl⁻ currents in 3T3 fibroblasts stably expressing CFTR (Anderson et al., 1991a).

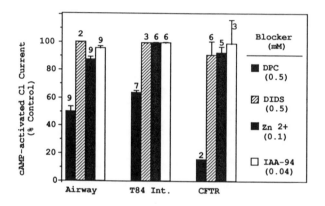

Figure 5. Effect of Cl⁻ channel blockers on cAMP-regulated Cl⁻ channels in the apical membrane of human airway epithelia (Airway) and T84 intestinal epithelia (T84), and in 3T3 fibroblasts stably expressing CFTR (CFTR). Blockers were added to the mucosal bath for epithelia and the extracellular bath for 3T3 fibroblasts.

Second, the time-dependent response to a voltage step was similar for Cl⁻ channels in the apical membrane and in cells expressing recombinant CFTR (Anderson et al., 1991a; 1991b). In both, the majority of the current was voltage-insensitive. However a small and variable component of slowly activating current was observed in the whole-cell studies.

Third, in the presence of symmetrical Cl⁻ concentrations, the current-voltage relationship was linear for endogenous apical membrane currents and current generated by recombinant CFTR.

Fourth, we measured the anion selectivity. Table 1 shows that the anion permeability sequence is Br⁻ > Cl⁻ > I⁻ in T84 intestinal epithelia, consistent with results in normal airway epithelia (Anderson et al., 1991c). cAMP-activated Cl⁻ currents generated by expression of CFTR with two different expression systems in two different cell types (3T3 fibroblasts and HeLa cells) (Table 1) had a similar anion selectivity. The conductance sequence was also similar for apical Cl⁻ channels and currents generated by expression of CFTR (Table 1).

TABLE 1

	n	Na⁺	Br⁻	P_X/P_{Cl} Cl⁻	I⁻	F⁻
T84 Epithelia						
Apical	(12)	0.16 ±0.01	1.21 ±0.06	1.00	0.56 ±0.03	0.11 ±0.03
3T3 Fibroblasts						
CFTR	(4)	0.01 ±0.01	1.11 ±0.03	1.00	0.59 ±0.07	0.30 ±0.01
HeLa Cells						
CFTR	(5)	0.09 ±0.03	1.24 ±0.09	1.00	0.57 ±0.08	
K95D	(4)	0.09 ±0.02	1.25 ±0.12	1.00	1.43 ±0.05	0.15 ±0.05
K335E	(5)	0.11 ±0.02	1.06 ±0.02	1.00	1.37 ±0.09	0.15 ±0.00
R347E	(4)	0.10 ±0.01	1.24 ±0.06	1.00	0.90 ±0.08	0.13 ±0.02
R1030E	(4)	0.12 ±0.01	1.46 ±0.07	1.00	0.81 ±0.10	0.33 ±0.05

Relative anion permeability of cAMP stimulated channels in the apical membrane and in cells expressing wild-type and mutant CFTR. Data are mean ± SEM of values calculated from currents in cAMP-stimulated individual cells. Permeability ratios, P_X/P_{Cl} where X is Na⁺, I⁻, Br⁻, or F⁻, were calculated from reversal potentials using the Goldman-Hodgkin-Katz equation.

Thus cAMP-activated Cl⁻ channels generated by expression of CFTR had properties similar to those of endogenous cAMP-activated Cl⁻ channels located in the apical membrane of secretory epithelia, ie., the channels that are defective in CF. These data supported, but did not prove the hypothesis that CFTR is itself a Cl⁻ channel. The notion that CFTR is a Cl⁻ channel had been difficult to accept, because CFTR does not resemble any known ion channels, but instead most resembles a family of energy-dependent transport proteins (Riordan et al., 1989; Ames et al., 19990; Hyde et al., 1990). This observation had lead some to speculate that CFTR is a Cl⁻ channel regulator.

The CFTR Transmembrane Domains Form Part of a Cl Channel: To test whether CFTR is a cAMP-regulated Cl⁻ channel, we used site-directed mutagenesis in an attempt to alter conductive properties of the channel pore.

Our rational for choosing the amino acids for mutation was as follows: ionic selectivity is determined by an interaction between the permeating ions and the amino acids that line the channel pore (Hille, 1984). We reasoned that if CFTR itself forms the Cl⁻ channel pore, then changing positively charged amino acids within the transmembrane domain of CFTR to negatively charged amino acids might alter its ionic selectivity.

We mutated amino acids within the putative membrane spanning sequences (M1 through M12) (Fig. 1) originally identified by Riordan and coworkers (Riordan et al., 1989). On the basis of their hydropathy and the prediction that they form alpha helices, these sequences are likely to span the lipid bilayer. Based on these considerations, we made the following mutations, converting basic residues to acidic residues: $Lys^{95} \rightarrow Asp^{95}$, $Lys^{335} \rightarrow Glu^{335}$, $Arg^{347} \rightarrow Glu^{347}$, and $Arg^{1030} \rightarrow Glu^{1030}$ (Anderson et al., 1991a).

cAMP reversibly activated whole-cell Cl⁻ currents in HeLa cells transfected with wild type or any of the mutant forms of CFTR (Table 1). As previously reported with the vaccinia virus-T7 hybrid expression system (Rich et al., 1990; Anderson et al., 1991b), 30 to 70% of the cells responded to cAMP. Wild type and mutated CFTR also showed similar responses to hyperpolarizing (-100 mV) and depolarizing (+100 mV) voltages: the majority of the current was voltage-insensitive.

In symmetrical Cl^- concentrations, CFTR generates Cl^- channels which display a linear current-voltage (I-V) relation. To maximize our ability to detect a change in the relative permeability of Na^+ to Cl^-, we used a NaCl concentration gradient. Similar permeabilities for Cl^- over Na^+ were observed in mutated and nonmutated CFTR, and in the apical membrane of T84 cells (Table 1).

Thus, we concluded that these mutations do not cause a general disruption of channel structure: cAMP-dependent channel regulation was intact, voltage-dependence was unchanged, and selectivity for Cl^- over Na^+ was preserved.

The anion permeability sequence of cAMP-regulated currents in wild-type CFTR and the apical membrane of T84 cells was: $Br^- \geq Cl^- > I^- > F^-$ (Table 1). Site-directed mutation altered the anion selectivity. When Lys^{95} or Lys^{335} were substituted with amino acids containing acidic side chains, the anion permeability sequence was converted to: $I^- > Br^- > Cl^- > F^-$ (Table 1). When Arg^{347} and Arg^{1030} were mutated, the permeability sequence was not different from that of wild-type CFTR, although the relative permeability of I^- to Cl^- increased (Table 1).

We also estimated the relative anion conductance of these mutated channels. Wild-type CFTR, $Arg^{347} \rightarrow Glu^{347}$, and $Arg^{1030} \rightarrow Glu^{1030}$ had cAMP-dependent channels with a relative conductance sequence that paralleled their permeability sequences: $Br^- \geq Cl^- > I^- \approx F^-$. The most dramatic change was in $Lys^{335} \rightarrow Glu^{335}$, where the conductivity sequence was: $Br^- > I^- > Cl^- \geq F^-$. Although the $Lys^{95} \rightarrow Asp^{95}$ mutation changed the permeability sequence, the relative conductivity sequence for the halides remained unchanged from wild-type.

These data indicate that CFTR is itself a cAMP-regulated Cl^- channel. The finding that several properties of apical membrane

Cl⁻ channels and of Cl⁻ channels induced by recombinant CFTR are the same is consistent with that conclusion. The finding that specific mutations in the membrane spanning sequences alter anion selectivity demonstrates that CFTR forms a cAMP-regulated anion pore and makes other interpretations unlikely.

cAMP Regulates the CFTR Cl Channel via the R Domain: CFTR Cl⁻ channels open upon phosphorylation by cAMP-dependent protein kinase (PKA) (Berger et al., 1991; Tabcharani et al., 1991). However, these studies did not identify the sites of phosphorylation nor the mechanism by which such phosphorylation opens the channel. In addition to the transmembrane domains which form at least part of a Cl⁻ channel pore, CFTR also contains two putative nucleotide binding domains (NBD1 and NBD2) and an R domain (Fig 1). The R domain has a number of potential phosphorylation sites for cAMP-dependent protein kinase (Riordan et al., 1989) Gregory et al. (1990) showed that CFTR can be phosphorylated by PKA. These observations suggested that the R domain might confer cAMP-dependence on the CFTR Cl⁻ channel.

To test this hypothesis, we examined the functional consequences of deleting the R domain (Rich et al., 1991b). We constructed a plasmid encoding CFTR in which amino acids 708-835 were deleted (CFTRΔR), expressed it in HeLa cells using the vaccinia virus/T7 hybrid expression system, and assessed cAMP-dependent Cl⁻ channel activity using the halide-sensitive fluorophore 6-methoxy-N-(3-sulfopropyl)-quinolinium (SPQ) assay or the whole-cell patch-clamp technique. In the SPQ assay, when I⁻ is replaced by NO_3^- fluorescence increases as I⁻ leaves the cell; an increase in halide permeability results in a more rapid increase in SPQ fluorescence.

Substitution of NO_3^- for I⁻ in cells expressing CFTR produced minimal changes in SPQ fluorescence (Fig. 6). A subsequent increase in intracellular cAMP, produced by addition of forskolin and 3-isobutyl-1-methylxanthine (IBMX), stimulated a rapid increase in fluorescence, indicating that cAMP increased anion permeability. In contrast, in unstimulated cells expressing CFTRΔR, substitution of I⁻ by NO_3^- caused an immediate, rapid increase in SPQ fluorescence (Fig. 6) which resembled the response observed in CFTR-expressing cells stimulated by cAMP. Subsequent addition of cAMP to cells expressing CFTRΔR further increased the rate of change in SPQ fluorescence.

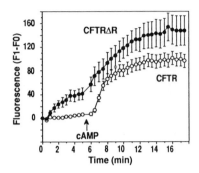

Figure 6. A) Change in fluorescence of SPQ-loaded HeLa cells expressing CFTR (open circles) or CFTRΔR (closed circles). NO3⁻ was substituted for I⁻ in the bathing medium at time zero. Five min. later cells were stimulated with 20 μM forskolin and 100 μM 3-isobutyl-1-methylxanthine (IBMX) to raise intracellular cAMP levels. Data are mean ± SEM of fluorescence at time t (Ft) minus the baseline fluorescence (Fo, average fluorescence measured in the presence of I⁻ for 2 min. prior to ion substitution.

Whole-cell patch-clamp studies also showed that cells expressing CFTRΔR had large basal currents, even in the absence of cAMP (Fig. 7A,B). In contrast, cAMP was required to stimulate Cl⁻ currents in cells expressing CFTR. cAMP-dependent stimulation produced a further increase in whole-cell Cl⁻ current in cells expressing CFTRΔR (Fig. 6B).

CFTRΔR-generated currents in unstimulated cells were similar to CFTR-generated currents in cAMP-stimulated cells: both currents were Cl⁻ selective; both were more permeable and conductive to Cl⁻ than to I⁻; both were inhibited by diphenylamine-2-carboxylate (DPC); and most of the current showed time-independent voltage effects. The preservation of these channel properties suggests that the R domain does not contribute in forming part of the channel pore. These results indicate that the R domain confers cAMP-dependence on the chloride channel. However additional sites

within CFTRΔR respond to cAMP, since Cl⁻ current increased further
with cAMP stimulation.

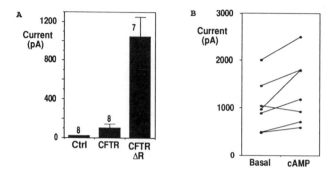

Figure 7.   Whole-cell patch clamp studies of CFTR and CFTRΔR
expressed in HeLa cells.   Basal current measured in untransfected
HeLa cells (Ctrl), and transfected with either CFTR or CFTRΔR.
Current measured in 7 cells expressing CFTRΔR before (Basal) and
during stimulation with cAMP.   Current increased upon addition of
either 10 μM forskolin and 100 μM IBMX (n+3) or 500 μM 888-(4-
chlorophenylthio) adenosine cyclic monophosphate (n=3).   Current
was measured during steps to +80 mV from a holding potential of -
60 mV.

Recent studies (Cheng et al., 1991) have shown that four serine
residues within the R domain are phosphorylated by PKA and that
mutation of these sites to alanine prevents cAMP-stimulated
activation of the channel.   Mutation of the one remaining PKA
phosphorylated site within CFTRΔR (S660N) abolishes the remaining
cAMP stimulation observed in CFTRΔR.   This combination of studies
demonstrates that the R domain is the site through which PKA
regulates the Cl⁻ channel.

FUTURE WORK

Despite this progress on the other domains, the function of the
two consensus nucleotide binding domains remains an enigma.

Understanding their function is particularly important because they are the site of the majority of CF-associated missense mutations. Some CF-associated NBD mutations may cause CF by preventing CFTR from reaching the correct cellular location (eg. ΔF508), (Cheng et al., 1991; Gregory et al, 1991). However, the mechanism by which other CF-associated NBD mutations, (which are processed normally), destroy channel activity is still unknown (Gregory et al., 1991).

REFERENCES

Ames, G.F., Mimura, C.S., and Shyamala, V. (1990). Bacterial periplasmic permeases belong to a family of transport proteins operating from Escherichia coli to human: Traffic ATPases. FEMS. Microbiol. Rev 6, 429-446

Anderson, M.P., Gregory, R.J., Thompson, S., Souza, D.W., Paul, S., Mulligan, R.C., Smith, A.E., and Welsh, M.J. (1991a). Demonstration that CFTR Is a chloride channel by alteration of its anion selectivity. Science 253, 202-205

Anderson, M.P., Rich, D.R., Gregory, R.J., Smith, A.E., and Welsh, M.J. (1991b). Generation of cAMP-activated chloride currents by expression of CFTR. Science 251, 679-682

Anderson, M.P. and Welsh, M.J. (1991c). Calcium and cAMP activate different chloride channels in the apical membrane of normal and cystic fibrosis epithelia. Proc Natl Acad Sci 88, 6003-6007

Berger, H.A., Anderson, M.P., Gregory, R.J., Thompson, S., Howard, P.W., Maurer, R.A., Mulligan, R., Smith, A.E., and Welsh, M.J. (1991) Identification and regulation of the CFTR-generated chloride channel. J. Clin. Invest. 88, 1422-1431

Cheng, S.H., Gregory, R.J., Marshall, J., Paul, S., Souza, D.W., White, G.A., O'Riordan, C.R., and Smith, A.E. (1990). Defective intracellular transport and processing of CFTR is the molecular basis of most cystic fibrosis. Cell 63, 827-834

Cheng, S.H., Rich, D.P., Marshall, J., Gregory, R.J., Welsh, M.J., and Smith, A.E. (1991) Phosphorylation of the R domain by cAMP-dependent protein kinase regulates the CFTR chloride channel. Cell 66, 1027-1036

Denning, G.M., Ostedgaard, L.S., Cheng, S.H., Smith a.E., and Welsh, M.J. (1991) Localization of cystic fibrosis transmembrane conductance regulator in chloride secretory epithelia. J. Clin. Invest. (In press)

Drumm, M.L., Pope, H.A., Cliff, W.H., Rommens, J.M., Marvin, S.A., Tsui, L-C., Collins, F.C., Frizzell, R.A., and Wilson, J.M. (1990). Correction of the cystic fibrosis defect in vitro by retrovirus-mediated gene transfer. Cell 62, 1227-1233

Elroy-Stein, O., Fuerst, T.R., and Moss, B. (1989). Cap-independent translation of mRNA conferred by

412

encephalomyocarditis virus 5' sequence improves the performance of the vaccinia virus/bacteriophage T7 hybrid expression system. Proc Natl Acad Sci U. S. A. 86, 6126-6130

Gregory, R.J., Cheng, S.H., Rich, D.R., Marshall, J., Paul, S., Hehir, K., Ostedgaard, L., Klinger, K.W., Welsh, M.J., and Smith, A.E. (1990). Expression and characterization of the cystic fibrosis transmembrane conductance regulator. Nature 347, 382-386

Gregory, R.J., Rich, D.P., Cheng, S.H., Souza, D.W., Paul, S., Manavalan, P., Anderson, M.P., Welsh, M.J., and Smith, A.E. (1991). Maturation and function of cystic fibrosis transmembrane conductance regulator variants bearing mutations in putative nucleotide-binding domains 1 and 2. Molec Cell Biol 11, 3886-3893

Hille, B. (1984) In: Ionic channels in excitable membranes, Sinauer Associates, Inc., Sunderland, MA, pp-420-426

Hyde, S.C., Emsley, P., Hartshorn, M.J., Mimmack, M.M., Gileadi, U., Pearce, S.R., Gallagher, M.P., Gill, D.R., Hubbard, R.E., and Higgins, C.F. (1990). Structural model of ATP-binding proteins associated with cystic fibrosis, multidrug resistance and bacterial transport [see comments]. Nature 346, 362-365

Kartner, N., Hanrahan, J.W., Jensen, T.J., Naismith, A.L., Sun, S., Ackerley, C.A., Reyes, E.F., Tsui, L-C., Rommens, J.M., Bear, C.E., and Riordan, J.R. (1991). Expression of the cystic fibrosis gene in non-epithelial invertebrate cells produces a regulated anion conductance. Cell 64, 681-691

Kerem, B., Rommens, J.M., Buchanan, J.A., Markiewicz, D., Cox, T.K., Chakravarti, A., Buchwald, M., and Tsui, L.C. (1989). Identification of the cystic fibrosis gene: genetic analysis. Science 245, 1073-1080

Madara, J.L., Stafford, J., Dharmsathaphorn, K., and Carlson, S. (1987) Structural analysis of a human intestinal epithelial cell line. Gastroenterology 92, 1133-1145

Ostedgaard, L.S., Shasby, D.M., and Welsh, M.J. (1991) Staphylococcus aureus alpha toxin permeabilizes the basolateral membrane of a Cl⁻ secreting epithelium. (Submitted)

Quinton, P.M. (1990). Cystic fibrosis: a disease in electrolyte transport. FASEB J 4, 2709-2717

Rich, D.P., Anderson, M.P., Gregory, R.J., Cheng, S.H., Paul, S., Jefferson, D.M., McCann, J.D., Klinger, K.W., Smith, A.E., and Welsh, M.J. (1990). Expression of cystic fibrosis transmembrane conductance regulator corrects defective chloride channel regulation in cystic fibrosis airway epithelial cells. Nature 347, 358-363

Rich, D.P., Gregory, R.J., Anderson, M.P., Manavalan, P., Smith, A.E., and Welsh, M.J. (1991). Effect of deleting the R domain on CFTR-generated chloride channels. Science 253, 205-207

Riordan, J.R., Rommens, J.M., Kerem, B., Alon, N., Rozmahel, R., Grzelczak, Z., Zielenski, J., Lok, S., Plavsic, N., Chou, J.L., Drumm, M.C., Ianuzzi, M.C., Collins, F.S., and Tsui, L-C. (1989). Identification of the cystic fibrosis gene: cloning and characterization of complementary DNA. Science 245, 1066-1073

Tabcharani, J.A., Chang, X-B., Riordan, J.R., and Hanrahan, J.W. (1991) Phosphorylation-regulated Cl⁻ channel in CHO cells stably expressing the cystic fibrosis gene. Nature 352, 628-631

THE 70-kDa PEROXISOMAL MEMBRANE PROTEIN

Takashi Hashimoto[1], Takehiko Kamijo[1,2], Ichiro Ueno[1,3], Keiju Kamijo[1,4] and Takashi Osumi[1,5]

[1]Department of Biochemistry; and [2]Department of Pediatrics, Shinshu University School of Medicine; [3]Central Clinical Laboratories, Shinshu University Hospital, Matsumoto, Nagano Matsumoto 390; [4]Division of Molecular Genetics, National Institute of Neuroscience, National Center of Neurology and Psychiatry, Kodaira 187; [5]Department of Life Science,Faculty of Science, Himeji Institute of Technology, Kamigori, Hyogo 678-12, Japan

SUMMARY: The 70 kDa peroxisomal membrane polypeptide is one of the major integral polyeptides and its content varies in accord with proliferation of peroxisomes. The structural analysis of the cDNA for this polypeptide and its topological study indicate that this membrane polypeptide anchors to the membrane via six transmembrane segments on amino-terminus, and the carboxy-terminal region having ATP-binding domain exposes to the cytosol. The possible role of this protein is discussed.

Peroxisomes are surrounded by a single membrane and filled with small granular matrix materials, and observed in almost all of eukaryotic cells. Peroxisomes in higher animals had been considerd to be a kind of fossil organelle, because the presence of important functions was not confirmed in this organelle except H_2O_2-forming oxidases and catalase. But, recently, several functions have been found in peroxisomes. Peroxisomal fatty acid oxidation system, for example, oxidizes very long-chain fatty acids which are not sufficiently metabolized by the mitochondrial β-oxidation system. The enzymes involved in the final steps of formation of bile acids are exclusively present in peroxisomes. The first three steps of plasmalogen biosynthesis are also located in peroxisomes (for review, see Fahimi & Sies,1987). The important functions of peroxisomes were strengthened by the recent discovery of a new class of the fatal hereditary diseases, peroxisomal diseases (for review, see Lazarow & Moser,1989).

A line of evidence for the roles of peroxisomal membrane is accumulating. Morphologically distinct peroxisomes are absent in the patient's cells with Zellweger syndrome which is the most severe disease among peroxisomal diseases. The cause of the absence of peroxisomes in this patients is supposed to be defects in localization of the peroxisomal proteins (Santos et al., 1988). The other interest is a specific mechanism that enables transport of the metabolites through the limiting membrane of peroxisomes.

Major peroxisomal membrane proteins: Rat liver peroxisomes were freeze-thawed, and sequentially washed with a phosphate buffer and a sodium carbonate solution. The integral membrane proteins were solubilized from the residue with a sodium dodecylsulfate (SDS) solution and purified by repeating SDS-polyacrylamide gel electrophoresis as described previously (Hashimoto et al.,1986). The major peroxisomal membrane polypeptides were 70 kDa, 26 kDa, and 22 kDa polypeptides (PMP70, PMP26, and PMP22). Further search for loosely membrane-bound proteins was carried out by using several methods such as the Triton X-114 solubilization and phase separation. The antibodies were raised against the candidate polypeptides. The localization of these polypeptides were examined by immunoelectron microscopic analysis. Presence of other major membrane polypeptide except long-chain acyl-CoA synthetase was not confirmed.

Induction of peroxisomal membrane proteins: The number and size of peroxisomes and the contents of the peroxisomal fatty acid oxidation enzymes of rat liver are markedly increased, when the rats were given with a peroxisome proliferator (for review, see Fahimi & Sies, 1987). The contents of the fatty acid oxidation enzymes reached the maximal levels about 2 weeks after the treatment, and returned to the control level about 1 week upon the withdrawal of the drug. The time-content curves of the peroxisomal membrane polypeptides were very similar to the change in the contents of the peroxisomal β-oxidation enzymes. Table 1 shows the induction of major membrane polypeptides, and the relation to the changes in the translatable mRNAs of these polypeptides.

Structure of PMP70: None of the functions of the major integral membrane proteins is known. Therefore, we carried out cloning of the cDNA for PMP70 first, since this polypeptide was abundant and its content changes in accord with proliferation of peroxisomes.

The cDNA clones for rat PMP70 (rPMP70) were isolated from the libraries prepared from livers of rats given DEHP (Kamijo et al., 1990). The cDNA clones of human PMP70 (hPMP70) were

Table 1. Effect of a peroxisome proliferator on the contents of major integral peroxisomal membrane polypeptides. Rats were fed a diet with and without containing 2% di(2-ethylhexyl)phthalate (DEHP), a peroxisome proliferator.

	Polypeptides (μ g/g liver)			mRNA (arbitrary unit)		
	Control A	DEHP B	B/A	Control A	DEHP B	B/A
PMP70	86	426	4.9	5.0	26.9	5.4
PMP26	26	350	13.5	3.4	38.4	11.4
PMP22	136	329	2.4	10.5	28.3	2.7

418

```
  1        10         20         30         40         50         60
MAAFSKYLTA RNSSLAGAAF LLFCLLHKRR RALGLHGKKS GKPPLQNNEK EGKKERAVVD
********** ********** ** ******* ********** ********** **********
MAAFSKYLTA RNSSLAGAAF LLLCLLHKRR RALGLHGKKS GKPPLQNNEK EGKKERAVVD
  1        10         20         30         40         50         60

 61        70         80         90        100        110        120
KVFLSRLSQI LKIMVPRTFC KETGYLILIA VMLVSRTYCD VWMIQNGTLI ESGIIGRSSK
*** *** ** ********** ****** *** ********** ********** ******** *
KVFFSRLIQI LKIMVPRTFC KETGYLVLIA VMLVSRTYCD VWMIQNGTLI ESGIIGRSRK
 61        70         80         90        100        110        120

121       130        140        150        160        170        180
DFKRYLFNFI AAMPLISLVN NFLKYGLNEL KLCFRVRLTR YLYEEYLQAF TYYKMGNLDN
****** *** ********** ********** ********** ********** **** *****
DFKRYLLNFI AAMPLISLVN NFLKYGLNEL KLCFRVRLTR YLYEEYLQAF TYYKKGNLDN
121       130        140        150        160        170        180

181       190        200        210        220        230        240
RIANPDQLLT QDVEKFCNSV VDLYSNLSKP FLDIVLYIFK LTSAIGAQGP ASMMAYLLVS
********** ********** ********** ********** ********** ****** **
RIANPDQLLT QDVEKFCNSV VDLYSNLSKP FLDIVLYIFK LTSAIGAQGP ASMMAYLVVS
181       190        200        210        220        230        240

241       250        260        270        280        290        300
GLFLTRLRRP IGKMTIMEQK YEGEYRFVNS RLITNSEEIA FYNGNKREKQ TIHSVFRKLV
********** ****** *** ****** *** ********** ********** * ********
GLFLTRLRRP IGKMTITEQK YEGEYRYVNS RLITNSEEIA FYNGNKREKQ TVHSVFRKLV
241       250        260        270        280        290        300

301       310        320        330        340        350        360
EHLHNFIFFR FSMGFIDSII AKYIATVVGY LVVSRPFLDL AHPRHLHSTH SELLEDYYQS
******* ** ********** *** ****** ********** ***** *** **********
EHLHNFILFR FSMGFIDSII AKYLATVVGY LVVSRPFLDL SHPRHLKSTH SELLEDYYQS
301       310        320        330        340        350        360

361       370        380        390        400        410        420
GRMLLRMSQA LGRIVLAGRE MTRLAGFTAR ITELMQVLKD LNHGKYERTM VSQQDKGIEG
********** ********** ********** ********** ********** **** *****
GRMLLRMSQA LGRIVLAGRE MTRLAGFTAR ITELMQVLKD LNHGKYERTM VSQQEKGIEG
361       370        380        390        400        410        420

421       430        440        450        460        470        480
AQASPLIPGA GEIINADNII KFDHVPLATP NGDILIQDLS FEVRSGANVL ICGPNGCGKS
 * ****** **** ***** ********** *** ** ** ********** **********
VQVIPLIPGA GEIIIADNII KFDHVPLATP NGDVLIRDLN FEVRSGANVL ICGPNGCGKS
421       430        440        450        460        470        480

481       490        500        510        520        530        540
SLFRVLGELW PLFGGHLTKP ERGKLFYVPQ RPYMTLGTLR DQVIYPDGKE DQKKKGISDQ
********** ***** **** ** ******* ********** ******** * *** *****
SLFRVLGELW PLFGGRLTKP ERRKLFYVPQ RPYMTLGTLR DQVIYPDGRE DQKRKGISDL
481       490        500        510        520        530        540

541       550        560        570        580        590        600
VLKGYLDNVQ LGHILEREGG WDSVQDWMDV LSGGEKQRMA MARLFYHKPQ FAILDECTSA
* * ****** ********** ********** ********** ********** **********
VQKEYLDNVQ LGHILEREGG WDSVQDWMDV LSGGEKQRMA MARLFYHKPQ FAILDECTSA
541       550        560        570        580        590        600

601       610        620        630        640        650        660
VSVDVEDYIY SHCRKVGITL FTVSHRKSLW KHHEYYLHMD GRGNYEFKKI TEDTVEFGS
***** ***  ********** ********** ********** ******** * *********
VSVDVEGYIY SHCRKVGITL FTVSHRKSLW KHHEYYLHMD GRGNYEFKQI TEDTVEFGS
601       610        620        630        640        650        660
```

Fig. 1. Amino acid sequences of rPMP70 and hPMP70.

isolated from a human hepatic cDNA library, using the rPMP70 cDNA as a probe. The composite nucleotide sequences of rPMP70 and hPMP70 contained the open reading frames with the same size of 1977 bp and their encoded amino acid sequences had 659 residues. The comparison between the coding sequences of rPMP70 and hPMP70 displays 90.6% and 95.0% identities at nucleotide and amino acid levels, respectively. (Fig. 1).

The hydropathy analysis (Kyte Doolittle, 1970) shows that PMP70 contains two domains. One is an amino-terminal hydrophobic membrane-associating region which presumably contains six transmembrane segments and the other is a hydrophilic region located in the carboxyl terminus.

A search of the NBRF data base using the entire deduced amino acid sequences of these PMP70 revealed a striking homology between the carboxy-terminal sequence of 200-250 amino acid residues of PMP70 and certain other proteins belonged to the ABC (ATP-binding cassette) superfamily (Higgins et al., 1986; Higgins et al., 1988; Hyde et al., 1990). Two sequences for ATP-binding folds (Walker et al., 1982) are highly conserved in PMP70. Hence we conclude that PMP70 is a newly identified member of the ATP-binding protein superfamily.

Topology of PMP70: Purified rat liver peroxisomes were incubated with proteinase K under the isotonic conditions. The mixture was centrifuged, separated into the supernatant and peroxisomal particles, and used for SDS-polyacrylamide gel electrophoresis and immunoblot analysis. The released polypeptide was of 24 kDa size which was derived from the carboxyl terminus by cleavage between the amino acid residues 457 and 458. The result indicates that a protrusion of PMP70 to cytosol is of 24 kDa or more of the carboxyl terminus which has the ATP-binding domain. The degradation of amino-terminal side by the treatment was not known, since the antibody preparation recognized only a extreme carboxyl terminus.

420

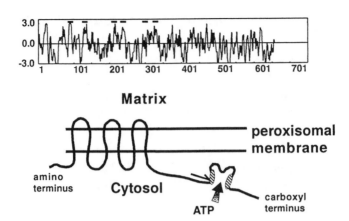

Fig. 2. Hydropathy profile and hypothetical model for the topology of PMP70. Cutting site by proteinase K is indicated by an arrow.

A hypothetical model of PMP70 in the peroxisomal membrane in analogy to the transmembrane proteins of ABC superfamily (Chen et al., 1986; Felmlee et al., 1985) is summarized in Fig. 2. Six transmembrane segments are in the amino-terminus and the the ATP-binding region is exposed to the cytosol.

Consideration of the function: Most of the members of the ABC superfamily mediate transport of small molecules, such as sugar, amino acids, inorganic ions, or of large molecules, such as proteins. From the transmembrane structure and homology to ABC transporters, PMP70 may involve in active transport across peroxisomal membrane.

Several observations on the roles of ATP in the peroxisomal functions have been reported. ATP stimulated the palmitoyl-CoA oxidation activity of isolated peroxisomes about 2-fold (Thomas et al., 1980). A latency of the activity of dihydroxyacetone phosphate acyltransferase, which located on the matrix side of the peroxisomal membrane, was released by additon of ATP, and the activity was increased about 4-fold, although ATP itself did not affect the enzyme activity (Wolvetang et al., 1990). These effect are assumed to be due to the presence of an ATP-dependent acyl-CoA carrier in the peroxisomal membrane.

Localization of the precursor of acyl-CoA oxidase into peroxisomes required ATP-hydrolysis (Imanaka et al., 1987). Therefore, we examined whether intactness of PMP70 is essential for import of the acyl-CoA oxidase precursor into isolated peroxisomes by our import assay method (Miyazawa et al., 1989).

Rat liver peroxisomes were preincubated with or without the presence of proteinase K, and then mixed with acyl-CoA oxidase synthesized from the cDNA by in vitro transcription/translation. When intact peroxisomes were used, a significant amount of the enzyme was associated to the particles, and about a half of this associated enzyme was resistant to the added proteinase K. When proteinase K-pretreated peroxisomes were used, the enzyme associated to the particles was decreased, and all of this component was digested by proteinase K.

The results suggest that the in vitro import assay method provides one of the useful tools for the study on the function of PMP70.

REFERENCES

Chen, C.J., Chin, J.E., Ueda, K., Clark, D.P., Pastan, I., Gottesman, M.M., and Robinson, I.B. (1986) Cell, 47, 381-389
Fahimi, H.D. and Sies, H., Ed. (1987) Peroxisomes in Biology and Medicine, Springer-Verlag, Berlin, 470 pp.
Felmlee, T., Pellet, S., and Welch, R.A. (1985) J. Bacteriol. 163, 94-105
Hashimoto, T., Kuwabara, T., Usuda, N.,and Nagata, T.(1986) J. Biochem. (Tokyo) 100, 301-310
Higgins, C.F., Hiles, I.D., Salmond, G.P.C., Downie, J.A., Evans, I.D., Holland, I.B., Gray, L., Buckel, S.D., Bell, A.W., and Hermodson, M.A. (1986) Nature, 323, 448-450
Higgins, C.F., Gallagher, M.P., Mimmack, M.L., and Pearce, S.R. (1988) BioEssays, 8, 111-116
Hyde, S.C., Emsley, P., Hartshorn, M.J., Mimmack, M.L, Gileadi, U., Pearce, S.R., Gallagher, M.P., Gill, D.R., Hubbard, R.E.,

422

and Higgins, C.F. (1990) Nature, 346, 362-365

Imanaka, T., Small, G.M., and Lazarow, P.B. (1987) J. Cell Biol. 105, 2915-2922

Kamijo, K., Taketani, S., Yokota, S., Osumi, T., and Hashimoto, T. (1990) J. Biol. Chem. 265, 4534-4540

Kyte, J. and Doolittle, R.F. (1970) J. Mol. Biol. 157, 105-132

Lazarow,P.B., and Moser,H.W. (1989) In: Metabolic Basis of Inherited Disease (C.R. Scriver, A.L. Beaudet, W. Sly, D. Valle, Ed), McGraw-Hill, New York, pp.1479-1561

Miyazawa, S., Osumi, T., Hashimoto, T., Ohno, K., Miura,S., and Fujiki, Y.(1989) Mol. Cell. Biol. 9, 83-91

Santos, M.J., Imanaka, T., Shio, H., Small, G.M., and Lazarow, P.B. (1988) Science, 239, 1536-1538

Thomas, J., Debeer, L.J., Schepper, P.J., and Mannaerts, G.P. (1980) Biochem. J. 190, 485-494

Walker, J.E., Saraste, M., Runswick, M.J., and Gay, N.J. (1982) EMBO J. 1, 945-951

Wolvetang, E.J., Tager, J.M., and Wanders R.J.A. (1990) Biochem. Biophys. Res. Commun. 170, 1135-1143

Adenine Nucleotides in Cellular Energy Transfer and Signal Transduction
S. Papa, A. Azzi & J.M. Tager (eds)
© 1992 Birkhäuser Verlag, Basel/Switzerland

LATENCY OF PEROXISOMAL ENZYMES IN DIGITONIN-PERMEABILIZED CELLS:
THE EFFECT OF ATP ON PEROXISOME PERMEABILITY

Ernst J. Wolvetang[1],   and Joseph M. Tager[2], Ronald J.A. Wanders[1]*

[1]Department of Pediatrics, University of Amsterdam, Academic
Medical Centre, Meibergdreef 9, 1105 AZ, Amsterdam, The Netherlands.
[2]Institute of Medical Biochemistry and Chemistry, University of
Bari, Piazza G. Cesare, 70124, Bari, Italy.

* Correspondence

SUMMARY: The activity of peroxisomal enzymes was studied in
cells which are selectively permeabilized with a low digitonin
concentration in a medium mimicking the composition of the
cytosol. In this assay system the cytosolic compartment of both
cultured human skin fibroblasts and isolated rat hepatocytes is
freely accessible to peroxisomal substrates and other low
molecular weight compounds. In selectively permeabilized human
skin fibroblasts peroxisomal palmitoyl-CoA ß-oxidation and
dihydroxyacetonephosphate acyltransferase (DHAP-AT) activity
display 50 and 70 % structure-linked latency, respectively. Upon
addition of Mg-ATP the latency of both DHAP-AT and C16-CoA
ß-oxidation is abolished in an ATP concentration dependent
manner. No effect of ATP is observed in sonicated fibroblast
preparations, where peroxisomal integrity is destroyed, or when
a non-hydrolyzable analogue of ATP is used. Preincubation with
inhibitors that inhibit ATPase activity in highly purified
peroxisomal fractions from rat liver partly prevent the
stimulatory effect of ATP on latent DHAP-AT activity in selectively permeabilized fibroblasts. Furthermore, the latency of
DHAP-AT is critically dependent on the GSH:GSSG ratio in the
medium. In selectively permeabilized isolated rat hepatocytes
urate oxidase and several other peroxisomal oxidases display
about 65 % structure-linked latency, which is not influenced by
omission of GSH from the medium or by addition of ATP. Furthermore we conclude that peroxisomes, when studied in situ, exhibit
a restricted permeability to at least some peroxisomal substrates and/or products, in contrast to peroxisomes in homogenates or after isolation. We suggest that a peroxisomal ATPase
is involved in the permeation of substrates (e.g. palmitoyl-CoA)
through the peroxisomal membrane in human skin fibroblasts.

INTRODUCTION

It has been known for many years that peroxisomes, unlike other cytoplasmic organelles, display a highly permeable character after isolation or in cell homogenates (Tolbert 1981 and Mannaerts and van Veldhoven 1987). Isolated peroxisomes have been found to be freely permeable to sucrose, cofactors, and the substrates and products of peroxisomal enzymes (de Duve and Baudhuin 1966, Baudhuin 1969, Beaufay et al. 1964, van Veldhoven et al. 1983 and 1987, Mannaerts and van Veldhoven 1987). Peroxisomes may even lose soluble enzymes during the isolation procedure (Thomas et al. 1980, Alexson et al. 1985). Enzymes in isolated peroxisomes therefore do not posses structure-linked latency. The question arises of whether the leakiness of isolated peroxisomes reflects the situation in the intact cell (Borst 1989). Recent studies on intact cells indicate that the peroxisomal membrane constitutes an effective permeability barrier to small compounds. Indeed peroxisomes in intact yeast cells posses an internal pH 1.2 units below the cytosolic pH, implying a restricted permeability of the peroxisomal membrane to protons in vivo (Nicolay et al. 1987, Waterham et al. 1990). Furthermore, pulse-labelling experiments in intact trypanosomes showed that the glycosomal membrane acts as a true permeability barrier to glycolytic intermediates in the cytosol (Visser et al. 1981). Prompted by these findings we set out to investigate the permeability properties of peroxisomes in situ, using cultured human skin fibroblasts and rat hepatocytes in which the permeability barrier imposed by the plasma membrane only was removed by treatment with a low concentration of digitonin.

MATERIALS AND METHODS

Cultured human skin fibroblasts were harvested by standard procedures. Hepatocytes were isolated from rat liver by collagenase perfusion (Berry and Friend 1969, Groen et al. 1982).

Cell homogenates were prepared by mechanical disruption according to standard procedures except that the medium described
below was used. Intact or sonicated cells were incubated for 2
min at 37 °C in a medium containing: 120 mM KCl, 25 mM NaCl, 1
mM EGTA, 0.3 mM CaCl$_2$, 15 mM GSH, 0-20 mM Mg-ATP (as indicated),
0.1 % (^{w}/$_v$) BSA, 50 mM MOPS (pH=7.2), 50 U/ml creatine kinase,
and 10 mM creatinephosphate. Depending on whether fibroblasts (2
mg protein/ml) or hepatocytes (0.5 mg protein/ml) were used 20
or 10 μg/ml digitonin respectively, was added to the medium.
During the subsequent 8 min either acyl-CoA:dihydroxy-
acetonephosphate acyltransferase (DHAP-AT), acyl-CoA:glycerol-3-
phosphate acyltransferase (G3P-AT) activity or peroxisomal
palmitoyl-CoA ß-oxidation was determined in the permeabilized
fibroblasts, as described previously (Wolvetang et al. 1990a,
Wanders et al. 1987). In rat hepatocytes the activity of urate
oxidase was determined either by measuring the disappearance of
urate (0.1 mM) from the medium, or the production of H$_2$O$_2$. In the
latter case 1 mM homovanillic acid, 1 mM sodium azide and 20
U/ml horseradish peroxidase were included in the medium. This
method was also used for measuring D-aspartate oxidase, glycolate oxidase and polyamine oxidase activities with 10 mM
D-aspartate, 10 mM sodium glycolate and 5 mM N-acetyl spermidine
as substrates, respectively. Latency of catalase, phosphoglucose
isomerase and lactate dehydrogenase activity was determined
after an 8 min incubation period as previously described
(Wolvetang et al. 1990a).

RESULTS AND DISCUSSION

In order to avoid the potentially deleterious effect of conventional isolation and homogenization procedures on peroxisomal
integrity, we selectively permeabilized the plasma membrane of
cells with digitonin in order to study the functional properties
of peroxisomes in situ. In Fig.1 we show that already at very
low digitonin concentrations (10-20 μg/ml) the latency of the

426

cytosolic enzymes phosphoglucose isomerase (◇) and lactate dehydrogenase (□) in cultured human skin fibroblasts (Panel A) as well as in freshly isolated rat hepatocytes (Panel B) is

FIG 1.
THE EFFECT OF DIGITONIN ON LATENCY OF PHOSPHOGLUCOSE ISOMERASE, LACTATE DEHYDROGENASE AND CATALASE IN HUMAN SKIN FIBROBLASTS (A) AND ISOLATED RAT HEPATOCYTES (B).

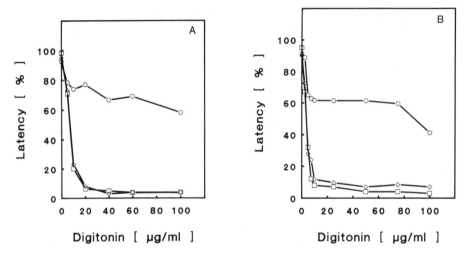

Cultured human skin fibroblasts (A) (1.5-2.0 mg protein/ml) or isolated rat hepatocytes (0.5 mg protein/ml) (B) were incubated in the cytosol mimicking medium described in Materials and Methods for 8 min at 37 °C in the presence of the indicated concentrations of digitonin. Subsequently, aliquots of the cell suspension were immediately transferred to assay mixtures for catalase (O), lactate dehydrogenase (□) and phosphoglucose isomerase (◊) activity measurement. Structure linked latency of the enzymes was calculated using the maximal activities obtained from incubations with sonicated cell preparations. The results are the values of the means of 3 separate experiments.

almost completely lost, indicating that the plasma membrane of these cells no longer imposes a permeability barrier to fructose-6-phosphate, pyruvate and NADH. Fig. 1 shows that the latency of catalase is lost in a biphasic manner, suggesting two distinct intracellular pools of catalase. First, at low concentrations of digitonin there is an initial drop in catalase latency which probably corresponds to the unmasking of cytosolic catalase. Only at concentrations of digitonin exceeding 100 µg/ml there is an unmasking of peroxisomal catalase. In the presence of low digitonin concentrations the latency of catalase remains high for at least 10-15 min in both cultured human skin fibroblasts and isolated rat hepatocytes (results not shown). In this time period, during which all experiments were performed, only a small amount of total cellular lactate dehydrogenase activity (5-15 %) is released from the permeabilized cells.

In the permeabilized fibroblasts the activity of dihydroxyacetonephosphate acyltransferase (DHAPAT), a membrane

TABLE I
STRUCTURE LINKED LATENCY OF ENZYMES IN DIGITONIN-PERMEABILIZED HUMAN SKIN FIBROBLSTS FROM CONTROL AND ZELLWEGER PATIENTS.

Activity measured	Type	Preparation	Activity (pmol/min.mg)	Latency (%)
DHAP-AT	Control	Permeabilization	21.2 ± 6.6 (4)	72
		Sonication	74.9 ± 5.3 (4)	
	Zellweger	Permeabilization	6.9 ± 1.3 (2)	33
		Sonication	10.3 ± 0.6 (2)	
G3P-AT	Control	Permeabilization	74.4 ± 15.0(4)	0
		Sonication	63.8 ± 14.2(4)	
	Zellweger	Permeabilization	92.6 (1)	0
		Sonication	85.6 (1)	
C16:0-CoA ß-oxidation	Control	Permeabilization	6.2 ± 0.6 (4)	50
		Sonication	12.0 ± 1.0 (4)	
	Zellweger	Permeabilization	1.2 ± 0.3 (2)	37
		Sonication	1.9 ± 0.4 (2)	

Cultured human skin fibroblasts from control subjects and Zellweger patients (1.5-2.0 mg protein/ml) were permeabilized with 20 µg/ml digitonin as described in Materials and Methods or sonicated, in both cases without addition of ATP. DHAP-AT, G3P-AT and peroxisomal palmitoyl-CoA (C16:0-CoA) ß-oxidation activity were determined as described in Materials and Methods during a sub-sequent 8 min incubation at 37 °C. Values are means ± SD with the number of experiments in parentheses.

bound peroxisomal enzyme with its catalytic site facing the peroxisomal interior, was studied. First we investigated whether
the substrates for DHAP-AT, palmitoyl-CoA and DHAP, could gain
free access to the cytosolic compartment. For this purpose the
activity of glycerol-3-phosphate acyltransferase (G3P-AT), an
enzyme closely resembling DHAP-AT both in function and specific
activity, but located at the outer face of the endoplasmic reticulum (Coleman et al. 1978), was investigated in selectively
permeabilized as well as sonicated fibroblasts. As shown in
Table I G3P-AT does not exhibit structure-linked latency, indicating that palmitoyl-CoA, G3P and by analogy also DHAP could
freely permeate into the cytosolic compartment. DHAP-AT, in
contrast to G3P-AT however, displays 70 % structure linked
latency under these conditions, strongly suggesting a restricted
permeability of the peroxisomal membrane to the substrates
and/or products of this enzyme, at least in the absence of ATP.
Peroxisomal palmitoyl-CoA ß-oxidation activity exhibits 50 %
structure linked latency in selectively permeabilized
fibroblasts. The observation that both DHAP-AT and palmitoyl-CoA
ß-oxidation activity were strongly reduced in fibroblasts from
patients with the Zellweger syndrome, which is characterized by
a general impairment of peroxisomal functions (Wanders et al.
1988), emphasizes that the acyltransferase and ß-oxidation
activities measured under the conditions used are solely peroxisomal. Table II shows that in selectively permeabilized isolated
rat hepatocytes urate oxidase, glycolate oxidase, D-aspartate
oxidase and polyamine oxidase exhibit about 60 % structure-
linked latency. This demonstrates that the membrane of rat liver
peroxisomes, when studied in situ, constitutes an effective permeability barrier to low molecular weight substrates and/or
products, in contrast to the situation in cell homogenates or
after isolation of the organelles.

TABLE II
STRUCTURE LINKED LATENCY OF PEROXISOMAL OXIDASES IN DIGITONIN
PERMEABILIZED RAT HEPATOCYTES.

Enzyme activity measured	Preparation	Activity (pmol/min.mg)	Latency (%)
Urate oxidase	Permeabilization	54 ± 9 (8)	63
	Sonication	148 ± 6 (8)	
D-aspartate oxidase	Permeabilization	43 ± 2 (4)	60
	Sonicatation	107 ± 13(4)	
Glycolate oxidase	Permeabilization	37 ± 4 (4)	60
	Sonicatation	92 ± 9 (4)	
Polyamine oxidase	Permeabilization	20 ± 1 (4)	62
	Sonication	52 ± 1 (4)	

Isolated rat hepatocytes (0.5 mg protein/ml) were permeabilized with 10 μg/ml digitonin or sonicated in the absence of ATP. Enzyme activities were determined as described in Materials and Methods during a subsequent 10 min incubation at 37 °C. Activities represent the means ± SD with the number of experiments in parentheses.

Subsequently it was found that upon addition of Mg-ATP to the medium the latency of both DHAP-AT (Fig 2a) and peroxisomal palmitoyl-CoA ß-oxidation (Fig 2b) in selectively permeabilized human skin fibroblasts was abolished in an ATP-dependent manner. No stimulatory effect of ATP was observed in sonicated fibroblast preparations, in which peroxisomal integrity was destroyed, indicating that the stimulatory effect was neither due to a direct effect on DHAP-AT or one or more peroxisomal ß-oxidation enzymes, nor due to an ATP-dependent resynthesis of palmitoyl-CoA lost by hydrolysis. We therefore suggest that the stimulatory effect of ATP should be attributed to an increased permeation of substrates (or products) through the peroxisomal membrane. Preliminary results indicate that other nucleotide triphosphates can substitute for ATP in stimulating latent DHAP-AT activity, albeit not as efficiently.

FIG 2
THE EFFECT OF ATP ON THE ACTIVITY OF ACYL-CoA:DIHYDROXYACETONE-
PHOSPHATE ACYLTRANSFERASE (A) AND PEROXISOMAL PALMITOYL-CoA
ß-OXIDATION (B) IN DIGITONIN-PERMEABILIZED OR SONICATED SUSPEN-
SIONS OF HUMAN SKIN FIBROBLASTS.

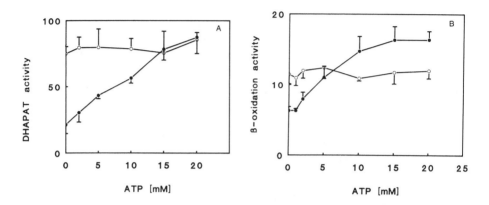

Cultured human skin fibroblasts (1.5-2.0 mg protein/ml) were
permeabilized as described in Materials and Methods in the
cytosol mimicking medium for 2 min at 37 °C or sonicated in the
presence of increasing concentrations of ATP. DHAP-AT activity
(A) and palmitoyl-CoA ß-oxidation (B) were measured during the
following 8 min as described in Materials and Methods. The
values are expressed as pmol/min.mg and represent the means ± SD
from 4 different experiments.

The fact that the rate of ß-oxidation in the presence of ATP in
permeabilized fibroblasts exceeds that in sonicated fibroblasts
may be due to the dilution of the peroxisomal ß-oxidation enzymes (which are soluble) and/or cofactors that occurs upon
sonication. DHAP-AT, which is a single enzyme strongly bound to
the peroxisomal membrane, would not suffer from this dilution
effect. An alternative explanation would be that substrates are
actively transported into peroxisomes, thus giving rise to an
increased intra-peroxisomal substrate concentration, which may
have a greater stimulatory effect on palmitoyl-CoA ß-oxidation
as compared to DHAP-AT.

TABLE III
THE EFFECT OF ATP, ATPASE INHIBITORS, DINITROPHENOL AND A NON-
HYDROLYZABLE ATP ANALOGUE ON THE ACTIVITY OF DHAP-AT IN
DIGITONIN-PERMEABILIZED OR SONICATED HUMAN SKIN FIBROBLASTS.

		DHAP-AT activity (pmol/min.mg)	
ATP (mM)	Additions	Digitonin treated fibroblasts	Sonicated fibroblasts
0	none	21.2 ± 6.6 (6)	74.9 ± 5.3 (4)
20	none	87.3 ± 22.3 (7)	85.4 ± 5.6 (4)
0	ATP-γ-S (20 mM)	25.3 ± 3.2 (2)	75.4 ± 5.7 (2)
20	DCCD (250 μM)	43.6 ± 2.8 (2)	100.6 ± 12.4 (2)
20	NEM (1000 μM)	39.8 ± 3.3 (2)	97.7 ± 8.9 (2)
20	Oligomycin (0.1 mg/ml)	56.9 ± 5.1 (2)	88.2 ± 4.1 (2)
20	Bafilomycin (250μM)	85.4 ± 4.0 (2)	89.6 ± 5.5 (2)
0	DNP (100 μM)	12.0 ± 3.0 (3)	78.0 ± 14.0 (3)
20	DNP (100 μM)	68.0 ± 8.0 (2)	79.0 ± 9.0 (3)

Digitonin permeabilized or sonicated human skin fibroblasts
(1.5-2.0 mg protein/ml) were incubated in the presence of the
indicated concentrations of ATPase inhibitors with or without 20
mM ATP or with 20 mM ATP- -S as indicated. When the effect of
DNP was invesstigated 10 μM atractyloside was included in the
medium. GSH was omitted from the preincubation when NEM was
used. DHAP-AT activity was measured as described in Materials
and Methods. Values represent the means ± SD with the number of
experiments between parentheses.

As shown in Table III addition of ATP-γ-S, an analogue of ATP
which can serve as substrate for some protein kinases but generally not for ATPases (Eckstein et al. 1985), failed to stimulate latent DHAP-AT activity, which leads us to conclude that
hydrolysis of ATP is necessary to overcome the peroxisomal
permeability barrier. This prompted us to investigate the effect
of ATPase inhibitors on the stimulatory effect of ATP on latent

DHAP-AT activity. Previously we (Wolvetang et al. 1990b) and others (del Valle 1988 et al., Malik et al. 1991), reported that the ATPase activity detected in highly purified peroxisomal fractions from rat liver was inhibited by DCCD, NEM and oligomycin at high concentrations. Similar properties have been reported for the ATPase activity in purified peroxisomal fractions from yeast (Douma et al. 1987). We now show (Table III) that exactly these ATPase inhibitors are able to inhibit, at least in part, the stimulatory effect of ATP on latent DHAP-AT activity in selectively permeabilized fibroblasts. It has been suggested that the peroxisomal ATPase in yeast is a proton translocating ATPase involved in the generation of the proton-gradient across the peroxisomal membrane (Douma et al. 1987) which has indeed been demonstrated in intact yeast cells (Nicolay et al. 1987, Waterham et al. 1990). Table III shows that the protonophore DNP had virtually no effect on the latency of DHAP-AT in selectively permeabilized fibroblasts. Furthermore, DNP did not affect the stimulatory effect of ATP on latent DHAP-AT activity, suggesting that at least in cultured human skin fibroblasts there is no direct involvement of a proton-gradient in the permeation of the substrates of DHAP-AT through the peroxisomal membrane. We rather suggest that a peroxisomal ATPase is directly responsible for the permeation of the substrates(or products) of DHAP-AT and palmitoyl-CoA β-oxidation through the peroxisomal membrane. Since DHAP-AT and peroxisomal β-oxidation share palmitoyl-CoA as common denominator, the effect of ATP may well be due to an enhanced rate of palmitoyl-CoA permeation through the peroxisomal membrane. The low affinity of this process for ATP compares well to the results of kinetic studies on the peroxisomal ATPase activities from yeast (Douma et al. 1987) and rat liver (Wolvetang et al. 1990b) showing relatively high apparent K_m values for ATP, which might be brought about by the loss of some regulatory factor upon cell-permeabilization and peroxisome isolation. It remains to be established whether the 70 kD peroxisomal membrane protein, which has been shown to be an ATP-binding transmembrane protein, displaying more than 40 % homology with members of the superfamily of multidrug resistance proteins (Kamijo et al. 1990), is involved in a ATP-driven transport of peroxisomal substrates (see also Hashimoto, this volume).

It was found that exchanging the cytosol mimicking medium for a buffered sucrose medium, which is routinely used for peroxisome isolation, leads to a strong reduction of structure-linked latency of DHAP-AT in permeabilized fibroblasts and of urate oxidase in permeabilized hepatocytes (results not shown). We therefore investigated whether or not DHAP-AT and urate oxidase display latency in cell homogenates which are prepared in the cytosol mimicking medium. This proved not to be the case. Possibly, perturbation of the cell ultrastructure inflicts irreversible damage to the peroxisomes. A peroxisomal reticulum in the cell, as has been demonstrated in regenerating rat liver (Yamamoto and Fahimi 1987a and 1987b), could be prone to damage by mechanical stress and might account for this behavior and the substantial loss of soluble enzymes from the peroxisomes during isolation (Thomas et al. 1980, Alexson et al. 1985). On the other hand it seems likely that as yet unknown factors, which are not present in the cytosol mimicking medium used in this study, are of importance for the functional intactness of peroxisomes and for detecting structure-linked latency of peroxisomal enzymes. In Fig 3 we show that one of the key factors for detecting latency of DHAP-AT in selectively permeabilized human skin fibroblasts is the GSH/GSSG ratio in the medium. Other thiol protecting compounds like ß-mercaptoethanol and dithiothreitol were found to be similarly effective in maintaining DHAP-AT latency, suggesting that a thiol group on the peroxisomal membrane is essential for the functional integrity of peroxisomes in fibroblasts. In permeablized hepatocytes, however, no significant loss of latency of peroxisomal oxidases was observed when GSH was omitted from the medium or when it was replaced by the oxidized form (GSSG) (results not shown).

434

FIG 3
THE EFFECT OF AN INCREASING GSH:GSSG RATIO ON DHAP-AT ACTIVITY
IN DIGITONIN PERMEABILIZED HUMAN SKIN FIBROBLASTS.

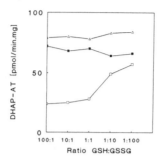

Cultured human skin fibroblasts (1.5-2.0 mg protein/ml) either
sonicated (Δ) or intact ($\blacksquare$, $\square$) were permeabilized as described
in Materials and Methods with 20 μg/ml digitonin at 37 °C for 2
min in the cytosol mimicking medium containing the indicated
GSH/GSSG ratio's (15 mM total) in the absence ($\square$) or presence
of 20 mM Mg-ATP ($\blacksquare$). DHAP-AT activity was measured during the
subsequent 10 min as described in Materials and Methods. The
result of a representative experiment is shown in Fig 3. Similar
results were obtained in 2 other experiments.

Furthermore, preliminary experiments reveal no stimulatory
effect of ATP on latent urate oxidase activity in permeabilized
rat hepatocytes. These observations might be interpreted to mean
that the transport of substrates across the peroxisomal membrane
is brought about by more than one transport system.
Alternatively, it might be that substrates is present in mammalian peroxisomes or that substantial differences between
peroxisomes as they occur in liver and fibroblast. Elucidation
of this question awaits investigation of the effect of ATP on
peroxisomal ß-oxidation and DHAP-AT activity in rat hepatocytes,
which is currently in progress.

REFERENCES
Alexson, S.E.H., Fujiki, Y., Shio, H. and Lazarow, P.B. (1985)
     J.Cell.Biol. 101, 294-305.
Baudhuin, P.(1969) Ann. N.Y. Acad.Sci. 168, 214-228.
     Beaufay, H., Jacques, P., Sellinger, O.Z., Berthet, J. and
     De Duve, C.(1964) Biochem.J. 92, 184-205.

Berry, M.N. and Friend, D.S. (1969) J.Cell.Biol. <u>43</u>, 506-520.

Borst, P. (1989) Biochim.Biophys.Acta <u>1008</u>, 1-13.

Coleman, R., and Bell, R.M. (1978) J. Cell. Biol. <u>76</u>, 245-253.

Douma, A.C., Veenhuis, M., Sulter, G.J. and Harder, W. (1987) Arch.Microbiol. <u>147</u>, 42-47.

Duve de, C. and Baudhuin, P.(1966) Physiol.Rev. <u>46</u>, 322-357.

Eckstein, F. (1985) Ann.Rev.Biochem. <u>54</u>, 367-402.

Groen, A.K., Sips, H.J., Vervoorn, R.C. and Tager, J.M. (1982) Eur.J.Biochem. <u>122</u>, 87-93.

Kamijo, K., Taketani, S., Yokota, S., Osumi, T. and Hashimoto, T. (1990) J.Biol.Chem. <u>265</u>, 4534-4540.

Malik, Z.A., Tappia, P.S., De Netto, L.A., Burdett, K., Sutton, R. and Connock, M.J. (1991) Comp.Biochem.Physiol. <u>99</u>, 295-300.

Mannaerts, G.P. and Van Veldhoven, P.P. (1987) Peroxisomes in Biology and Medicine (H.D. Fahimi and H. Sies eds.) Springer-Verlag Berlin Heidelberg, pp 169-176.

Nicolay, K., Veenhuis, M., Douma, A.C. and Harder, W. (1987) Arch.Microbiol. <u>147</u>, 37-41.

Thomas, J., Debeer, L.J., De Schepper, P.J. and Mannaerts, G.P.(1980) Biochem.J. <u>190</u>, 485-494.

Tolbert, N.E. (1981) Annu.Rev.Biochem. <u>50</u>, 133-157.
    Valle del, R., Soto, U., Necochea, C. and Leighton, F. (1988) Biochem.Biophys.Res.Commun. <u>156</u>, 1353-1359.

Veldhoven van, P., Debeer, L.J. and Mannaerts, G.P. (1983) Biochem.J. <u>210</u>, 685-693.

Veldhoven van, P., Just, W.W. and Mannaerts, G.P. (1987) J. Biol.Chem. <u>262</u>, 4310-4318.

Visser, N., Opperdoes, F.R., Borst, P. (1981) Eur. J.Biochem. <u>118</u>, 521-526.

Wanders RJA, van Roermund CWT, van Wijland MJA, Schutgens RBH, Schram AW, van den Bosch H and Tager JM (1987) Biochim. Biophys. Acta <u>919</u>, 21-25.

Wanders, R.J.A, Heymans, H.S.A., Schutgens, R.B.H., Barth, P.G., Van den Bosch, H. and Tager, J.M. (1988) J.Neurol.Sci. <u>88</u>, 1-39.

Waterham, H.R., Keizer-Gunnink, I., Goodman, J.M., Harder, W. and Veenhuis, M. (1990) FEBS Lett. <u>262</u>, 17-19.

Wolvetang EJ, Tager JM and Wanders RJA (1990a) Biochem Biophys Res Comm <u>170</u>, 1135-1143.

Wolvetang, E.J., Wanders, R.J.A., Schutgens, R.B.H., Berden, J.A. and Tager, J.M. (1990b) Biochim.Biophys.Acta <u>1035</u>, 6-11.

Wolvetang, E.J., Tager, J.M. and Wanders, R.J.A. (1991) Biochim. Biophys.Acta <u>1095</u>, 122-126.

Yamamoto, K. and Fahimi, H.D. (1987a) J.Cell.Biol. <u>105</u>, 713-722.

Yamamoto, K. and Fahimi, H.D. (1987b) Eur.J.Cell Biol. <u>43</u>, 293-300.

Adenine Nucleotides in Cellular Energy Transfer and Signal Transduction
S. Papa, A. Azzi & J.M. Tager (eds)
© 1992 Birkhäuser Verlag, Basel/Switzerland

DUAL ACTION OF FLUORESCEIN DERIVATIVES ON K_{ATP} CHANNELS

Jan R. de Weille, Michele Müller and Michel Lazdunski
Institut de Pharmacologie Moléculaire et Cellulaire,
660 route des Lucioles, Sophia Antipolis, 06560 Valbonne, France.

SUMMARY

Fluorescein derivatives are known to bind to nucleotide binding sites on transport ATPases. Their effect on ATP-sensitive K^+ (K_{ATP}) channel activity has been studied in insulinoma cells using $^{86}Rb^+$-efflux and patch-clamp techniques. Fluorescein derivatives have two opposite effects. Firstly, like ATP, they can inhibit active K_{ATP} channels. Secondly, they are able to reactivate K_{ATP} channels subjected to run-down in the absence of cytoplasmic ATP. K_{ATP} channel reactivation clearly does not require channel phosphorylation as is commonly believed. Either irreversible inhibition or activation of the K_{ATP} channel was obtained with the sulphydryl reagent eosin-5-maleimide. The irreversibly activated channel could still be inhibited by 2mM ATP. The results indicate the existence of two nucleotide-binding sites, one activator site and one inhibitor site. After activation by fluorescein derivatives K_{ATP} channels become resistant to a selective blocker of this channel, the sulfonylurea glibenclamide, suggesting allosteric interactions between nucleotide-binding sites and the sulfonylurea receptor.

INTRODUCTION

Potassium channels that are sensitive to intracellular variations of ATP (K_{ATP}) are important for the regulation of insulin secretion from pancreatic ß-cells. In the absence of extracellular glucose, the ß-cell is electrically silent. When glucose is raised to physiological concentrations (10 mM), intracellular ATP augments, K_{ATP} channels close and the cell membrane depolarizes, giving rise to generation of action potentials and Ca^{2+} ion influx (Arkhammer et al., 1986; Cook & Hales, 1984). The ensuing Ca^{2+}-stimulated insulin release by ß-cells (Yap Nelson et al., 1987) ultimately leads to a reduction of blood glucose levels. The K_{ATP} channel thus forms a link in a negative feedback loop that controls the ATP concentration in the pancreatic ß-cell.

Treatment of RINm5F insulinoma cells with oligomycin and 2-deoxy-glucose, to block both oxidative phosphorylation and glycolysis, leads to an increase of $^{86}Rb^+$ efflux from these cells due to the opening of the K_{ATP} channel (Schmid-Antomarchi et al., 1987). $^{86}Rb^+$ efflux reaches a half-maximum value ($K_{0.5}$) at 0.8 mM ATP. The $K_{0.5}$ contrast sharply with those ($K_{0.5}=10-60\mu M$) found by the voltage-clamp technique using membrane patches excised from ß-cells (Cook & Hales, 1984; Findlay, 1988a; Kakei et al., 1986; Ribalet & Ciani, 1987). Part of this difference may be due to the fact that the K_{ATP} channel is also blocked, although less effectively, by other nucleotides such as ADP, AMP, ATPγS, AMP-PNP, AMP-PCP, guanidine phosphates and pyridine nucleotides (Dunne et al., 1988, Petersen & Dunne, 1989). Probably the ATP/ADP ratio predicts more adequately the K_{ATP} channel behavior under physiological conditions than the ATP concentration alone (Ribalet & Ciani, 1987; Petersen & Dunne, 1989). Much evidence indicates that cytoplasmic nucleotides control the K_{ATP} channel via interaction with two different sites on the channel. Firstly, ATP and other nucleotides inhibit the K_{ATP} channel by a mechanism that does not require the presence of Mg^{2+} ions and which is also produced by non-hydrolysable ATP analogues. Secondly, ATP in combination with Mg^{2+} ions is essential to avoid rundown of the K_{ATP} channel. In excised patches with an ATP-free solution on the cytoplasmic side, K_{ATP} channel activity declines within tens of seconds to minutes until patches become completely silent. However, channel activity reappears after transient exposure of the intracellular face of the patch to ADP or ATP in combination with Mg^{2+}. Reactivation of the channel after rundown is either absent or much less effective if Mg^{2+} is omitted or if non-hydrolysable adenosine analogues are used such as ADPßS, AMP-PNP or AMP-PCP (Trube & Hescheler, 1984; Ohno-Shosaku et al., 1987; Findlay, 1988b; Petersen & Dunne, 1989). It has been suggested that K_{ATP} channel reactivation by Mg-ATP implies channel phosphorylation (Lederer & Nichols, 1989; Petersen & Dunne, 1989; Nelson et al., 1990).

Apart from their regulation by nucleotides, K_{ATP} channels are modulated by various drugs. Sulphonylureas, drugs that restore insulin secretion in patients affected by non insulin-dependent diabetes mellitus, have been shown to block the K_{ATP} channel selectively (Sturgess et al., 1985; Schmid-Antomarchi et al., 1987). The sulphonylurea receptor is very likely located on the K_{ATP} channel since the channel in the excised patches is consistently blocked by sulphonylureas.

Fluoresceine and its derivatives (FD's) have been used to label nucleotide binding sites. 4-benzoyl(benzoyl)-1-amidofluorescein has been used to label mitochondrial F_1-ATPase (Pal & Coleman, 1990) and eosin was found to bind to the ATP binding site of the Na^+/K^+-ATPase (Skou & Esmann, 1981). The action of FD's on K_{ATP} channel activity and sulfonylurea binding to HIT-T15 insulinoma cels will be evaluated.

METHODS

HIT-T15 ß-cells were cultured as previously described (Praz et al., 1983). $^{86}Rb^+$ efflux experiments were carried out by preincubating cells in Krebs solution, supplemented with 0.1 µCi/ml $^{86}RbCl$ (250 µl/well) at 37°C. After 120-150 min of incubation, efflux experiments were started by replacing KCl for $^{86}RbCl$.

Efflux was stopped by washing the cells and remaining radio-activity was counted. Glibenclamide-binding experiments were carried out according to Schmid-Antomarchi et al. (1987). Membranes were incubated with 1 nM (^{3}H)-glibenclamide in the presence of increasing amounts of bengale rose or phloxin-B for 1 h on ice. 0.5 ml samples were vacuum-filtered and washed at 4°C. Filters were incubated with 5 ml Biofluor (NEN) and counted for radioactivity.

Current-clamp experiments were carried out using the whole-cell suction-pipette technique (Hamill et al., 1981). Single-channel currents were recorded from inside-out and outside-out membrane patches and their membrane potentials were voltage-clamped at 0 mV. In all experiments, the intracellular solution contained 150 mM KCl, 1 mM MgCl$_2$, 2 mM EGTA, 10 mM HEPES-KOH, pH 7.2. The extra-cellular solution was: 140 mM NaCl, 5mM KCl, 2mM MgCl$_2$, 2 mM CaCl$_2$, 10 mM HEPES-NaOH, pH 7.3.

RESULTS AND DISCUSSION

Membrane potentials were recorded from HIT-T15 cells that were intracellularly dialyzed with a 150 mM KCl solution containing 0.5 mM ATP, 2mM ADP and 2mM MgCl. A few minutes after breaking into the whole-cell configuration, cells hyperpolarized to values close to the potassium equilibrium potential, due to the opening of K$_{ATP}$ channels. It was found that extracellular application of 100 μM of the FD, phloxin-B, induced membrane depolarization that was accom-panied by an increase of membrane resistance, suggesting the closure of K$^+$ channels (Fig. 1). In outside-out membrane patches, again with 0.5 ATP and 2mM ADP in the pipette, 100 μM phloxin-B induced closure of glibenclamide-sensitive K$^+$ channels. Hence, phloxin-B inhibits K$_{ATP}$ channels. The FD, eosin (250 μM), also inhibited K$_{ATP}$ channel activity in outside-out patches (Fig. 1). As these experiments showed that FDs may inhibit K$_{ATP}$ channel activity when applied extracellularly, it was possible to quantify the effect of FDs by using the ^{86}Rb$^+$ tracer-flux technique. Fig. 2

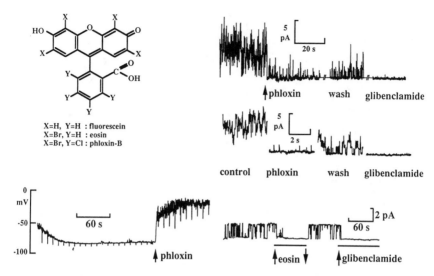

X=H, Y=H : fluorescein
X=Br, Y=H : eosin
X=Br, Y=Cl : phloxin-B

Fig. 1. Membrane potentials of HIT-T15 insulinoma cells were recorded using the suction-pipette technique (bottom, left). Cells were intracellularly dialyzed with a solution containing 2 mM ADP, 0.5 mM ATP and 2.5 μg/ml oligomycin. After passing into the whole-cell configuration, cells hyperpolarized, due to opening of K_{ATP} channels. Downward deflections in this trace were induced by injection of 300 ms, 5 pA current pulses indicating variations in membrane resistance. Upon extracellular application of 100 μM phloxin-B, cells depolarized and membrane resistance increased. K_{ATP} channel activity in outside-out membrane patches, excised with a pipette containing 2 mM ADP and 0.5 mM ATP, was rapidly inhibited by 100 μM phloxin-B (right, upper traces) or 250 μM eosin (lower trace). Channel activity that reappeared upon washing, was subsequently blocked by 20 μM glibenclamide. Upward deflections correspond to channel openings.

shows that treatment of insulinoma cells with 2.5 μg/ml oligomycin induced an important increase in $^{86}Rb^+$ efflux, which was antagonized dose-dependently by various FDs. Assumedly, the hydrophobic FDs penetrate the membrane and act on a nucleotide binding site located on the intracellular domain of the K_{ATP} channel. Indeed it was found that phloxin-B, when applied to the intracellular face of membrane patches, excised in an ATP-free solution, reversibly inhibited K_{ATP} channel activity (Fig. 3).

In a series of 34 such experiments it was noted that upon application of phloxin-B, channel activity either immediately subsided or

442

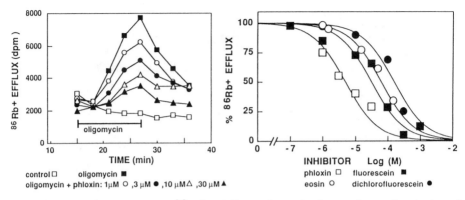

Fig. 2. The increase of $^{86}Rb^+$ efflux through K_{ATP} channels induced by application oligomycin, is dose-dependently inhibited by preincubation of the HIT-T15 cells with phloxin-B (left panel). Dose-inhibition curves thus obtained with phloxin-B, fluorescein, eosin and dichlorofluorescein are shown in the right panel.

increased transiently before subsiding. An analog situation occurs when subjecting certain voltage-controlled channels to a step in membrane potential. During a depolarizing step, voltage-controlled channels may either inactivate immediately (one does not observe channel openings) or pass into the inactivated state via one or more open states (a few openings are observed before the channel inactivates). An analogous model, describing K_{ATP} channel inhibition by FDs, is shown in Fig. 3. Two binding sites for FDs are assumed: one inhibitory site, which occupation entails inactivation, and one activatory site, which occupation leads to opening of the K_{ATP} channel. If both sites are occupied, the channel is in the inactivated state. Merely depending on chance, the activatory site may be occupied before the inhibitory site is, thus resulting in an transient increase of channel activity. Interestingly, K_{ATP} channel activity, if it had disappeared completely after excision in the ATP-free solution, was restored upon washout of phloxin-B (right lower trace in Fig. 3). Hence, reactivation of rundown channels may be induced by FDs and clearly does not require phosphorylation.

It has been demonstrated that 1 mM of the sulphydryl reagent N-

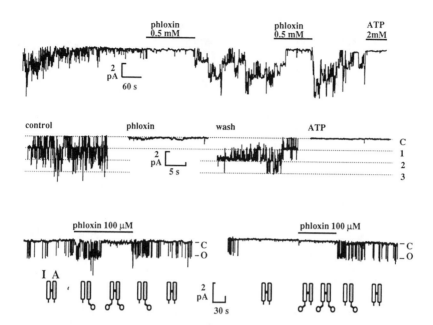

Fig. 3. Inside-out membrane patches were excised in a K^+-rich medium by a pipette containing a Na^+-rich and ATP-free solution and voltage-clamped at 0 mV. K_{ATP} channel activity was reversibly inhibited by application of 500 µM phloxin-B to the intracellular face of the membrane (Upper traces). During application of 100 µM phloxin-B, K_{ATP} channel activity either increased transiently or remained absent (lower traces). Run-down channel activity was restored following exposure to phloxin (right most record). The cartoons underneath the lower records represent the interpretation of events based on the two nucleotide-binding site model.

ethylmaleimide irreversibly inhibits K_{ATP} channel activity in skeletal muscle that can be prevented if 1 mM ATP is present, suggesting that a functional SH-group is located near the ATP-binding site(s) (Weik & Neumcke, 1989). If two nucleotide binding sites are present on the channel as proposed above, either of the two sites might be covalently modified by sulphydryl reagents, resulting in either permanent activation or permanent inhibition of the K_{ATP} channel. In Fig. 4 results obtained with eosin-5-maleimide (EMA) applied to inside-out patches at a low concentration of 25 µM and for a short period of 2 min are presented. In 7 out of 23 patches only reversible effects were observed, probably because exposure was to short to induce irreversible cross-linking.

444

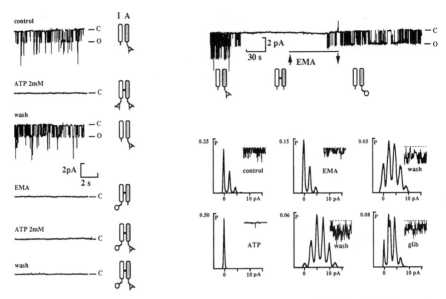

Fig. 4. Inside-out membrane patches were treated with 25 µM eosin-maleimide (EMA) and channel activity was recorded at 0 mV. Channel activity in control patches was completely suppressed by exposure of the intracellular face of the membrane to 2 mM ATP. Upon wash-out of ATP, channel activity reappeared. In 9 out of 23 inside-out patches, exposure of the intracellular face of the membrane resulted in irreversible K_{ATP} channel inhibition since 2 mM of ATP was unable to restore channel activity (traces on the left). In 7 out of 23 patches irreversible K_{ATP} channel activation by EMA was observed (top right). Current-density histograms from a single inside-out membrane patch show that after EMA-treatment channel activity was increased. Channels was fully blocked by 2 mM of ATP, but had become less sensitive to glibenclamide (100 nM). The absices in the histograms are composed of 100 bins of 0.2 pA width. The ordinates, representing probability per bin, are scaled as indicated. The peaks around 0 pA represent the closed channel probabilities, other peaks indicate 1, 2 or more channels open.

In 9 patches K_{ATP} channel activity disappeared and could not be reactivated by exposure to ATP. Hence, channels were irreversibly inactivated by EMA (Fig. 4, leftmost traces). In the remaining 7 patches, EMA induced channel activity that did not subside after washout of the sulphydryl reagent, but could still be inhibited by 2 mM of ATP, indicating that only the activatory nucleotide-binding site was modified. The reason to believe that EMA-induced activation was in fact irreverible was given by the observation

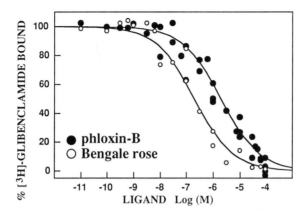

Fig. 5. Binding of 1 nM [³H]glibenclamide to insulinoma cell membranes was inhibited by increasing amounts of Bengale rose and phloxin-B. K_i's are 0.2 ± 0.1 and 2 ± 1 µM respectively. The zero level was determined by data points obtained with 100 µM unlabelled glibenclamide.

that 100 nM of glibenclamide, a concentration that completely inhibits K_{ATP} channel activity in control patches, did not inhibit the EMA-modified channel (Fig 4, bottom, rightmost panel). Hence, the K_{ATP} channel is controlled by two FD-sensitive nucleotide-binding sites containing essential SH-groups. Sulphydryl groups are also known to be essential components in the interaction of ATP with regulatory binding-sites in transport ATPases such as the (Na/K)ATPase (Schuurmans Stekhoven & Bonting, 1981) and they are readily modified by N-ethylmaleimide.

The results indicate that the K_{ATP} channel becomes insensitive to sulfonylureas if the activatory nucleotide-binding site is occupied by FDs. Fig. 5 shows that binding of radiolabelled glibenclamide to insulinoma cell membranes is inhibited dose-dependently by the Fds phloxin-B and Bengale rose, which is possibly due to a negative allosteric interaction between ATP-binding site(s) and the sulfonylurea receptor. This finding is consistent with observations that intracellular ADP interferes with the ability of tolbutamide to inhibit the pancreatic K_{ATP} channel (Zünkler et al. 1988) and glibenclamide to inhibit the cardiac K_{ATP} channel (Venkatesh and Weiss, 1991).

446

REFERENCES

Arkhammer, P., Nilsson,T., Rorsman, P., and Berggren, P.-O. (1986) J. Biol. Chem. 261, 5448-5454

Cook, D. L., and Hales, N. (1984) Nature 311, 271-273

Dunne, M. J., Findlay, I., and Petersen, O. H. (1988) J. Membrane Biol. 102, 205-216

Findlay, I. (1988a) Pflügers Arch. 412, 37-41

Findlay, I. (1988b) J. Membr. Biol. 101, 83-92

Hamill, O. P., Marty, A., Neher, E., Sakmann, B., and Sigworth, F. J. (1981) Pflügers Arch. 351,85-100.

Kakei, M., Kelly, R. P., Ashcroft, S. J. H., and Ashcroft, F. M. (1986) FEBS Lett. 208, 63-66

Lederer, W., J. & Nichols, C. G. (1989) J. Physiol. 419, 193-211

Nelson, M. T., Patlak, J. B., Worley, J. F. and Standen, N. B. (1990) Am. Phys. Soc. 259, C3-C18

Ohno-Shosaku, T., Zünkler, B. J., and Trube, G. (1987) Pflügers Arch. 408, 133-138

Pal, P. K., and Coleman, P. S. (1990) J. Biol. Chem. 265, 14996-15002

Praz, G. A., Kalban, P. A., Wollheim, C. B., Blondel, B., Strauss, A. J. and Reynold, A. E. (1983) Biochem. J. 210, 345-352

Petersen, O. H., and Dunne, M. J. (1989) Pflügers Arch. 414, S115-S120

Ribalet, B., and Ciani, S. (1987) Proc. Natl. Acad. Sci. U. S. A. 84, 1721-1725

Schmid-Antomarchi, H., de Weille, J. R., Fosset, M., and Lazdunski, M. (1987) J. Biol. Chem. 262, 15840-15844

Schuurmans Stekhoven, F. and Bonting, S.I. (1981) Phys. Rev. 61, 1-76

Skou, J. C., and Esmann M. (1981) Biochim. Biophys. Acta 674,232-240

Sturgess, N., Ashford, M. L. J., Cook, D. L., and Hales, C. N. (1985) Lancet ii 8453, 474-475

Trube, G., and Hescheler, J. (1984) Pflügers Arch. 401, 178-184

Venkatesh, N. and Weiss, J. N. (1991) Clinical Res. 1, 89A

Weik R., and Neumcke B. (1990) J. Membrane Biol. 110, 217-226

de Weille, J. R., and Lazdunski, M. (1990) in Ion Channels (Narahashi, T., ed.) Vol. 2, pp 205-222, Plenum Press, New York

Yap Nelson, T., Gaines, K. L., Rajan, A. S., Berg, M., and Boyd, A. E. (1987) J. Biol. Chem. 262, 2608-2612

Zünkler, B. J., Lins, S., Ohno-Shosaku, T, Trube, G. and Panten, U. (1988) FEBS lett. 239, 241-244

CONTROL OF SYNTHESIS OF UNCOUPLING PROTEIN AND ATPase IN ANIMAL
AND HUMAN BROWN ADIPOSE TISSUE

Josef Houstek[1], Jan Kopecky[1], Stanislav Pavelka[1], Petr Tvrdik[1],
Marie Baudysova[1], Karel Vizek[2], Jana Hermanska[1] and Dagmar
Janikova[1]

[1]Institute of Physiology, Czechoslovak Academy of Sciences, 142
20 Prague; [2]Research Institute for Mother and Child, 147 10
Prague, Czechoslovakia

SUMMARY: Biogenesis of two energy-converting and adenine
nucleotide-binding proteins of brown adipose tissue mitochondria
- the uncoupling protein (UCP) and ATP synthase (ATPase) has been
investigated in situ and in cell cultures. Highly differentiated
synthesis of these proteins was found to result from distinct
regulatory events. Tissue-specific, heat-producing UCP is
transcriptionally regulated by a cAMP-dependent and
triiodothyronine-dependent mechanism which is operated through β3
adrenoreceptors and involves de-novo synthesis of intracellular
iodothyronine 5'-deiodinase. UCP gene can be rapidly turned on
and the rate of its expression can differ by several orders of
magnitude. In contrast, the synthesis of ATPase which is highly
depressed in brown adipose tissue appears to be
posttranscriptionally controled and does not exhibit apparent
sensitivity to catecholamines.

INTRODUCTION

In all mammalian tissues the primary role of mitochondrial energy
conversion is to produce ATP by means of H^+-translocating
ATP-synthase (ATPase). The only exception are thermogenic
mitochondria of brown adipose tissue (BAT) which generate heat
instead of ATP due to uncoupling of oxidative phosphorylation
(for review see Nicholls and Locke, 1984). Thermogenesis in BAT
is based on existence of regulated H^+-channel of the inner
mitochondrial membrane called uncoupling protein (UCP). UCP is
encoded by single nuclear gene and has Mr of 32 kDa (for review
see Ricquier & Bouillaud, 1986), the functional unit is a dimer

with one regulatory nucleotide-biding site. Although UCP belongs to a family of structurally related mitochondrial transport proteins, it is strictly specific for BAT. Similarly as is ATPase the key enzyme of oxidative phosphorylation, the UCP serves as the key and rate-limiting component of heat production (Sundin et al., 1987).

Thermogenesis in BAT is a powerful way of defence against cold in mammalian newborns, rodents and hibernators (Nicholls et al., 1986) and the tissue thermogenic potential must be subtly accommodated to developmental, thermal and dietetic conditions (Houstek et al., 1989). One of the major regulatory mechanisms are fast and pronounced changes in composition of mitochondrial proteins which appear to be under control of catecholamines and thyroid hormones. In this report we summarize recent studies on biogenesis of mitochondrial proteins in BAT, in particular of the UCP and ATPase.

RESULTS AND DISCUSSION

Asynchronous synthesis of mitochondrial proteins in BAT: The content of UCP and ATPase may vary substantially in BAT. The best example is spontaneous perinatal recruitment of BAT, which occurs either prenatally, as in mouse and rat, or postnatally, as in hamster (Sundin et al., 1981; Houstek et al., 1988, 1990a). During this process rapid growth and differentiation of adipocytes is accompanied by increase in number of mitochondria and changes of their enzymic apparatus. In the initial stage of recruitment BAT contains significant amount of cytochrome oxidase (COX) which is in relation to ATPase, but no UCP is yet detectable. BAT mitochondria are nonthermogenic and principally do not differ from phosphorylating mitochondria of other tissues. The UCP gene starts to be expressed rather abruptly, 2 days before the birth of mouse and 7 days after the birth of hamster and specific content of UCP then rapidly increases. Similarly increases also the content of COX but the amount of ATPase either levels off or even decreases. The typical thermogenic mitochondria that are formed within few days are defined by high content of UCP (up to 10% of mitochondrial protein) which parallels high oxidative capacity and "replaces ATPase", both functionally and structurally. This transformation of BAT mitochondria illustrates how different and asynchronous can be the synthesis of mitochondrial proteins in BAT. Similar recruitment process occurs also during cold exposure, or as a result of certain diet (Himms-Hagen 1986; Rothwell & Stock 1986). It has been long realized that the long-term, biosynthetic processes in BAT including differential expression of UCP and ATPase genes (Houstek et al., 1990b) are initiated and to some extent controled by norepinephrine released from sympathetic nerves in the tissue (Cannon et al., 1989). For the study of cell biology of these processes the cell cultures of BAT represent an excellent and well controled in vitro model system.

BAT cell cultures: Despite several previous attempts to perform primary cell cultures of BAT, the cultivating conditions were elaborated enough only very recently to allow complete differentiation and maturation of BAT cells (Kopecky et al., 1990; Rehnmark et al., 1990). This is apparent as the ability to express the UCP gene which is the late marker of BAT differentiation. In a typical cultivation experiment the cell population consists after 1 day mainly of undifferentiated interstitial cells and preadipocytes, after 3 days the cells become polyhedral and show rapid proliferation. The confluence is reached around day 6 when almost all cells develop the typical appearance of multilocular adipocytes. As assessed by Western blotting (Kopecky et al., 1990) and immunofluorescence microscopy (Houstek et al., 1990b) the amount of COX increases during cell differentiation and at about confluence it reaches the amount of enzyme present in cells of original tissue. Similarly spontaneous is the synthesis of ATPase which can be determined using the anti F_1-ATPase antibodies. In contrast, the UCP antigen (Fig. 1 and Kopecky et al., 1990) and UCP mRNA signal (Rehnmark et al., 1990; Houstek et al., 1991a) are completely absent or very low in

450

control confluent cells. After the addition of norepinephrine the
UCP mRNA and UCP antigen content rapidly increase, but no change
is observed in the content of ATPase (Fig. 1) and COX antigens.
Changes in mRNA levels (Rehnmark et al., 1990) shortly precede
the synthesis of UCP protein (Kopecky et al., 1990), which lasts
for at least 24 hours and results in accumulation and
incorporation to mitochondrial fraction of near physiological
amounts of UCP. Induction of UCP synthesis thus exerts both in
vivo and in vitro remarkable selectivity with respect to other
mitochondrial proteins and the control of UCP synthesis is
apparently attained at transcriptional level. These cell culture
experiments are also in good agreement with in vivo changes in
UCP and UCP mRNA in response to cold or to catecholamine
administration (Bouillaud et al., 1984; Jacobsson et al., 1986).

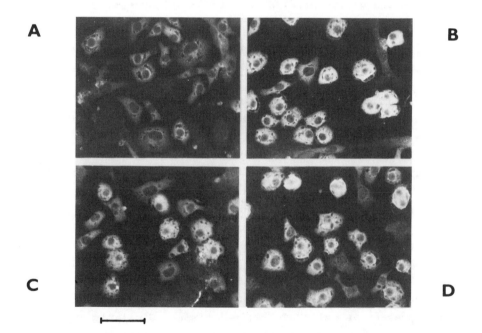

Fig. 1: Immunofluorescence microscopy of UCP (A,B) and ATPase
(C,D) in control (A,C) and 0.1 μM norepinephrine-supplemented
(24h; B,D) cultured BAT cells. Bar 50 μm.

Adrenergic receptors involved in control of UCP gene expression:
Both the acute and long-term thermogenic responses of BAT seem to
be regulated via β-adrenergic pathway (Cannon et al., 1989;
Houstek et al., 1989). BAT cells are endowed with $\alpha 1$ and numerous
$\beta 1$ and $\beta 2$ receptors (Svoboda et al., 1979; Rothwell et al.,
1985). In addition, they contain another type of atypical
adrenergic receptor called $\beta 3$ (Arch 1989). With the help of
cultured BAT cells it became possible to analyze in detail the
regulatory effects on UCP synthesis that are due to direct
interaction of hormones with BAT cell receptors (Kopecky et al.,
1990; Rehnmark et al., 1990; Houstek et al., 1990b). Both
norepinephrine ($\alpha 1$-, β-agonist) and isoprenaline (pure β-agonist)
have been found similarly effective in stimulating UCP and UCP
mRNA synthesis. The stimulation was largely prevented by
β-adrenergic antagonist propranolol while $\alpha 1$-selective antagonist
prazosin was much less effective. Interestingly, the optimal
norepinephrine concentration (0.1 μM) is identical with that
required for acute thermogenic response (Mohell et al., 1983;
norepinephrine-stimulated respiration), but the dose-response
curve of UCP synthesis has nonsaturating character. Pronounced
increase of UCP synthesis could also be induced by modulating
intracellular concentration of cAMP. Forskolin, an activator of
adenylate cyclase increased the UCP synthesis to more than 75%
and dibutyryl- or bromo-cAMP to 95% of the norepinephrine value.
These results clearly demonstrate that the character of
stimulation is preferentially β-adrenergic and cAMP-mediated.
Furthermore, both the UCP content and UCP mRNA levels exert
near-maximal stimulation by $\beta 3$-adrenergic agonists CGP-12177 and
BRL-37344 (Rehnmark et al., 1990; Bronnikov et al., 1991). Since
CGP-12177 is also a potent $\beta 1$-antagonist (Staehelin et al., 1983)
it must be concluded that induction of UCP gene expression is
achieved via $\beta 3$ rather than $\beta 1$ receptors.

The role of thyroxine 5'-deiodinase in control of UCP gene:
Thyroid hormones are known to stimulate biosynthesis of
mitochondrial proteins in general and in some cases also quite

selectively via triiodothyronine (T_3)-responsive nuclear genes (Nelson 1990). The cold-induced activation of BAT thermogenic potential can be best mimicked by simultaneous administration of norepinephrine and T_3 (Himms-Hagen 1986). In vivo studies also indicate that adrenergic stimulation of UCP mRNA and UCP synthesis, as well as of several other proteins require active 5'-deiodination of thyroxine (T_4) directly in BAT (Bianco & Silva, 1987). Similarly to brain, pituitary and pineal glands BAT contains type II thyroxine 5'-deiodinase (5'D) which is insensitive to propylthiouracil and has low Km values for T_4 and rT_3 (Leonard et al., 1983; Visser et al., 1983). 5'D of BAT is largely stimulated during cold exposure (Kopecky et al., 1986) and by sympathetic stimulation (Silva & Larsen, 1983). It also increases during perinatal recruitment of BAT. In developing hamster more than 10-fold increase in specific activity correlated closely with the increase of UCP antigen (Table I) and

Table I: UCP content and thyroxine 5'D activity in perinatal BAT

BAT	UCP content (pmol/mg protein)	5'D activity (fmol/h/mg protein)
hamster (a)		
8th day	12 + 6	10 + 7
11th day	60 + 14	43 + 12
14th day	129 + 27	84 + 15
17th day	155 + 20	116 + 25
human newborn (b)		
25th gestational week	28 + 9	146 + 47
32nd gestational week	65 + 14	425 + 65
40th gestational week	63 + 11	350 + 70

Data are Mean + SD, n = 5-7 (a), or 3-4 (b). Measurements were performed as described in (Houstek et al., 1990a, 1991b).

similar correlation was found in rat and bovine (Iglesias et al.,
1987; Giralt et al., 1989). Also in human BAT (Table I) obtained
at autopsies of pathological newborns the increase in specific
content of UCP between 25th and 32nd gestational weeks was
accompanied by similar increase in 5'D specific activity.

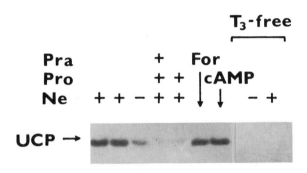

Fig. 2: Western blot analysis of UCP induced in cultured BAT
cells. Norepinephrine (Ne, 0.1 μM), propranolol (Pro, 50 μM),
prazosin (Pra, 50 μM), forskolin (For, 1 μM) and dibutyryl-cAMP
(cAMP 1 mM) were added for 24 h as indicated. From day 3 cells
were cultured for 4 days in serum-free medium supplemented or
devoid of triiodothyronine (T$_3$, 0.1 nM).

The most direct evidence for involvement of 5'D in control of
UCP synthesis comes again from cell culture experiments. As
depicted in Fig. 2, the omission of thyroid hormone from
serum-free cultivation medium prevented induction of UCP
synthesis by norepinephrine and similar effect was noted at the
level of UCP mRNA (Rehnmark et al., 1990). BAT cell cultures were
also found to contain 5'D activity when properly stimulated
(Houstek et al., 1990c). Incubation of confluent BAT cells with
norepinephrine and dibutyryl-cAMP caused up to 7- and 17-fold
activation, respectively, of 5'D due to de-novo synthesis of the
enzyme and the effect of norepinephrine was abolished by
propranolol but not by prazosin. Also the stimulation by
α_1-agonist was very weak (Table II). Since similar activation is
induced by norepinephrine, isoprenaline and CGP-12177 (Table II),

Table II: Induction of thyroxine 5'D activity in BAT cells cultured for 7 days.

additions	5'D activity (%)	UCP content (%)
control	18 ± 9	14 ± 9
1 μM norepinephrine	100 ± 26	100 ± 15
1 μM isoprenaline	87 ± 20	83 ± 14
1 μM oxymetazoline	22 ± 8	21 ± 10
1 μM CGP-12177	80 ± 15	80 ± 12

UCP content was determined 24 h and 5'D activity 8 h after addition of agonists as described in (Houstek et al., 1990c). Data are Mean ± SD, n = 3-4.

it can be concluded that the induction of 5'D in BAT is mainly regulated via β3-adrenergic receptors, rather than via α1-receptors as originally proposed by Silva and Larsen (1983).

The striking similarity in the adrenergic control of UCP and 5'D genes is further illustrated by the fact that UCP and 5'D can be well induced only during short time window between day 6 and 8 of cell cultivation. These data strongly support the view that adrenergic control of UCP synthesis is attained via cAMP-dependent activation of 5'D and UCP genes which could be of "tandem character". Increase in cAMP thus would affect UCP gene only when 5'D becomes sufficiently activated to allow for necessary local production of T_3. As the thyroid hormones themselves cannot induce the synthesis of UCP, the effect of T_3 is of essential, but permissive character. It has been recently shown that transcriptional control of UCP gene consists of rapid and short-lived increase of nascent UCP gene transcripts and much longer-lasting mRNA processing and stabilization (Rehnmark 1991). Interestingly the activation of 5'D in cultured BAT cells exhibits similar time course as primary UCP gene transcripts since stimulation of 5'D reaches the maximum after 8-12 hours and declines afterwards, while UCP synthesis remains activated for at least another 12 hours.

Control of synthesis of ATPase: In thermogenically active BAT the high oxidative capacity is not paralleled by phosphorylating activity and the content of ATPase is exceptionally low. This concerns both the catalytic F_1-part (Cannon & Vogel, 1977; Houstek et al., 1978) and the membrane bound F_o-part of the enzyme (Svoboda et al., 1981). The mechanism of selective reduction of ATPase in BAT is not known, however it may reflect depressed synthesis of the enzyme due to reduced expression of nuclear genes which encode 11 of 13 ATPase subunits (Walker et al., 1987).

In order to compare the expression of ATPase and COX genes we have screened the total RNA from BAT and other tissues of mouse with specific cDNA probes (Houstek et al., 1991a) for the content of transcripts of nuclearly encoded ATPase β-subunit and COX VIb subunit. As apparent from Table III, the steady state levels of both types of mRNA were much higher in BAT than in

Table III: Relative levels of β-ATPase and COX VIb mRNAs in mouse tissues.

tissue	β-ATPase mRNA		COX VIb mRNA	
	/mg tissue	/ATPase	/mg tissue	/COX
brown adipose	25.87	7.84	30.27	1.54
heart	10.75	0.70	10.75	0.70
liver	2.78	0.37	2.20	0.29
brain	2.65	0.56	3.05	1.22

The Mean of mRNA values from three Northern blots and ATPase and COX values from four Western blots were used for calculations (data from Houstek et al., 1991a)

456

heart, liver and brain. Similar tissue profiles were obtained for mRNAs of mitochondrially-encoded ATPase 6 and COX I genes. Resulting βF_1/COX VIb mRNA ratio has been found equal in all tissues tested and excludes any possibility that the low content of ATPase in BAT is caused by limited transcription of ATPase genes.

The β-subunit of ATPase is one of the few mitochondrial proteins that seem to be hormonally regulated both at transcriptional and translational levels, at least in developing liver (Izquierdo et al., 1990). In BAT, however, no change in the levels of β-ATPase mRNA accompanied norepinephrine-induced 100-fold increase of UCP mRNA in cultured BAT cells, and similarly ineffective was cold exposure for 3 days of warm acclimated animals (Houstek et al., 1991a). These results are in accordance with Western blot data on the synthesis of UCP, ATPase and COX in cultured BAT cells (Kopecky et al., 1990) and suggest that suppression of the synthesis of ATPase in BAT is due to posttranscriptional regulation which might involve translation, transport or assembly of ATPase subunits.

Acknowledgements: This work was supported by grant No 51146 from the Czechoslovak Academy of Sciences.

REFERENCES

Arch, J. (1989) Proc. Nutr. Soc. 48, 215-223
Bianco, A.C. and Silva, J.E. (1990) J. Clin. Invest. 79, 295-300
Bouillaud, F., Ricquier, D., Mory, G., and Thibault, J. (1984) J. Biol. Chem. 259, 11583-11586
Bronnikov, G., Houstek, J., and Nedergaard, J. (1991) J. Biol. Chem., in press
Cannon, B., Rehnmark, S., Nechad, M., Herron, D., Jacobsson, A., Kopecky, J., Obregon, M.J., and Nedergaard, J. (1989) In: Living in the Cold II (A. Malan, B. Canguilhem, Eds), John Libbey Eurotext Ltd, London, pp. 359-366
Cannon, B. and Vogel, G. (1977) FEBS Lett. 76, 284-289
Giralt, M., Casteilla, L., Vinas, O., Mampel, T., Iglesias, R., Robelin, G., and Villarroya, F. (1989) Biochem. J. 259, 555-559
Himms-Hagen, J. (1986) In: Brown Adipose Tissue (P. Trayhurn, D.G. Nicholls, Eds), Edward Arnold Ltd., London, pp. 214-268
Houstek, J., Janikova, D., Bednar, J., Kopecky, J., Sebestian, J., and Soukup, T. (1990a) Biochim. Biophys. Acta 1015, 441-449
Houstek, J., Kopecky, J., Baudysova, M., Janikova, D., Pavelka, S., and Klement, P. (1990b) Biochim. Biophys. Acta 1018,

457

243-247

Houstek, J., Kopecky, J., and Drahota, Z. (1978) Comp. Biochem. Physiol. 608, 209-214

Houstek, J., Kopecky, J., Janikova, D., Bednar, J., Holub, M., Soukup, T., Mikova, M., and Vizek, K. (1989) In: Molecular Basis of Membrane-Associated Diseases (A. Azzi, Z. Drahota, S. Papa, Eds), Springer Verlag, pp. 265-284

Houstek, J., Kopecky, J., Rychter, Z., and Soukup, T. (1988) Biochim. Biophys. Acta 935, 16-25

Houstek, J., Pavelka, S., Baudysova, M., and Kopecky, J. (1990c) FEBS Lett. 274, 185-188

Houstek, J., Tvrdik, P., Pavelka, S., and Baudysova, M. (1991a) FEBS Lett. in press

Houstek, J., Vizek, K., Pavelka, S., Kopecky, J., Cermakova, M., and Hermanska, J. (1991b) in preparation

Iglesias, R., Fernandez, J.A., Mampel, T., Obregon, M.J., and Villarroya, F. (1987) Biochim. Biophys. Acta 923, 233-240

Izquierdo, J.M., Luis, A.M., and Cuezva, J.M. (1990) J. Biol. Chem. 265, 9090-9097

Jacobsson, A., Nedergaard, J., and Cannon, B. (1986) Biosci. Rep. 6, 621-631

Kopecky, J., Baudysova, M., Zanotti, F., Janikova, D., Pavelka, S., and Houstek, J. (1990) J. Biol. Chem. 265, 22204-22209

Kopecky, J., Sigurdson, L., Park, I.R.A., and Himms-Hagen, J. (1986) Am. J. Physiol. 251, E1-E7

Leonard, J.L., Mellen, S.A., and Larsen, P.R. (1983) Endocrinology 112, 1153-1155

Mohell, N., Nedergaard, J., and Cannon, B. (1983) Eur. J. Pharmacol. 93, 183-193

Nelson, B.D. (1990) Biochim. Biophys. Acta 1018, 275-277

Nicholls, D.G., Cunningham, S.A., and Rial, E. (1986) In: Brown Adipose Tissue (P. Trayhurn, D.G. Nicholls, Eds) Edward Arnold, London, pp. 52-85

Nicholls, D.G. and Locke, R.M. (1984) Physiol. Rev. 64, 1-64

Rehnmark, S. (1991) Doctor Dissertation Thesis, Wenner-Gren Institute, Stockholm University, Sweden

Rehnmark, S., Nechad, M., Herron, D., Cannon, B., and Nedergaard, J. (1990) J. Biol. Chem. 265, 16464-16471

Ricquier, D. and Bouillaud, F. (1986) In: Brown Adipose Tissue (P. Trayhurn, D.G. Nicholls, Eds) Edward Arnold, London, pp. 86-104

Rothwell, N.J. and Stock, M.J. (1986) In: Brown Adipose Tissue (P. Trayhurn, D.G. Nicholls, Eds) Edward Arnold, London, pp. 269-338

Rothwell, N., Stock, M., and Sudeza, D. (1985) Am. J. Physiol. 248, E397-E402

Silva, J.E. and Larsen, P.R. (1985) J. Clin. Invest, 76, 2296-2305

Staehelin, M., Simons, P., Jaeggki, K., and Wigger, N. (1983) J. Biol. Chem. 258, 3496-3502

Sundin, U., Herron, D., and Cannon, B. (1981) Biol. Neonate, 39, 141-149

Sundin, U., Moore, G., Nedergaard, J., and Cannon, B. (1987) Am. J. Physiol. 252, R822-R832

Svoboda, P., Houstek, J., Kopecky, J., and Drahota, Z. (1981)

458

Biochim. Biophys. Acta 634, 321–330

Svoboda, P., Svartengren, J., Snochowski, M., Houstek, J., and Cannon, B. (1979) Eur. J. Biochem. 102, 203–210

Visser, T.J., Kaplan, M.M., Leonard, J.L., and Larsen, P.R. (1983) J. Clin. Invest. 71, 992–1002

Walker, J.E., Cozens, A.L., Dyer, M.R., Fearnley, I.M., Powell, S.J., and Runswick, M.J. (1987) Chem. Scr. 27B, 97–105

Adenine Nucleotides in Cellular Energy Transfer and Signal Transduction
S. Papa, A. Azzi & J.M. Tager (eds)
© 1992 Birkhäuser Verlag, Basel/Switzerland

DETECTION OF THE UNCOUPLING PROTEIN (A NUCLEOTIDE-MODULATED H+- TRANSLOCATOR) AND ITS mRNA IN HUMAN DISEASES

GABRIELLA GARRUTI [1], RICCARDO GIORGINO [1] and DANIEL RICQUIER [2]

[1]Istituto di Clinica Medica, Endocrinologia e Malattie Metaboliche, Università di Bari, Policlinico, 70124 Bari, Italy; [2]Centre De Recherche sur l'Endocrinologie Moléculaire et le Developpement, (CNRS) 92190 Meudon-Bellevue, France.

SUMMARY
The mitochondrial inner membrane contains several transport systems for metabolites. Among them there is the Uncoupling Protein (UCP), which is a 33KDa protein whose activity is inhibited by purine nucleotides (GDP, GTP, ADP and ATP) and is stimulated by free fatty acids following stimulation of cells by Norepinephrine in rodents. UCP is a H+ translocator which allows the regulated uncoupling of respiration and is uniquely expressed in a specialised adipose tissue, the brown adipose tissue (BAT). Since BAT is a thermogenic organ in rodents and UCP dissipates the gradient generated by the respiratory chain, the presence of UCP in the inner membrane of brown adipocyte mitochondria might account for BAT thermogenic capacity. Human UCP whose primary structure has already been determined by gene sequencing was recently detected in perirenal adipose deposits of human subjects with Pheochromocytoma (Pheo), a catecholamine-secreting tumor; thus, it was argued that catecholamines might play a crucial role in the expression of UCP in humans as well as in rodents. Furthermore, a high prevalence of BAT seems to be associated with cancer-induced cachexia and increased BAT activity has been found in some malignant diseases. However, no identification of UCP and its mRNA was performed in the last two studies. Hence, we investigated whether UCP and its mRNA are detectable in adipose deposits of adults with pathological conditions distinct from Pheo, with or without malignant diseases. In parallel assays we studied a group of subject with malignant or non-malignant Pheochromocytoma. All patients with Neoplasms did not show any sign of cancer-induced cachexia. By utilizing Western and Northern blotting we confirm that UCP and its mRNA are detectable in subjects with Pheo. However we also demonstrated the presence of human UCP in fat deposits of people with both endocrine and non-endocrine diseases distinct from Pheo, either in presence or in absence of malignancy.

460

INTRODUCTION

Brown adipose tissue (BAT) is a specialised adipose tissue in mammals. BAT function is that of maintaining body temperature because BAT is mainly devoted to dissipate oxidation energy as heat (Thermogenesis).

However BAT represents only a small component of mammals adipose mass, whose major part is constituted by white adipose tissue which stores energy as triglycerides and might determine the development of obesity.

Between 1961 and 1965 it was demonstrated that BAT accounted for thermogenesis in rodents. However at this time noone knew which biochemical mechanism was responsible for BAT heat production. Further studies demonstrated the existence of a H+-translocator in the inner mitochondrial membrane of brown adipocytes (Nicholls et al. 1984, Kopecky et al. 1984) which allowed proton re-entry into the mitochondrial matrix without the obbligatory proton coupling to respiration. Since this mitochondrial carrier is strongly inhibited by purine nucleotides (GDP, GTP, ADP and ATP), brown adipocyte activity has been longer characterized by measuring the high-affinity binding of GDP to brown adipocyte mitochondria.

In successive studies the primary structure of UCP has been characterized in various species by both gene sequencing (Bouillaud et al.,1986, Cassard et al 1990), cDNA and amino-acid sequence analysis (Aquila H. et al. 1985) and the unequivocal demonstration was given that UCP is a unique proton conductance pathway present in brown adipocyte mitochondria.

In animal models both UCP expression and activity are strongly activated by norepinephrine (NE) and animal BAT is richly innervated by sympathetic nervous fibres (Stock et al. 1986). Recent studies have demonstrated that some novel B-adrenergic agonists having a slimming effect in animals, stimulate UCP synthesis (Muzzin et al., 1989) and that NE activates the transcription of the UCP gene "in vivo" (Ricquier et al., 1986) and in brown adipocytes differentiated in culture (Renmark S. et al. , 1990) thus suggesting that, at least in animals, UCP expression is mainly modulated by B-adrenergic receptors.

The occurrence of BAT in humans has been demonstrated by different authors but the physiological role of human BAT and its hormonal regulation are not clear at all.

BAT seems to be important in newborns and infants to regulate body temperature, but it seems to disappear in adult life.

In some elegant experiments Cunningham et al. (1985) studied the proton current generated by the respiratory chain in adipocytes isolated from perinephric fat deposits of 22 adult subjects. The authors concluded that the respiratory capacity of human BAT was limited as compared to that of animals (Cunningham et al.,1985).

The association between adrenal Pheochromocytoma (Pheo), a catecholamine-secreting tumor, and BAT was firstly reported by Melicow (quoted by Heaton, 1972) who based the diagnosis of BAT on the histological detection of a tissue showing a multilocular appearence and possessing a large number of mitochondria. However, these criteria seem inadequate whilst the detection of UCP with molecular biological techniques allows the unequivocal demonstration of BAT.

The human UCP gene has been sequenced (Cassard et al. 1990) and the H-UCP-0.5 genomic probe detected a 1.8Kb UCP mRNA in perinephric deposits obtained from six subjects with Pheo (Bouillaud et al, 1988). This led to the suggestion that the large amount of catecholamines produced by the tumor might account for the stimulation of human UCP expression.

An interesting relationship between cancer-induced cachexia (CIC) and BAT have been recently reported (Brooks et al. 1981). CIC is a common manifestation of advanced cancer which is characterized by physical disability and mortality.

In an animal model of cancer-induced cachexia (CIC) an increased BAT activity has been found which strongly correlates to the profound weight loss (Brooks et al. 1981).

In a study on human subjects a high prevalence of BAT has been recently found in adults with CIC (Shellock et al. 1986). An increased BAT activity has also been measured in children with malignant diseases as compared to children operated on for urological non-neoplastic diseases (Bianchi et al. 1989).

However no identification of UCP or of its mRNA was performed in the latter two human studies. We have investigated whether UCP and its mRNA are detectable in situations other than Pheo by utilizing molecular biology techniques (Western and Northern blotting). In our study we included subjects with malignant tumors and subjects with non-malignant tumors in order to assess whether BAT UCP detection was associated with the occurence of a malignancy.

In parallel experiments we studied a group of subjects operated on for both malignant and non-malignant Pheo in which we expected to detect characteristic human UCP and UCP mRNA.

SUBJECTS

Specimens of perirenal and periadrenal adipose tissue were obtained from 3 groups of adult humans (men=13, women=17) undergoing urological surgery. Group A included 11 subjects with Pheo (Malignant n=1, Non-malignant n=10). Group B included subjects with non-malignant adrenal tumors distinct from Pheo (n=9). In Group C we considered adults operated on for non-endocrine urological diseases (Malignant n=3, Non-malignant n=7).

All subjects with neoplasms did not show any sign of cancer-induced cachexia. The clinical details of our patients are reported in Table I and II.

TABLE I. Diagnosis and UCP detection in adults of the groups A, B and C

GROUP A	UCP	GROUP B	UCP	GROUP C	UCP
H11 Pheo	-	H1 Primary Aldosteronism	+	H23 P.U.O	+
H24 "	+	H7 Renin-ST	+	H10 Small Kidney	+
H25 "	+	H8 Primary Aldosteronism	+	H26 Hydronephr.	-
H33 "	+	H9 " "	-	H27 Kd HMGLP	+
H34 "	+	H13 " "	+	H31 Kd Adenocr	-
H35 "	-	H28 " "	-	H32 " "	-
H36 "	-	H37 " "	-	H38 P.U.O	+
H43 "	+	H41 " "	+	H39 P.U.O.	-
H50 "	+	H47 Cushing's Syndrome	-	H48 Kd Adenocr	+
H51 "	+			H53 Kd Infarction	+
H52 "	+				

Pheo= Pheochromocytoma; Renin-ST= Renin secreting-tumor; P.U.O= Pelvi-Ureteric obstruction; Kd=Kidney; HMGLP= Hemangiolipoma; Adenocr= Adenocarcinoma; (+)= Present; (-)= Absent. The subject H25 had a malignant Pheochromocytoma.

TABLE II. Increase in Catecholamine and Aldosterone levels in the adults of the groups A, B and C

GROUP A	↑Cat	↑Aldo	GROUP B	↑Cat	↑Aldo	GROUP C	↑Cat	↑Aldo
H11	+	-	H1	NA	+	H23	NA	NA
H24	+	NA	H7	NA	+	H10	NA	+
H25	+	+	H8	NA	+	H26	NA	NA
H33	+	+	H9	NA	+	H27	NA	NA
H34	+	+	H13	-	+	H31	NA	NA
H35	+	+	H28	NA	+	H32	NA	NA
H36	+	+	H37	NA	+	H38	NA	NA
H43	+	+	H41	-	+	H39	NA	NA
H50	+	NA	H47	NA	-	H48	NA	-
H51	+	NA				H53	NA	+
H52	+	NA						

↑Cat= Increased urinary and plasma catecholamine levels; ↑Aldo= Increased urinary and plasma aldosterone levels; NA= Non Available (+)= Present; (-)= Absent.

METHODS

Western blot analysis was used to detect UCP in mitochondria preparations, as described previously (Ricquier et al., 1991).

464

Mitochondrial proteins were electrophoresed in a polyacrylamide gel and transferred to a nitrocellulose fitted by an electric device. Antibodies were prepared as reported elsewhere (Bouillaud et al. 1983). In Northern analysis the human genomic probe (H-UCP-0.5) [Cassard et al., 1990] was used to detect UCP mRNA. It has
already been demonstrated that the above mentioned genomic fragment is able to hybridize a 1.8Kb UCP mRNA which corresponds to the human UCP mRNA isolated from a human hibernoma, which is a lipoma whose cytology reproduces that of BAT (Bouillaud et al.,1988).

RESULTS

We demonstrated the presence of UCP in 8, 5 and 6 cases in the groups A, B and C respectively. The H-UCP-0.5 genomic probe detected a typical human UCP mRNA in most of the subjects of the three groups (Table I; Fig.1 and 2).

Fig 1. Western blot analysis of samples from different fat deposits of adult subjects

1. 37-day infant; 2. empty space; 3. H23 Pelvi-ureteric obstruction (periadrenal fat); 4. H24 Pheochromocytoma (perirenal fat); 5. H24 Pheochromocytoma (periadrenal fat); 6. H25 Pheochromocytoma (periadrenal fat); 7. H27 Kidney Hemangiolipoma (peridranal fat).

<u>Fig 2.</u> Northern blot analysis of total RNA from fat samples of human subjects undergoing urological surgery.

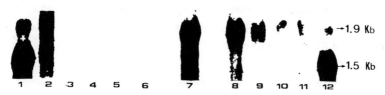

1. BAT of Djungarian Hamster; 2. H1, Primary aldosteronism (periadrenal fat); 3. H1, Primary Aldosteronism (perirenal fat); 4. Empty space; 5. H26 Hydronephrosis (periadrenal fat); 6. H32 Kidney adenocarcinoma (periadrenal fat) ; 7. H23 Pelvi-ureteric obstruction (periadrenal fat) ; 8. H10 Small Kidney (periadrenal fat) ; 9. H7 Renin-secreting tumor (periadrenal fat) ; 10. H13, Primary aldosteronism (periadrenal fat); 11. H13, Primary aldosteronism (perirenal fat)   12. BAT of Djungarian Hamster

DISCUSSION

Our results confirm the presence of UCP in the perirenal and periadrenal adipose deposits of adults with Pheo (Bouillaud et al. 1983, Ricquier et al. 1982). They also demonstrate that a typical BAT UCP can be found not only in subjects with adrenal tumors distinct from Pheo, but also in several subjects operated on for non-endocrine non-neoplastic diseases. Unexpectedly a common feature of most of the subjects in which we detected UCP was the presence of increased urinary and circulating levels of aldosterone. No increase in catecholamine levels has been usually reported in the diseases included in the groups B and C. Interestingly in one subject with Pheo (H36 in Group A) and in 2 patients with Primary Aldosteronism (H9 and H37 in Group B) who have been treated for long periods with beta-blocker therapy we did not detect UCP despite the high circulating levels of norepinephrine and aldosterone. Since in animals UCP expression is stimulated by the activation of beta-(atypical)-adrenergic receptors it seems likely that either catecholamines or other hormones (aldosterone) might modulate the expression of human UCP through the activation of a beta-adrenergic receptor.

REFERENCES:

Aquila H, Link T A, Klingeberg M, The uncoupling protein   from brown fat mitochondria is related to the ADP/ATP   carrier. Analysis of sequence homologies and folding of the   protein in   the membrane. EMBO J, 1985, 4:2369- 2376

Bianchi A, Bruce J, Cooper AL, Childs C, Kohli M, Morris-Jones   P, Rothwell NJ. Increased brown adipose tissue activity in   children with malignant disease. Horm Metab Res,1989,   21:640-641.

Bouillaud F, Combes-George M, Ricquier D. Mitochondria of adult human brown adipose tissue contain a 32000Mr uncoupling   protein.   Biosci Rep. 1983, 3:775-80.

Bouillaud F, Weissenbach J, Ricquier D, Complete cDNA-derived   amino-acid sequence of rat brown adipose tissue uncoupling   protein. J Biol Chem, 1986, 261: 1487-1490.

Bouillaud F, Villarroya F, Hentz H, Raimbault S, Cassard AM,   Ricquier D. Detection of brown adipose tissue uncoupling   protein mRNA   in adult patients by a human genomic probe.   Clin Sci, 1988, 75:21- 27.

Brooks S L, Neville A M, Rothwell N J, Stock M J, Wilson S,   Sympathetic activation of brown adipose tissue thermogenesis   in cachexia. Biosci Rep. 1981, 1: 509-17.

Cassard A M, Bouillaud F, Mattei M G, Hentz E, Raimbault   S., Thomas M, Ricquier D., Human uncoupling protein gene: structure,   comparison with Rat gene and assignement to the   long arm of the   chromosome 4. J Cell Biochem, 1990, 43:   255-264.

Cunningham S, Leslie P, Hopwood D, Illingworth P, Jung RT,   Nicholls DG, Peden N, Rafael J, Rial E. The   characterization   and the energetic potential of   brown adipose tissue in man.   Clin. Sci. 1985, 69:343-348.

Heaton J M. The distribution of brown adipose tissue in the   human. J Anat, 1972, 112:35-39.

Kopecky J, Guerrieri F, Jezek P, Drahota Z, Housteck J   Molecular mechanisms of   uncoupling in brown adipose tissue   mitochondria. The non-identity of proton   and chloride   conducting pathways. FEBS, 1984, 170:186-190.

Muzzin P, Ravelli J P, Ricquier D, Meier M K, Assimacoupoulus- Jeannet F. and Giacobino J P. The novel thermogenic B- adrenergic agonist Ro16-8714 increases interscapular   brown fat B-receptor-adenylate cyclase and the uncoupling   protein mRNA level   in obese (fa/fa) Zucker rats, Biochem J, 1989,261: 721-724.

Nicholls D G and Locke R M. Thermogenic mechanisms in   brown fat. Physiol Rev 1984, 64, 1-64

468

Ricquier D, Néchad M, Mory G. Ultrastructural and biochemical
    characterization of human brown adipose tissue in   pheochromocytoma. J
    Clin Endocrinol Metab, 1982, 13:501-520.

Ricquier D, Bouillaud F,   Toumelin P, Mory G, Bazin R, Arch J,   Pénicaud L.
    Expression of uncoupling mRNA in thermogenic or   weakly thermogenic brown
    adipose tissue. Evidence for a   rapid beta-adrenoceptor-mediated step
    during activation of   Thermogenesis. J Biol Chem, 1986, 261:13905-13910.

Ricquier D, Casteilla L and Bouillaud F. Molecular studies of   the
    uncoupling   protein. FASEB J, 1991, 5:2237-2242.

Rehnmark S., Néchad M., Herron D., Cannon B, and Nedergaard J.,   Alpha and
    beta-adrenergic induction of the expression of   the uncoupling protein
    thermogenin in brown adipocytes   differentiated in culture. J Biol Chem,
    1990, 265:16464- 16471

Shellock F G, Riedinger MS, Fishbein MC. Brown adipose tissue   in cancer
    patients: possible cause of cancer induced   cachexia. J Cancer Res Clin
    Oncol, 1986, 111: 82-85.

Stock M J & Rothwell NJ, (1986) Brown adipose tissue and diet-   induced
    tharmogenesis. In Brown adipose tissue (Trayhurn P   and Nicholls DG eds)
    pp269-298, Edward Arnold , London.

ACKNOWLEDGEMENTS

The International Symposium on "Adenine Nucleotides in Cellular Energy Transfer and Signal Transduction", held in Fasano, Bari, from 7th till 10th of October 1991, has been an activity of the International Biomedical Institute (IBMI Symposium No. 11).

The Symposium and the publication of the present book was supported by financial contributions from the UNESCO Global Network for Molecular and Cell Biology and from Biomed S.p.A.

The Symposium was also supported by the International Union of Biochemistry and Molecular Biology (IUBMB Symposium No. 218), the International Union of Pure and Applied Biophysics, the National Research Council of Italy, the Region of Puglia, the Commune of Fasano and the following companies: Imperial Chemical Industries (ICI Pharma Italia), Kontron Instruments Italia, Amity S.r.l., Beckman Analytical, Pharmacia.

IBMI

The International Biomedical Institute (IBMI) is a non-profit
organization founded in Bari in 1986, recognized by Regione
Puglia as Ente Morale, and by the UNESCO Global Network for
Molecular and Cell Biology as a member institution.

IBMI objectives are to promote and develop scientific, clinical
and technical research in selected areas of biological science
and medicine.

In 1988 IBMI joined the consortium "Biomed S.p.A." set up by
the Chamber of Commerce of Bari, Farmitalia-Carlo Erba, Kontron Instruments Italia, SVIM Service and other companies to
finance: the building of a non-profit research institute in
Puglia, the promotion of research programmes in its own research laboratories or in other research institutes, the
organization of courses and scientific meetings and the
realization of technical-scientific publications.

SUBJECT INDEX